FUNDAMENTALS
OF PROBABILITY

FUNDAMENTALS OF PROBABILITY

Saeed Ghahramani
Towson State University

PRENTICE HALL, Upper Saddle River, New Jersey 07458

Library of Congress Cataloging-in-Publication Data

Ghahramani, Saeed.
    Fundamentals of probability / Saeed Ghahramani.
      p.   cm.
    Includes index.
    ISBN 0-13-065798-0
    1. Probabilities.   I. Title.
QA273.G464   1996
519.2—dc20   95-16473
                                        CIP

Acquisitions editor: *Ann Heath*
Production editor: *Barbara Grasso Mack*
Managing editor: *Jeanne Hoeting*
Director of production and manufacturing: *David W. Riccardi*
Cover design: *Jayne Conte*
Cover photo: *Authenticated News International*
Manufacturing buyer: *Alan Fischer*

© 1996 by Prentice-Hall, Inc.
Simon & Schuster / A Viacom Company
Upper Saddle River, NJ 07458

Printed in the United States of America

10 9 8 7 6 5 4 3

ISBN 0-13-065798-0

PRENTICE-HALL INTERNATIONAL (UK) LIMITED, LONDON
PRENTICE-HALL OF AUSTRALIA PTY. LIMITED, SYDNEY
PRENTICE-HALL CANADA INC., TORONTO
PRENTICE-HALL HISPANOAMERICANA, S.A., MEXICO
PRENTICE-HALL OF INDIA PRIVATE LIMITED, NEW DELHI
PRENTICE-HALL OF JAPAN, INC., TOKYO
SIMON & SCHUSTER ASIA PTE. LTD., SINGAPORE
EDITORA PRENTICE-HALL DO BRASIL, LTDA., RIO DE JANEIRO

To Lili, Adam, and Andrew

CONTENTS

12   Simulation   428

Appendix   454

Answers to Odd-Numbered Exercises   457

Index   I-1

PREFACE

This book is designed for a course in basic probability to be taken by mathematics, physics, engineering, statistics, actuarial science, operations research, and computer science majors. It assumes a second course in calculus, and hence can also be used in basic probability courses taken by students who have completed two calculus courses. The aim of this one-semester textbook is to present probability in the most natural way: through a great number of attractive and instructive examples and exercises that motivate the definitions, theorems, and methodology of the theory. Examples and exercises have been very carefully designed to arouse the reader's curiosity and hence motivate the student to delve into the theory with enthusiasm.

As one mathematician wrote, when writing a book, authors are always faced with two opposing forces. One is the natural tendency to want to put too much in the book, because *everything* is important and *everything* has to be said the author's way! On the other hand, authors also have to keep in mind a clear definition of the focus, the level, and the audience of the book, thereby choosing carefully what is "in" and what is "out." I think that this book is a resolution of the tension generated by these opposing forces.

Instructors will enjoy the versatility of the book. They can choose their favorite problems and exercises from a collection of almost 1200, and if necessary omit some sections and/or theorems to teach it at an appropriate level.

Exercises for most sections are divided into two categories: A and B. Problems of category A are routine and those of category B are somewhat challenging. However, exercises of category B are not uniformly challenging. Some of these exercises were included because the students found them somewhat difficult. A number of my colleagues and I have taught probability using this book for several years. For 3-credit-hour classes, we have been able to omit many sections without difficulty. The book is designed to be challenging and yet instructive for students with different levels of abilities.

I have tried to maintain an approach that is mathematically rigorous and at the same time closely matches the historical development of probability. In particular, whenever appropriate, I have included historical remarks. I have also included discussions of a number of probability problems published in recent years in journals such as *Mathematics Magazine* and *American Mathematical Monthly*. These are interesting and instructive problems that deserve discussion in classrooms.

Chapter 12 concerns computer simulation. This chapter is divided into several sections, presenting algorithms that are used to find approximate solutions to complicated probabilistic problems. These sections can be discussed independently when relevant materials from earlier chapters are being taught, or they can be discussed all at the same time toward the end of the semester. Although I believe that the emphasis should remain on concepts, methodology, and the mathematics of the subject, I also think that students should be asked to read the material on simulation and perhaps do some projects. Computer simulation is an excellent means to acquire insight into the nature of a problem, its functions, its magnitude, and the characteristics of the solution.

Some of the other advantages of this book are:

- The historical roots and applications of many of the theorems and definitions are presented in detail and suitable counterexamples are given.
- Examples and exercises in each section do not refer to problems and exercises in other chapters or sections—a style that often frustrates students and instructors.
- Whenever a new concept is introduced, its relation with previous concepts and theorems is explained.
- Although the usual analytic proofs are given, simple probabilistic arguments have been presented to promote deeper understanding of the subject.
- The book begins with discussions on probability and its definition rather than with combinatorics. I believe that combinatorics should be taught after students have learned the preliminary concepts of probability. The advantage of this approach is that the need for methods of counting will

occur naturally to students, and the connection between the two areas becomes clear from the beginning. Moreover, combinatorics becomes more interesting and enjoyable.

- Students who begin learning probability have a tendency to think that sample spaces always have a finite number of sample points. To reduce such tendencies, I have introduced the concept of *random selection of a point from an interval* in Chapter 1 and have applied it where appropriate throughout the book. Moreover, since the basis of simulating indeterministic problems is selection of random points from $(0, 1)$, in order to understand simulations, students need to be thoroughly familiar with this concept.

- Conditional probabilities, independence, and conditional independence are discussed comprehensively from different directions with appropriate examples and exercises.

- The concepts of expectation and variance are introduced early, as I believe these extremely important concepts should be defined and used as soon as possible. One benefit of doing this is that when random variables such as Poisson and normal are studied, the associated parameters will be understood immediately rather than remaining ambiguous until expectation and variance are introduced. Therefore, from the beginning, students will have a natural feeling about such parameters.

- Special attention is paid to the Poisson distribution; it is made clear that this distribution is frequently applicable, for two reasons: first, because it approximates the binomial distribution, and second, it is the mathematical model of an enormous class of phenomena. The comprehensive presentation of the Poisson process and its applications can be understood by junior- and senior-level students.

- Students often have difficulties understanding functions or quantities such as the density function of a continuous random variable and the formula for mathematical expectation. For example, they might wonder why $\int x f(x)\, dx$ is the appropriate definition for $E(X)$ and why correction for continuity is necessary. I have explained the reason behind such definitions, theorems, and concepts and have demonstrated why they are the natural extensions of discrete cases.

- In the first six chapters, many examples and exercises concerning the selection of random points from intervals are introduced. Consequently, in Chapter 7, when discussing uniform random variables, I have been able to calculate the distribution and (by differentiation) the density function of X, a random point from an interval (a, b). In this way the concept of a uniform random variable and the definition of its density function are readily motivated.

- In Chapters 7 and 8 the usefulness of uniform densities is shown using many examples. In particular, applications of uniform density in *geometric probability theory* are emphasized.

- Normal density, arguably the most important density function, is easily motivated by De Moivre's theorem. In Section 7.4 I have introduced the standard normal density, the elementary version of the central limit theorem, and the normal density just as they were developed historically. My experience shows that this is a good pedagogical approach. When teaching this approach, the normal density becomes natural and does not look like a strange function appearing out of the blue.

- Exponential random variables naturally occur as *times between consecutive events of Poisson processes*. The time of occurrence of the nth event of a Poisson process has a gamma distribution. For these reasons I have motivated exponential and gamma distributions by Poisson processes. In this way we can obtain many examples of exponential and gamma random variables from the abundant examples of Poisson processes that we already know. Another advantage is that it helps us visualize memoryless random variables by looking at the interevent times of Poisson processes.

- Joint distributions and conditioning are often trouble areas for students. A detailed explanation and many applications concerning these concepts and techniques have made these materials somewhat easier for students to understand.

- The concepts of covariance and correlation are motivated thoroughly.

- A complete solution manual is provided for instructors, and the answers to odd-numbered exercises are included at the end of the book.

Acknowledgments

While writing this book, many people helped me either directly or indirectly. Lili, my beloved wife, deserves an accolade for her patience and encouragement; so do my children. I am thankful to my parents and mother-in-law for their moral support.

My colleagues and friends at the Towson State University have read or taught from various revisions of the manuscript and have offered useful advice. In particular, I am grateful to Professors Mostafa Aminzadeh, Raouf Boules, James Coughlin, Geoffrey Goodson, Sharon Jones, Sayeed Kayvan, Bill Rose, Martha Siegel, Houshang Sohrab, and Eric Tissue. I want to thank Houshang Sohrab especially for his kindness and generosity with his time spent very carefully reading the entire manuscript three times, Bill

Rose for many organizational and other constructive suggestions, and James Coughlin for comments that have improved the final product.

Thanks are also due to the students who have used preliminary versions of the manuscript. Some of these students, especially my friend Vincent Dobbs, read the text meticulously and made many useful comments.

I am also grateful to the following reviewers for their valuable suggestions and constructive criticisms: Jay Devore, California Polytechnic Institute—San Luis Obispo; Sergey Fomin, Massachusetts Institute of Technology; D. H. Frank, Indiana University of Pennsylvania; M. Lawrence Glasser, Clarkson University; Paul T. Holmes, Clemson University; Joe Kearney, Davenport College; Philippe Loustaunau, George Mason University; John Morrison, University of Delaware; Elizabeth Papousek, Fisk University; Richard J. Rossi, California Polytechnic Institute—San Luis Obispo; James R. Schott, University of Central Florida; Kyle Siegrist, University of Alabama—Huntsville; Olaf Stackelberg, Clemson University. In particular, Dr. Devore's excellent comments improved the manuscript substantially.

The preliminary manuscript of this book was put onto the computer using a grant by the Faculty Development of the Towson State University. The figures of the same version were produced by Carolyn Westbrook from the Graphic Services of this university. It is a pleasure to acknowledge the help of Carolyn and Towson State University.

Special thanks are due to Prentice Hall's former and present editors, particularly Jerome Grant and Ann Heath, for their encouragement and assistance in seeing this effort through. I also appreciate very much the excellent job Barbara Mack has done as production editor.

Last but not least, I want to express my gratitude for all technical help I have received from my good friend and colleague Professor Howard Kaplon. Over the years Howard has assisted me a lot with computers; this book has also benefited from his endless help.

Saeed Ghahramani

Chapter 1

AXIOMS OF PROBABILITY

1.1 Introduction

In search of natural laws that govern a phenomenon, science often faces "events" that may or may not occur. The event of *disintegration of a given atom of radium* is one such example because, in any given time interval, such an atom may or may not disintegrate. The event of *finding no defect in inspection of a microwave oven* is another example, since an inspector may or may not find defects in the microwave oven. The event that *an orbital satellite in space is at a certain position* is a third example. In any experiment, an event that may or may not occur is called *random*. If the occurrence of an event is inevitable, it is called *certain*, and if it can never occur, it is called *impossible*. For example, the event that an object travels faster than light is impossible and the event that in a thunderstorm flashes of lightning precede any thunder echoes is certain.

Knowing that an event is random determines only that the existing conditions under which the experiment is being performed do not guarantee its occurrence. Therefore, the knowledge obtained from randomness itself is hardly decisive. It is highly desirable to determine the exact value or an estimate of the chance of the occurrence of a random event quantitatively. The theory of probability has emerged from attempts to deal with this problem. In many different fields of science and technology, it has been observed that under a long series of experiments, the proportion of the time that an event occurs may appear to approach a constant. It is these constants that probability theory (and statistics) aims at predicting and describing as quantitative measures of the chance of occurrence of events. For example, if a fair coin is tossed repeatedly, the proportion of the heads approaches 1/2. Hence probability theory postulates that the number 1/2 be assigned to the event of *getting heads in a toss of a fair coin.*

Historically, from the dawn of civilization, human beings have been interested in games of chance and gambling. However, the advent of probability as a mathematical discipline is relatively recent. Ancient Egyptians, in about 3500 B.C., were using astragali, a four-sided die-shaped bone found in the heels of some animals, to play a game now called *Hounds and Jackals.* The ordinary six-sided die was made about 1600 B.C. and since then has been a major instrument in all kinds of games. The ordinary deck of playing cards, which is probably the most popular tool in games and gambling, is much more recent than dice. Although it is not known where and when it originated, there are reasons to believe that it was invented in China sometime between the seventh and tenth centuries. Clearly, through gambling and games of chance people have gained intuitive ideas about the frequency of occurrence of certain events and, hence, about probabilities. But surprisingly, scholarly studies of the chances of events were not begun until the fifteenth century. The Italian scholars Luca Paccioli (1445–1514), Niccolò Tartaglia (1499–1557), Girolamo Cardano (1501–1576), and especially Galileo Galilei (1564–1642) were among the first prominent mathematicians who calculated probabilities concerning many different games of chance. They also tried to construct a mathematical foundation for probability. Cardano even published a handbook on gambling with sections discussing methods of cheating. Nevertheless, real progress started in France in 1654, when Blaise Pascal (1623–1662) and Pierre de Fermat (1601–1665), two giants of mathematics, exchanged several letters in which they discussed general methods for the calculation of probabilities. In 1655, the famous Dutch scholar Christian Huygens (1629–1695) joined them, and their collaboration was very fruitful. In 1657 Huygens published the first book on probability, *De Ratiocinates in Aleae Ludo (On Calculations in Games of Chance).* This book marked the real birth of probability. Scholars who read it realized that they were faced with a deep theory. Discussions of solved and unsolved problems and the many new ideas generated interested readers in this challenging new field.

After the work of Pascal, Fermat, and Huygens, the book written by James Bernoulli (1654–1705) in 1713 and that by Abraham de Moivre (1667–1754) in 1730 were major breakthroughs. In the eighteenth century, studies of great mathematicians such as Pierre-Simon Laplace (1749–1827), Siméon Denis Poisson (1781–1840), and Karl Friedrich Gauss (1777–1855) started to grow probability and its applications very rapidly and in many different directions. In the nineteenth century, prominent Russian mathematicians Pafnuty Chebyshev (1821–1894), Andrei Markov (1856–1922), and Aleksandr Lyapunov (1857–1918) advanced the works of Laplace, De Moivre, and Bernoulli considerably. By the early twentieth century, probability was already a grown and developed theory, but its foundation was not firm. A major goal was to put it on firm mathematical grounds. Until then, among other interpretations perhaps the *relative frequency interpretation* of probability was the most satisfactory. According to this interpretation, to define p, the probability of the occurrence of an event A of an experiment, we look at a series of sequential or simultaneous performances of the experiment and observe that the proportion of times that A occurs approaches a constant. Then we count $n(A)$, the number of times that A occurs in n of these performances of the experiment and define $p = \lim_{n \to \infty} n(A)/n$. This definition is mathematically problematic and cannot be the basis of a rigorous probability theory. Some of the difficulties that this definition creates are:

1. In practice, $\lim_{n \to \infty} n(A)/n$ cannot be computed since it is impossible to repeat an experiment infinitely many times. Moreover, if for a large n, $n(A)/n$ is taken as an approximation for the probability of A, there is no way to analyze the error.

2. There is no reason to believe that the limit of $n(A)/n$ as $n \to \infty$ exists. Also, if the existence of this limit is accepted as an axiom, many dilemmas arise that cannot be solved. For example, there is no reason to believe that in a different series of experiments and for the same event A, this ratio approaches the same limit. Hence the uniqueness of the probability of the event A is not guaranteed.

3. By this definition, probabilities that are based on our personal belief and knowledge are not justifiable. Thus statements such as "the probability that the price of oil will be raised in the next six months is 60%"; "the probability that the 50,000th decimal figure of the number π is 7 exceeds 10%"; "the probability that it will snow next Christmas is 30%"; and "the probability that Mozart was poisoned by Salieri is 18%" would be meaningless.

In 1900, at the International Congress of Mathematicians in Paris, David Hilbert (1862–1943) proposed 23 problems whose solutions were, in his opinion, crucial to the advancement of mathematics. One of these problems was the axiomatic treatment of the theory of probability. In his lecture, Hilbert quoted Weierstrass, who had said: "The final object, always to be kept in mind, is to arrive at a correct understanding of the foundations of the science." Hilbert added that a thorough understanding of special theories of a science is necessary for successful treatment of its foundation.

Probability had reached that point and was studied enough to warrant the creation of a firm mathematical foundation. Some work toward this goal has been done by Émile Borel (1871–1956), Serge Bernstein (1880–1968), and Richard von Mises (1883–1953), but it was not until 1933 that Andrei Kolmogorov (1903–1987), a prominent Russian mathematician, successfully axiomatized the theory of probability. In Kolmogorov's work, which is now universally accepted, three self-evident and indisputable properties of probability (discussed later) are taken as *axioms* and the entire theory of probability is developed rigorously based on these axioms. In particular, the existence of a constant p, as the limit of the proportion of the number of times that the event A occurs when the number of experiments increases to ∞, in some sense, is shown. Subjective probabilities based on our personal knowledge, feelings, and beliefs may also be modeled and studied by this axiomatic approach.

In this book we present a study of the mathematics of probability based on the axiomatic approach. Since in this approach the concepts of *sample space* and *event* play a central role, we explain these concepts next in detail.

1.2 Sample Space and Events

If the outcome of an experiment is not certain but all of its possible outcomes are predictable in advance, then the set of all these possible outcomes is called the *sample space* of the experiment and is usually denoted by S. Therefore, the sample space of an experiment consists of all possible outcomes of the experiment. These outcomes are sometimes called *sample points* or simply *points* of the sample space. In the language of probability, certain subsets of S are referred to as *events*. So events are sets of points of the sample space. Some examples follow.

Example 1.1 For the experiment of *tossing a coin once*, the sample space S consists of two points (outcomes), "heads" (H) and "tails" (T). Thus $S = \{H, T\}$. ♦

Example 1.2 Suppose that an experiment consists of two steps. First a coin is flipped. If the outcome is tails, a die is tossed. If the outcome is heads, the coin is flipped again. The sample space of this experiment is $S = \{T1, T2, T3, T4, T5, T6, HT, HH\}$. For this experiment, the event of *heads in the first flip of the coin* is $E = \{HT, HH\}$ and the event of *an odd outcome* when the die is tossed is $F = \{T1, T3, T5\}$. ♦

Example 1.3 Consider the experiment of measuring the lifetime of a light bulb. Since any nonnegative real number can be considered as the lifetime of the light bulb (in hours), the sample space is $S = \{x : x \geq 0\}$. In this experiment, $E = \{x : x \geq 100\}$ is the event that *the light bulb lasts at least 100 hours*, $F = \{x : x \leq 1000\}$ is the event that *it lasts at most 1000 hours*, and $G = \{505.5\}$ is the event that *it lasts exactly 505.5 hours*. ◆

Example 1.4   Suppose that a study is being done on all families with one, two, or three children. Let the outcomes of the study be the genders of the children in descending order of their ages. Then

$$S = \{b, g, bg, gb, bb, gg, bbb, bgb, bbg, bgg, ggg, gbg, ggb, gbb\}.$$

Here the outcome b means that the child is a boy and g means that it is a girl. The events $F = \{b, bg, bb, bbb, bgb, bbg, bgg\}$ and $G = \{gg, bgg, gbg, ggb\}$ represent families with the eldest child a boy and families with exactly two girls, respectively. ◆

Example 1.5   A bus with a capacity of 34 passengers stops at a station sometime between 11 and 11:40 every day. The sample space of the experiment consisting of counting the number of passengers on the bus and measuring the arrival time of the bus is

$$S = \left\{ (i, t) : 0 \leq i \leq 34, \ \ 11 \leq t \leq 11\frac{2}{3} \right\}, \tag{1.1}$$

where i represents the number of passengers and t the arrival time of the bus in hours and fractions of hours. The subset of S defined by $F = \{(27, t) : 11\frac{1}{3} < t < 11\frac{2}{3}\}$ is the event that the bus arrives between 11:20 and 11:40 with 27 passengers. ◆

REMARK: Different manifestations of outcomes of an experiment might lead to different representations for the sample space of the same experiment. For instance, in Example 1.5, the outcome that *the bus arrives at t with i passengers* is represented by (i, t), where t is expressed in hours and fractions of hours. By this representation, (1.1) is the sample space of the experiment. Now if the same outcome is denoted by (i, t), where t is the number of minutes after 11 that the bus arrives, then the sample space takes the form

$$S' = \{(i, t) : 0 \leq i \leq 34, \ 0 \leq t \leq 40\}.$$

To the outcome that *the bus arrives at 11:20 with 31 passengers*, in S the corresponding point is $(31, 11\frac{1}{3})$, while in S' it is $(31, 20)$. ◆

Example 1.6   (Round-off Error)   Suppose that each time Jay charges an item to his credit card, he will round the amount to the nearest dollar in his records. Therefore, the round-off error, which is the true value charged minus the amount recorded, is random with the sample space

$$S = \{0, 0.01, 0.02, \ldots, 0.49, -0.50, -0.49, \ldots, -0.01\},$$

where we have assumed that for any integer amount a, Jay rounds $a.50$ to $a + 1$. The event of rounding off at most 3 cents in a random charge is given by $\{0, 0.01, 0.02, 0.03, -0.01, -0.02, -0.03\}$.   ◆

If the outcome of an experiment belongs to an event E, we say that the event E has *occurred*. For example, if we draw two cards from an ordinary deck of 52 cards and observe that one is a spade and the other a heart, all of the events $\{sh\}$, $\{sh, dd\}$, $\{cc, dh, sh\}$, $\{hc, sh, ss, hh\}$, and $\{cc, hh, sh, dd\}$ have occurred because sh, the outcome of the experiment, belongs to all of them. However, none of the events $\{dh, sc\}$, $\{dd\}$, $\{ss, hh, cc\}$, and $\{hd, hc, dc, sc, sd\}$ have occurred because sh does not belong to any of them.

In the study of probability theory the relations between different events of an experiment play a central role. In the remainder of this section we study these relations. In all of the following definitions the events belong to a fixed sample space S.

An event E is said to be a *subset* of the event F if whenever E occurs F also occurs. This means that all of the sample points of E are contained in F. Hence considering E and F solely as two sets, E is a subset of F in the usual set-theoretic sense: that is, $E \subseteq F$.

Events E and F are said to be *equal* if the occurrence of E implies the occurrence of F, and vice versa, that is, if $E \subseteq F$ and $F \subseteq E$, hence $E = F$.

An event is called the *intersection of two events* E and F if it occurs only whenever E and F occur simultaneously. In the language of sets this event is denoted by EF or $E \cap F$ because it is the set containing exactly the common points of E and F.

An event is called the *union* of two events E and F if it occurs whenever at least one of them occurs. This event is $E \cup F$ since all of its points are in E or F or both.

An event is called the *complement* of the event E if it only occurs whenever E does not occur. The complement of E is denoted by E^c. It is clear that $E^c = S - E$, where $A - B = A \cap B^c$.

An event is called *certain* if its occurrence is inevitable. Thus the sample space is a certain event. An event is called *impossible* if there is certainty in its nonoccurrence. Therefore, the empty set $\emptyset$, which is S^c, is an impossible event.

If the joint occurrence of two events E and F is impossible, we say that E and F are *mutually exclusive*. Thus E and F are mutually exclusive if the occurrence of E precludes the occurrence of F, and vice versa. Since the event representing the joint occurrence of E and F is EF, their intersection, E and F are mutually exclusive if $EF = \emptyset$. A set of events $\{E_1, E_2, \ldots\}$ is called *mutually exclusive* if the joint occurrence of any two of them is impossible, that is, if $\forall i \neq j$, $E_i E_j = \emptyset$. Thus $\{E_1, E_2, \ldots\}$ is mutually exclusive if and only if every pair of them is mutually exclusive.

The events $\bigcup_{i=1}^{n} E_i$, $\bigcap_{i=1}^{n} E_i$, $\bigcup_{i=1}^{\infty} E_i$, and $\bigcap_{i=1}^{\infty} E_i$ are defined in a similar way to $E_1 \cup E_2$ and $E_1 \cap E_2$. For example, if $\{E_1, E_2, \ldots, E_n\}$ is a set of events, by $\bigcup_{i=1}^{n} E_i$ we mean the event in which at least one of the events E_i, $1 \leq i \leq n$, occurs. By $\bigcap_{i=1}^{n} E_i$ we mean an event that occurs only when all of the events E_i, $1 \leq i \leq n$, occur.

Example 1.7 At a busy international airport arriving planes land on a first come, first served basis. Let E, F, and H be the events that there are at least five, at most three, and exactly two planes waiting to land, respectively. Then

1. E^c is the event that at most four planes are waiting to land.
2. F^c is the event that at least four planes are waiting to land.
3. E is a subset of F^c; that is, if E occurs, then F^c occurs. Therefore, $EF^c = E$.
4. H is a subset of F; that is, if H occurs, then F occurs. Therefore, $FH = H$.
5. E and F are mutually exclusive; that is, $EF = \emptyset$. E and H are also mutually exclusive since $EH = \emptyset$.
6. FH^c is the event that the number of planes waiting to land is zero, one, or three. $\blacklozenge$

Sometimes Venn diagrams are used for a practical representation of the relations among events of a sample space. The sample space S of the experiment is usually shown as a large rectangle, and inside S, circles or other geometrical objects are drawn to indicate the events of interest. Figure 1.1 presents Venn diagrams for EF, $E \cup F$, E^c, and $(E^c G) \cup F$. The shaded regions are the indicated events.

Unions, intersections, and complementations satisfy many useful relations between events. A few of these relations are as follows: $(E^c)^c = E$, $E \cup E^c = S$, and $EE^c = \emptyset$:

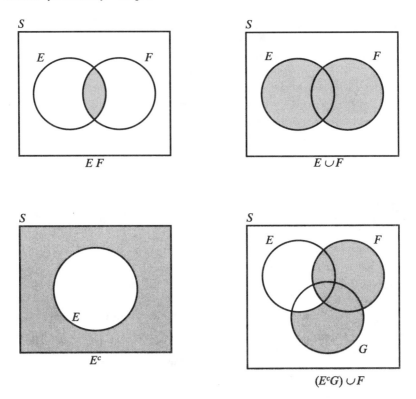

Figure 1.1   Venn diagrams of the events specified.

Commutative laws: $E \cup F = F \cup E, EF = FE$.
Associative laws: $E \cup (F \cup G) = (E \cup F) \cup G, E(FG) = (EF)G$.
Distributive laws:

$$(EF) \cup H = (E \cup H)(F \cup H), \quad (E \cup F)H = (EH) \cup (FH).$$

De Morgan's first law:

$$(E \cup F)^c = E^c F^c, \quad \left(\bigcup_{i=1}^{n} E_i\right)^c = \bigcap_{i=1}^{n} E_i^c, \quad \left(\bigcup_{i=1}^{\infty} E_i\right)^c = \bigcap_{i=1}^{\infty} E_i^c.$$

De Morgan's second law:

$$(EF)^c = E^c \cup F^c, \quad \left(\bigcap_{i=1}^{n} E_i\right)^c = \bigcup_{i=1}^{n} E_i^c, \quad \left(\bigcap_{i=1}^{\infty} E_i\right)^c = \bigcup_{i=1}^{\infty} E_i^c.$$

These and similar identities are usually proved by the *elementwise method*. The idea is to show that the events on both sides of the equation are formed of the same sample points. To use this method, we prove set inclusion in both directions. That is, sample points belonging to the event on the left also belong to the event on the right, and vice versa. An example follows.

Example 1.8 Prove De Morgan's first law: For E and F, two events of a sample space S, $(E \cup F)^c = E^c F^c$.

PROOF: First we show that $(E \cup F)^c \subseteq E^c F^c$, then we prove the reverse inclusion $E^c F^c \subseteq (E \cup F)^c$. To show that $(E \cup F)^c \subseteq E^c F^c$, let x be an outcome that belongs to $(E \cup F)^c$. Then x does not belong to $E \cup F$, meaning that x is neither in E nor in F. So x belongs to both E^c and F^c and hence to $E^c F^c$. To prove the reverse inclusion, let $x \in E^c F^c$. Then $x \in E^c$ and $x \in F^c$, implying that $x \notin E$ and $x \notin F$. Therefore, $x \notin E \cup F$ and thus $x \in (E \cup F)^c$. ♦

Note that Venn diagrams are excellent means to give intuitive justification for the validity or to create counterexamples and show invalidity of relations. However, they are not appropriate to prove relations. This is because of the large number of cases that must be considered (particularly if more than two events are involved). For example, suppose that by means of Venn diagrams, we want to prove the identity $(EF)^c = E^c \cup F^c$. First we must draw appropriate representations for all possible ways that E and F can be related: cases such as $EF = \emptyset$, $EF \neq \emptyset$, $E = F$, $E = \emptyset$, $F = S$, and so on. Then in each particular case we should find the regions that represent $(EF)^c$ and $E^c \cup F^c$ and observe that they are the same. Even if these two sets have different representations in only one case, the identity would be false.

EXERCISES

A

1. A box contains three red and five blue balls. Define a sample space for the experiment of recording the colors of three balls that are drawn from the box one by one, with replacement.

2. Define a sample space for the experiment of choosing a number from the interval $(0, 20)$. Describe the event that such a number is an integer.

3. Define a sample space for the experiment of putting three different books on a shelf in random order. If two of these three books are a

two-volume dictionary, describe the event that these volumes stand in increasing order side by side (i.e., Volume I precedes Volume II).

4. Two dice are rolled. Let E be the event that the sum of the outcomes is odd and F be the event of at least one 1. Interpret the events EF, E^cF, and E^cF^c.

5. Define a sample space for the experiment of drawing two coins from a purse that contains two quarters, three nickels, one dime, and four pennies.   For the same experiment describe the following events: drawing 26 cents; drawing more than 9 but less than 25 cents; drawing 29 cents.

6. A telephone call from a certain person is received sometime between 7:00 A.M. and 9:10 A.M. every day.   Define a sample space for this phenomenon and describe the event that the call arrives within 15 minutes of an hour.

7. Let E, F, and G be three events; explain the meaning of the relations $E \cup F \cup G = G$ and $EFG = G$.

8. A limousine that carries passengers from an airport to three different hotels just left the airport with two passengers. Describe the sample space of the stops and the event that both of the passengers get off at the same hotel.

9. Find the simplest possible expression for the following events.
   (a) $(E \cup F)(F \cup G)$.
   (b) $(E \cup F)(E^c \cup F)(E \cup F^c)$.

10. At a certain university, every year eight to 12 professors are granted University Merit Awards. This year among the nominated faculty are Drs. Jones, Smith, and Brown. Let A, B, and C denote the events that these professors will be given awards, respectively. In terms of A, B, and C, find an expression for the event that the award goes to (a) only Dr. Jones; (b) at least one of the three; (c) none of the three; (d) exactly two of them; (e) exactly one of them; (f) Drs. Jones or Smith but not both.

11. Prove that the event B is impossible if and only if for every event A, $A = (B \cap A^c) \cup (B^c \cap A)$.

12. Let E, F, and G be three events. Determine which of the following statements are correct and which are incorrect. Justify your answers.
   (a) $(E - EF) \cup F = E \cup F$.
   (b) $F^cG \cup E^cG = G(F \cup E)^c$.
   (c) $(E \cup F)^cG = E^cF^cG$.
   (d) $EF \cup EG \cup FG \subset E \cup F \cup G$.

B

13. Prove De Morgan's second law, $(AB)^c = A^c \cup B^c$, (a) by elementwise proof; (b) by applying De Morgan's first law to A^c and B^c.

14. Let A and B be two events. Prove the following relations by the elementwise method.
    (a) $(A - AB) \cup B = A \cup B$.
    (b) $(A \cup B) - AB = AB^c \cup A^c B$.

15. Let $\{A_n\}_{n=1}^{\infty}$ be a sequence of events. Prove that for every event B:
    (a) $B(\bigcup_{i=1}^{\infty} A_i) = \bigcup_{i=1}^{\infty} BA_i$.
    (b) $B \bigcup (\bigcap_{i=1}^{\infty} A_i) = \bigcap_{i=1}^{\infty} (B \cup A_i)$.

16. Define a sample space for the experiment of putting seven different books on a shelf in random order. If three of these seven books are a three-volume dictionary, describe the event that these volumes stand in increasing order side by side (i.e., Volume I precedes Volume II and Volume II precedes Volume III).

17. Let $\{A_1, A_2, A_3, \ldots\}$ be a sequence of events. Find an expression for the event that "infinitely many of the A_i's occur."

18. Let $\{A_1, A_2, A_3, \ldots\}$ be a sequence of events of a sample space S. Find a sequence $\{B_1, B_2, B_3, \ldots\}$ of mutually exclusive events such that for all $n \geq 1$, $\bigcup_{i=1}^{n} A_i = \bigcup_{i=1}^{n} B_i$.

1.3   Axioms of Probability

In mathematics, the goals of researchers are to obtain new results and prove their correctness, create simple proofs for already established results, discover or create connections between different fields of mathematics, construct and solve mathematical models for real-world problems, and so on. To discover new results, mathematicians use trial and error, instinct and inspired guessing, inductive analysis, studies of special cases, and other methods. But when a new result is discovered, its validity remains subject to skepticism until it is rigorously proven. Sometimes attempts to prove a result fail and contradictory examples are found. Such examples which invalidate a result are called *counterexamples*. No mathematical proposition is settled unless it is either proven or refuted by a counterexample. If a result is false, a counterexample exists to refute it. Similarly, if a result is valid, a proof must be found for its validity, although in some cases it might take years, decades, or even centuries to find it.

Proofs in probability theory (and virtually any other theory) are done in the framework of the *axiomatic method*. By this method, if we want to convince any rational person, say Sonya, that a statement L_1 is correct, we will show her how L_1 can be deduced logically from another statement L_2 that might be acceptable to her. However, if Sonya does not accept L_2, we should demonstrate how L_2 can be deduced logically from a simpler statement L_3. If she disputes L_3, we must continue this process until somewhere along the way we reach a statement that, without further justification, is acceptable to her. This statement will then become the basis of our argument. Its existence is necessary since otherwise the process continues ad infinitum without any conclusions. Therefore, in the axiomatic method, first we adopt certain simple, indisputable, and consistent statements without justifications. These are *axioms* or *postulates*. Then we agree on how and when one statement is a logical consequence of another one, and finally using the terms which are already clearly understood, axioms, and definitions, we obtain new results. New results found in this manner are called *theorems*. Theorems are statements that can be proved. Upon establishment, they are used for discovery of new theorems and the process continues and a theory evolves.

In this book, our approach is based on the axiomatic method. There are three axioms upon which probability theory is based, and except for them, everything else needs to be proved. We will now explain these axioms.

Let S be the sample space of an experiment. To each event A of S a number, called the *probability* of A and denoted by $P(A)$, is associated with A and satisfies the following axioms.

Axiom 1 $P(A) \geq 0$.

Axiom 2 $P(S) = 1$.

Axiom 3 *If $A_1, A_2, A_3, \ldots$ is a sequence of mutually exclusive events (i.e., the joint occurrence of every pair of them is impossible: $A_i A_j = \emptyset$ when $i \neq j$), then*

$$P\left(\bigcup_{i=1}^{\infty} A_i\right) = \sum_{i=1}^{\infty} P(A_i).$$

Axiom 1 states that the probability of the occurrence of an event is always nonnegative. Axiom 2 guarantees that the probability of the occurrence of the event S which is certain is 1. Axiom 3 states that for a sequence of mutually exclusive events the probability of the occurrence of at least one of them is equal to the sum of their probabilities.

Axiom 2 is merely a convenience to make things definite. It would be equally reasonable to have $P(S) = 100$ and interpret probabilities as percents (which we frequently do).

Let S be the sample space of an experiment. Let A and B be events of S. We say that A and B are *equally likely* if $P(A) = P(B)$. Let ω_1 and

ω_2 be sample points of S. We say that ω_1 and ω_2 are *equally likely* if the events $\{\omega_1\}$ and $\{\omega_2\}$ are equally likely, that is, if $P(\{\omega_1\}) = P(\{\omega_2\})$.

We will now prove some immediate implications of the axioms of probability.

THEOREM 1.1 *The probability of the empty set $\emptyset$ is 0. That is, $P(\emptyset) = 0$.*

PROOF: Let $A_1 = S$ and $A_i = \emptyset$ for $i \geq 2$; then $A_1, A_2, A_3, \ldots$ is a sequence of mutually exclusive events. Thus by Axiom 3,

$$P(S) = P\left(\bigcup_{i=1}^{\infty} A_i\right) = \sum_{i=1}^{\infty} P(A_i) = P(S) + \sum_{i=2}^{\infty} P(\emptyset),$$

implying that $\sum_{i=2}^{\infty} P(\emptyset) = 0$. This is possible only if $P(\emptyset) = 0$. ◆

THEOREM 1.2 *Let $\{A_1, A_2, \ldots, A_n\}$ be a mutually exclusive set of events, then*

$$P\left(\bigcup_{i=1}^{n} A_i\right) = \sum_{i=1}^{n} P(A_i).$$

PROOF: For $i > n$, let $A_i = \emptyset$. Then $A_1, A_2, A_3, \ldots$ is a sequence of mutually exclusive events. From Axiom 3 and Theorem 1.1 we get

$$P\left(\bigcup_{i=1}^{n} A_i\right) = P\left(\bigcup_{i=1}^{\infty} A_i\right) = \sum_{i=1}^{\infty} P(A_i)$$

$$= \sum_{i=1}^{n} P(A_i) + \sum_{i=n+1}^{\infty} P(A_i) = \sum_{i=1}^{n} P(A_i) + \sum_{i=n+1}^{\infty} P(\emptyset)$$

$$= \sum_{i=1}^{n} P(A_i). ◆$$

Letting $n = 2$, Theorem 1.2 implies that if A_1 and A_2 are mutually exclusive, then

$$P(A_1 \cup A_2) = P(A_1) + P(A_2). \tag{1.2}$$

The intuitive meaning of (1.2) is that if an experiment can be repeated indefinitely, then for two mutually exclusive events A_1 and A_2, the proportion of times that $A_1 \cup A_2$ occurs is equal to the sum of the proportion of times that A_1 occurs and the proportion of times that A_2 occurs. For example, for the experiment of tossing a fair die, $S = \{1, 2, 3, 4, 5, 6\}$ is the sample space. Let A_1 be the event that the outcome is 6 and A_2 be the event that the outcome is odd. Then $A_1 = \{6\}$ and $A_2 = \{1, 3, 5\}$. Since all sample points are equally likely to occur (by the definition of a fair die) and the number of sample points of A_1 is 1/6 of the number of sample points of S, we expect that $P(A_1) = 1/6$. Similarly, we expect that $P(A_2) = 3/6$. Now $A_1 A_2 = \emptyset$ implies that the number of sample points of $A_1 \cup A_2$ is $(1/6 + 3/6)$th of the number of sample points of S. Hence we should expect that $P(A_1 \cup A_2) = 1/6 + 3/6$, which is the same as $P(A_1) + P(A_2)$. This and many other examples suggest that (1.2) is a reasonable relation to be taken as Axiom 3. However, if we do this, difficulties arise when a sample space contains infinitely many sample points, that is, when the number of possible outcomes of an experiment is not finite. For example, in successive throws of a die let A_n be the event that the first 6 occurs on the nth throw. Then we would be unable to find the probability of $\bigcup_{n=1}^{\infty} A_n$ which represents the event that eventually a 6 occurs. For this reason, Axiom 3, which is the infinite analog of (1.2), is required. It by no means contradicts our intuitive ideas of probability and one of its great advantages is that Theorems 1.1 and 1.2 are its immediate consequences.

A significant implication of (1.2) is that for any event A, $P(A) \leq 1$. To see this, note that $P(A \cup A^c) = P(A) + P(A^c)$. Now by Axiom 2, $P(A \cup A^c) = P(S) = 1$; therefore, $P(A) + P(A^c) = 1$. This and Axiom 1 imply that $P(A) \leq 1$. Hence *the probability of the occurrence of an event is always some number between 0 and 1*.

REMARK: Let S be the sample space of an experiment. The set of all subsets of S is denoted by $\mathcal{P}(S)$ and is called the *power set* of S. Since the aim of probability theory is to associate a number between 0 and 1 to every subset of the sample space, probability is a function P from $\mathcal{P}(S)$ to $[0, 1]$. However, in theory, there is one exception to this: If the sample space S is not countable, not all of the elements of $\mathcal{P}(S)$ are events. There are elements of $\mathcal{P}(S)$ that are not in some sense (defined in more advanced probability texts) *measurable*. These elements are not events. Since in real-world problems we are only dealing with those elements of $\mathcal{P}(S)$ that are measurable, we are not concerned with these exceptions. In general, if the domain of a function is a collection of sets, it is called a *set function*. Hence probability is a real-valued, nonnegative set function. ♦

Example 1.9 A coin is called *unbiased* or *fair* if whenever it is flipped, the probability of obtaining heads equals that of tails. Suppose that in an experiment an unbiased coin is flipped. The sample space of such an experiment is $S = \{T, H\}$. Since the events $\{ H \}$ and $\{ T \}$ are equally likely to occur, $P(\{T\}) = P(\{H\})$, and since they are mutually exclusive,

$$P(\{T, H\}) = P(\{T\}) + P(\{H\}).$$

Hence Axioms 2 and 3 imply that

$$1 = P(S) = P(\{H, T\}) = P(\{H\}) + P(\{T\})$$
$$= P(\{H\}) + P(\{H\}) = 2P(\{H\}).$$

This gives that $P(\{H\}) = 1/2$ and $P(\{T\}) = 1/2$. Now suppose that an experiment consists of flipping a biased coin where the outcome of tails is twice as likely as heads, then $P(\{T\}) = 2P(\{H\})$. Hence

$$1 = P(S) = P(\{H, T\}) = P(\{H\}) + P(\{T\})$$
$$= P(\{H\}) + 2P(\{H\}) = 3P(\{H\}).$$

This shows that $P(\{H\}) = 1/3$, thus $P(\{T\}) = 2/3$. ◆

Example 1.10 Sharon has made five loaves of bread that are identical except that one of them is underweight. Sharon's husband chooses one of these loaves at random. Let B_i, $1 \leq i \leq 5$, be the event that he chooses the ith loaf. Since all five loaves are equally likely to be drawn, we have

$$P(\{B_1\}) = P(\{B_2\}) = P(\{B_3\}) = P(\{B_4\}) = P(\{B_5\}).$$

But the events $\{B_1\}, \{B_2\}, \{B_3\}, \{B_4\}, \{B_5\}$ are mutually exclusive and the sample space is $S = \{B_1, B_2, B_3, B_4, B_5\}$. Therefore, by Axioms 2 and 3,

$$1 = P(S) = P(\{B_1\}) + P(\{B_2\}) + P(\{B_3\}) + P(\{B_4\}) + P(\{B_5\})$$
$$= 5 \cdot P(\{B_1\}).$$

This gives $P(\{B_1\}) = 1/5$ and hence $P(\{B_i\}) = 1/5$, $1 \leq i \leq 5$. Therefore, the probability that Sharon's husband chooses the underweight loaf is $1/5$. ◆

From Examples 1.9 and 1.10 it should be clear that if a sample space contains N points which are equally likely to occur, then the probability of each outcome (sample point) is $1/N$. In general, this can be shown as follows. Let $S = \{s_1, s_2, \ldots, s_N\}$ be the sample space of an experiment; then if all of the sample points are equally likely to occur, we have

$$P(\{s_1\}) = P(\{s_2\}) = \cdots = P(\{s_N\}).$$

But $P(S) = 1$ and the events $\{s_1\}, \{s_2\}, \ldots, \{s_N\}$ are mutually exclusive. Therefore,

$$
\begin{aligned}
1 = P(S) &= P(\{s_1, s_2, \ldots, s_N\}) \\
&= P(\{s_1\}) + P(\{s_2\}) + \cdots + P(\{s_N\}) \\
&= NP(\{s_1\}).
\end{aligned}
$$

This shows that $P(\{s_1\}) = 1/N$. Thus $P(\{s_i\}) = 1/N$ for $1 \le i \le N$.

One simple consequence of the axioms of probability is that if the sample space of an experiment contains N points which are all equally likely to occur, then the probability of the occurrence of any event A is equal to the number of points of A, say $N(A)$, divided by N. Historically, until the introduction of the axiomatic method by A. N. Kolmogorov in 1933, this fact was taken as the definition of the probability of A. It is now called *the classical definition of probability*. The following theorem, which shows that the classical definition is a simple result of the axiomatic approach, is also an important tool for computation of probabilities of events of experiments with finite sample spaces.

THEOREM 1.3 *Let S be the sample space of an experiment. If S has N points that are all equally likely to occur, then for any event A of S,*

$$P(A) = \frac{N(A)}{N},$$

where $N(A)$ is the number of points of A.

PROOF: Let $S = \{s_1, s_2, \ldots, s_N\}$, where each s_i is an outcome (a sample point) of the experiment. Since the outcomes are equiprobable, $P(\{s_i\}) = 1/N$ for all i, $1 \le i \le N$. Now let $A = \{s_{i_1}, s_{i_2}, \ldots, s_{i_{N(A)}}\}$, where $s_{i_j} \in S$ for all i_j. Since $\{s_{i_1}\}, \{s_{i_2}\}, \ldots, \{s_{i_{N(A)}}\}$ are mutually exclusive, Axiom 3 implies that

$$
\begin{aligned}
P(A) &= P(\{s_{i_1}, s_{i_2}, \ldots, s_{i_{N(A)}}\}) \\
&= P(\{s_{i_1}\}) + P(\{s_{i_2}\}) + \cdots + P(\{s_{i_{N(A)}}\}) \\
&= \underbrace{\frac{1}{N} + \frac{1}{N} + \cdots + \frac{1}{N}}_{N(A) \text{ terms}} = \frac{N(A)}{N}. \quad \blacklozenge
\end{aligned}
$$

Example 1.11 Let S be the sample space of "flipping a fair coin three times" and A be the event of "at least two heads"; then

$$S = \{HHH, HTH, HHT, HTT, THH, THT, TTH, TTT\}$$

and $A = \{HHH, HTH, HHT, THH\}$. So $N = 8$ and $N(A) = 4$. Therefore, the probability of at least two heads in flipping a fair coin three times is $N(A)/N = 4/8 = 1/2$. ◆

Example 1.12 An elevator with two passengers stops at the second, third, and fourth floors. If it is equally likely that a passenger gets off at any of the three floors, what is the probability that the passengers get off at different floors?

SOLUTION: Let a and b denote the two passengers and a_2b_4 mean that a gets off at the second floor and b gets off at the fourth floor with similar representations for other cases. Let A be the event that the passengers get off at different floors. Then

$$S = \{a_2b_2, a_2b_3, a_2b_4, a_3b_2, a_3b_3, a_3b_4, a_4b_2, a_4b_3, a_4b_4\}$$

and $A = \{a_2b_3, a_2b_4, a_3b_2, a_3b_4, a_4b_2, a_4b_3\}$. So $N = 9$ and $N(A) = 6$. Therefore, the desired probability is $N(A)/N = 6/9 = 2/3$. ◆

Example 1.13 A number is selected at random from the set of natural numbers $\{1, 2, 3, 4, \ldots, 1000\}$. What is the probability that it is divisible by 3?

SOLUTION: Here the sample space contains 1000 points, so $N = 1000$. Let A be the set of all numbers between 1 and 1000 which are divisible by 3. Then $A = \{3m : 1 \leq m \leq 333\}$. So $N(A) = 333$. Therefore, the probability that a random natural number between 1 and 1000 is divisible by 3 equals $333/1000$. ◆

Example 1.14 A number is selected at random from the set $\{1, 2, \ldots, N\}$. What is the probability that it is divisible by k, $1 \leq k \leq N$?

SOLUTION: Here the sample space contains N points. Let A be the event that the outcome is divisible by k. Then $A = \{km : 1 \leq m \leq [N/k]\}$, where $[N/k]$ is the greatest integer less than or equal to N/k (to compute $[N/k]$, just divide N by k and round down). So $N(A) = [N/k]$ and $P(A) = [N/k]/N$. ◆

REMARK: As explained in the remark following Example 1.5, different manifestations of outcomes of an experiment might lead to different representations for the sample space of the same experiment. Because of this, different sample points of a representation might not have the same probability of occurrence. For example, suppose that a study is being done on families with three children. Let the outcomes of the study be the number of girls and the number of boys in a randomly selected family. Then

$$S = \{bbb, bgg, bgb, bbg, ggb, gbg, gbb, ggg\}$$

and

$$\Omega = \{bbb, bbg, bgg, ggg\}$$

are both reasonable sample spaces for the genders of the children of the family. In S, for example, bgg means that the first child of the family is a boy, the second child is a girl, and the third child is also a girl. In Ω, bgg means that the family has one boy and two girls. Therefore, in S all sample points occur with the same probability, namely 1/8. In Ω, however, probabilities associated to the sample points are not equal: $P\{bbb\} = P\{ggg\} = 1/8$, but $P\{bbg\} = P\{bgg\} = 3/8$. ◆

1.4   Basic Theorems

THEOREM 1.4   *For any event A, $P(A^c) = 1 - P(A)$.*

PROOF: Since $AA^c = \emptyset$, A and A^c are mutually exclusive. Thus $P(A \cup A^c) = P(A) + P(A^c)$. But $A \cup A^c = S$ and $P(S) = 1$, so $1 = P(S) = P(A \cup A^c) = P(A) + P(A^c)$. Therefore, $P(A^c) = 1 - P(A)$. ◆

This theorem states that the probability of nonoccurrence of the event A is 1 minus the probability of its occurrence. For example, consider $S = \{(i, j) : 1 \le i \le 6, \ 1 \le j \le 6\}$, the sample space of tossing two fair dice. If A is the event of getting a sum of 4, then $A = \{(1, 3), (2, 2), (3, 1)\}$ and $P(A) = 3/36$. Theorem 1.4 states that the probability of A^c, the event of not getting a sum of 4, which is harder to count, is $1 - 3/36 = 33/36$. As another example, consider the experiment of selecting a random number from the set $\{1, 2, 3, \ldots, 1000\}$. By Example 1.13, the probability that the number selected is divisible by 3 is 333/1000. Thus by Theorem 1.4, the probability that it is not divisible by 3, a quantity harder to find directly, is $1 - 333/1000 = 667/1000$.

THEOREM 1.5   *If $A \subseteq B$, then $P(B - A) = P(BA^c) = P(B) - P(A)$.*

PROOF: $A \subseteq B$ implies that $B = (B - A) \cup A$ (see Figure 1.2). But $(B - A)A = \emptyset$. So the events $B - A$ and A are mutually exclusive and

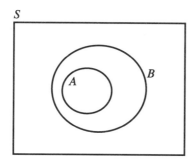

Figure 1.2 $A \subseteq B$ implies that $B = (B - A) \cup A$.

$P(B) = P((B - A) \cup A) = P(B - A) + P(A)$. This gives $P(B - A) = P(B) - P(A)$. ◆

COROLLARY   *If $A \subseteq B$, then $P(A) \leq P(B)$.*

PROOF: By Theorem 1.5, $P(B - A) = P(B) - P(A)$. Since $P(B - A) \geq 0$, we have that $P(B) - P(A) \geq 0$. Hence $P(B) \geq P(A)$. ◆

This corollary says that, for instance, it is less likely that a computer has one defect than it has at least one defect. Note that in Theorem 1.5, the condition of $A \subseteq B$ is necessary. The relation $P(B - A) = P(B) - P(A)$ is not true in general. For example, in rolling a fair die, let $B = \{1, 2\}$ and $A = \{3, 4, 5\}$, then $B - A = \{1, 2\}$. Therefore, $P(B - A) = 1/3$, $P(B) = 1/3$, and $P(A) = 1/2$. Hence $P(B - A) \neq P(B) - P(A)$.

THEOREM 1.6 $P(A \cup B) = P(A) + P(B) - P(AB)$.

PROOF: $A \cup B = A \cup (B - AB)$ (see Figure 1.3) and $A(B - AB) = \emptyset$, so A and $B - AB$ are mutually exclusive events and

$$P(A \cup B) = P(A \cup (B - AB)) = P(A) + P(B - AB). \qquad (1.3)$$

Now since $AB \subseteq B$, Theorem 1.5 implies that $P(B - AB) = P(B) - P(AB)$. Therefore, (1.3) gives $P(A \cup B) = P(A) + P(B) - P(AB)$. ◆

Example 1.15   Suppose that in a community of 400 adults, 300 bike or swim or do both, 160 swim, and 120 swim and bike. What is the probability that an adult selected at random from this community bikes?

SOLUTION: Let A be the event that the person swims and B be the event that he or she bikes, then $P(A \cup B) = 300/400$, $P(A) = 160/400$, $P(AB) =$

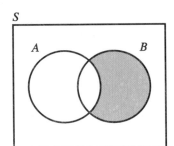

Figure 1.3 $A \cup B = A \cup (B - AB)$.

120/400. Hence the relation $P(A \cup B) = P(A) + P(B) - P(AB)$ implies that

$$P(B) = P(A \cup B) + P(AB) - P(A)$$

$$= \frac{300}{400} + \frac{120}{400} - \frac{160}{400} = \frac{260}{400} = 0.65. \quad \blacklozenge$$

Example 1.16   A number is chosen at random from the set of numbers $\{1, 2, 3, \ldots, 1000\}$. What is the probability that it is divisible by 3 or 5 (i.e., either 3 or 5 or both)?

SOLUTION: The number of integers between 1 and N that are divisible by k is computed by dividing N by k and then rounding down (see Examples 1.13 and 1.14). Therefore, if A is the event that the outcome is divisible by 3 and B is the event that it is divisible by 5, then $P(A) = 333/1000$ and $P(B) = 200/1000$. Now AB is the event that the outcome is divisible by both 3 and 5. Since a number is divisible by 3 and 5 if and only if it is divisible by 15 (3 and 5 are prime numbers), $P(AB) = 66/1000$ (divide 1000 by 15 and round down to get 66). Thus the desired probability is computed as follows:

$$P(A \cup B) = P(A) + P(B) - P(AB)$$

$$= \frac{333}{1000} + \frac{200}{1000} - \frac{66}{1000} = \frac{467}{1000}. \quad \blacklozenge$$

Theorem 1.6 gives a formula to calculate the probability that at least one of A and B occurs. We may also calculate the probability that at least one of the events $A_1, A_2, A_3, \ldots,$ and A_n occurs. For three events $A_1, A_2,$ and A_3,

$$P(A_1 \cup A_2 \cup A_3) = P(A_1) + P(A_2) + P(A_3) - P(A_1 A_2)$$

$$- P(A_1 A_3) - P(A_2 A_3) + P(A_1 A_2 A_3).$$

For four events,

$$P(A_1 \cup A_2 \cup A_3 \cup A_4) = P(A_1) + P(A_2) + P(A_3) + P(A_4) - P(A_1A_2)$$
$$- P(A_1A_3) - P(A_1A_4) - P(A_2A_3) - P(A_2A_4)$$
$$- P(A_3A_4) + P(A_1A_2A_3) + P(A_1A_2A_4)$$
$$+ P(A_1A_3A_4) + P(A_2A_3A_4) - P(A_1A_2A_3A_4).$$

In general, *to calculate $P(A_1 \cup A_2 \cup \cdots \cup A_n)$, first find all of the possible intersections of events from $A_1, A_2, \ldots, A_n$ and calculate their probabilities. Then add the probabilities of those intersections that are formed of an odd number of events and subtract the probabilities of those formed of an even number of events.* (This procedure is called the *inclusion–exclusion principle.*)

Example 1.17 Suppose that 25% of the population of a city read newspaper A, 20% read newspaper B, 13% read C, 10% read both A and B, 8% read both A and C, 5% read B and C, and 4% read all three. If a person from this city is selected at random, what is the probability that he or she does not read any of these newspapers?

SOLUTION: Let E, F, and G be the events that the person reads A, B, and C, respectively. The event that the person reads at least one of the newspapers A, B, and C is $E \cup F \cup G$. Therefore, $1 - P(E \cup F \cup G)$ is the probability that he or she reads none of them. Since

$$P(E \cup F \cup G) = P(E) + P(F) + P(G) - P(EF) - P(EG)$$
$$- P(FG) + P(EFG)$$
$$= 0.25 + 0.20 + 0.13 - 0.10 - 0.08 - 0.05 + 0.04 = 0.39,$$

the desired probability equals $1 - 0.39 = 0.61$. ◆

Example 1.18 Dr. Grossman, an internist, has 520 patients, of which (1) 230 are hypertensive, (2) 185 are diabetic, (3) 35 are hypochondriac and diabetic, (4) 25 are all three, (5) 150 are none, (6) 140 are only hypertensive, and finally, (7) 15 are hypertensive and hypochondriac but not diabetic. Find the probability that Dr. Grossman's next appointment is hypochondriac but neither diabetic nor hypertensive. Assume that appointments are all random. This implies that even hypochondriacs do not make more visits than others.

SOLUTION: Let T, C, and D denote the events that the next appointment of Dr. Grossman is hypertensive, hypochondriac, and diabetic, respectively.

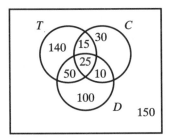

Figure 1.4   Venn diagram of Example 1.18.

The Venn diagram of Figure 1.4 shows that the number of patients with only hypochondria is 30. Therefore, the desired probability is $30/520 \approx 0.06$.   ◆

EXERCISES

A

1. Gottfried Wilhelm Leibniz (1646–1716), the German mathematician, philosopher, statesman, and one of the supreme intellects of the seventeenth century, believed that in a throw of a pair of fair dice, the probability of obtaining the sum 11 is equal to that of obtaining the sum 12. Do you agree with Leibniz? Explain.

2. Suppose that 33% of the people have O$^+$ blood and 7% have O$^-$. What is the probability that the next president of the United States has type O blood?

3. A company has only one position with three highly qualified applicants: John, Barbara, and Marty. However, because the company has only a few women employees, Barbara's chance to be hired is 20% higher than John's and 20% higher than Marty's. Find the probability that Barbara will be hired.

4. In a psychiatric hospital, the number of patients with schizophrenia is three times the number with psychoneurotic reactions, twice the number with alcohol addictions, and 10 times the number with involutional psychotic reaction. If a patient is selected from the list of all patients with one of these four diseases randomly, what is the probability that he or she suffers from schizophrenia? Assume that none of these patients has more than one of these four diseases.

5. Let A and B be two events. Prove that

$$P(AB) \geq P(A) + P(B) - 1.$$

6. A card is drawn from an ordinary deck of 52 cards at random. What is the probability that it is, (a) a black ace or a red queen; (b) a face or a black card; (c) neither a heart nor a queen?

7. Which of the following statements is true? If a statement is true, prove it. If it is false, give a counterexample.
   (a) If $P(A) + P(B) + P(C) = 1$, then the events A, B, and C are mutually exclusive.
   (b) If $P(A \cup B \cup C) = 1$, then A, B, and C are mutually exclusive events.

8. Suppose that in the Baltimore metropolitan area 25% of the crimes occur during the day and 80% of the crimes occur in the city. If only 10% of the crimes occur outside the city during the day, what percent occur inside the city during the night? What percent occur outside the city during the night?

9. Let A, B, and C be three events. Prove that

$$P(A \cup B \cup C)$$
$$= P(A) + P(B) + P(C) - P(AB) - P(AC) - P(BC) + P(ABC).$$

10. Eleven chairs are numbered 1 through 11. Four girls and seven boys sit on these chairs at random. What is the probability that chair 5 is occupied by a boy?

11. A ball is thrown at a square that is divided into n^2 identical squares. The probability that the ball hits the square of the ith column and jth row is p_{ij}, where $\sum_{i=1}^{n} \sum_{j=1}^{n} p_{ij} = 1$. In terms of p_{ij}'s, find the probability that the ball hits the jth horizontal strip.

12. Among 33 students in a class, 17 of them earned A's on the midterm exam, 14 earned A's on the final exam, and 11 did not earn A's on either examination. What is the probability that a randomly selected student from this class earned an A on both exams?

13. From a small town 120 persons were selected at random and asked the following question: Which of the three shampoos A, B, or C do you use? The following results were obtained: 20 use A and C, 10 use A and B but not C, 15 use all three, 30 use only C, 35 use B but not C, 25 use B and C, and 10 use none of the three. If a person is selected at random from this group, what is the probability that he or she uses (a) only A; (b) only B; (c) A and B? (Draw a Venn diagram.)

14. The coefficients of the quadratic equation $x^2 + bx + c = 0$ are determined by tossing a fair die twice (the first outcome is b, the second one is c). Find the probability that the equation has real roots.

15. Two integers m and n are called relatively prime if 1 is their only common positive divisor. Thus 8 and 5 are relatively prime, whereas 8 and 6 are not. A number is selected at random from the set $\{1, 2, 3, \ldots, 63\}$. Find the probability that it is relatively prime to 63.

16. A number is selected randomly from the set of natural numbers $\{1, 2, \ldots, 1000\}$. What is the probability that (a) it is divisible by 3 but not by 5; (b) it is divisible neither by 3 nor by 5?

17. The secretary of a college has calculated that from the students who took calculus, physics, and chemistry last semester, 78% passed calculus, 80% physics, 84% chemistry, 60% calculus and physics, 65% physics and chemistry, 70% calculus and chemistry, and 55% all three. Show that these numbers are not consistent and therefore the secretary has made a mistake.

B

18. From an ordinary deck of 52 cards, we draw cards at random and without replacement until only cards of one suit are left. Find the probability that the cards left are all spades.

19. A number is selected at random from the set of natural numbers $\{1, 2, \ldots, 1000\}$. What is the probability that it is divisible by 4 but neither by 5 nor by 7?

20. For a Democratic candidate to win an election, she must win districts I, II, and III. Polls have shown that the probability of winning I and III is 0.55, losing II but not I is 0.34, and losing II and III but not I is 0.15. Find the probability that this candidate wins all three districts. (Draw a Venn diagram.)

21. Two numbers are successively selected at random and with replacement from the set $\{1, 2, \ldots, 100\}$. What is the probability that the first one is greater than the second?

22. Let $A_1, A_2, A_3, \ldots$ be a sequence of events of a sample space. Prove that

$$P\left(\bigcup_{n=1}^{\infty} A_n\right) \le \sum_{n=1}^{\infty} P(A_n).$$

This is called *Boole's inequality.*

23. Let $A_1, A_2, A_3, \ldots$ be a sequence of events of an experiment. Prove that

$$P\left(\bigcap_{n=1}^{\infty} A_n\right) \geq 1 - \sum_{n=1}^{\infty} P(A_n^c).$$

HINT: Use Boole's inequality, discussed in Exercise 22.

1.5   Continuity of Probability Function

Let (here and everywhere else throughout the book) $\mathbf{R}$ denote the set of all real numbers. We know from calculus that a function $f: \mathbf{R} \to \mathbf{R}$ is called continuous at a point $c \in \mathbf{R}$ if $\lim_{x \to c} f(x) = f(c)$. It is called continuous on $\mathbf{R}$ if it is continuous at all points $c \in \mathbf{R}$. We also know that this definition is equivalent to the sequential criterion $f: \mathbf{R} \to \mathbf{R}$ *is continuous on* $\mathbf{R}$ *if and only if, for every convergent sequence* $\{x_n\}_{n=1}^{\infty}$ *in* $\mathbf{R}$,

$$\lim_{n \to \infty} f(x_n) = f(\lim_{n \to \infty} x_n). \tag{1.4}$$

This property, in some sense, is shared by the probability function. To explain this, we need to introduce some definitions. But first recall that probability is a set function from $\mathcal{P}(S)$, the set of all possible events of the sample space S, to $[0, 1]$.

A sequence $\{E_n, n \geq 1\}$ of events of a sample space is called *increasing* if

$$E_1 \subseteq E_2 \subseteq E_3 \subseteq \cdots \subseteq E_n \subseteq E_{n+1} \cdots;$$

it is called *decreasing* if

$$E_1 \supseteq E_2 \supseteq E_3 \supseteq \cdots \supseteq E_n \supseteq E_{n+1} \supseteq \cdots.$$

For an increasing sequence of events $\{E_n, n \geq 1\}$, by $\lim_{n \to \infty} E_n$ we mean the event that at least one E_i, $1 \leq i < \infty$ occurs. Therefore,

$$\lim_{n \to \infty} E_n = \bigcup_{n=1}^{\infty} E_n.$$

Similarly, for a decreasing sequence of events $\{E_n, n \geq 1\}$, by $\lim_{n \to \infty} E_n$ we mean the event that every E_i occurs. Thus in this case

$$\lim_{n \to \infty} E_n = \bigcap_{n=1}^{\infty} E_n.$$

The following theorem will express the property of probability function that is analogous to (1.4).

THEOREM 1.7   (Continuity of Probability Function)   *For any increasing or decreasing sequence of events, $\{E_n, n \geq 1\}$:*

$$\lim_{n \to \infty} P(E_n) = P(\lim_{n \to \infty} E_n).$$

PROOF: For the case where $\{E_n, n \geq 1\}$ is increasing, let $F_1 = E_1, F_2 = E_2 - E_1, F_3 = E_3 - E_2, \ldots, F_n = E_n - E_{n-1}, \ldots$. Clearly, $\{F_i, i \geq 1\}$ is a mutually exclusive set of events that satisfies the following relations:

$$\bigcup_{i=1}^{n} F_i = \bigcup_{i=1}^{n} E_i = E_n, \qquad n = 1, 2, 3, \ldots,$$

$$\bigcup_{i=1}^{\infty} F_i = \bigcup_{i=1}^{\infty} E_i.$$

Hence

$$P(\lim_{n \to \infty} E_n) = P\left(\bigcup_{i=1}^{\infty} E_i\right) = P\left(\bigcup_{i=1}^{\infty} F_i\right) = \sum_{i=1}^{\infty} P(F_i) = \lim_{n \to \infty} \sum_{i=1}^{n} P(F_i)$$

$$= \lim_{n \to \infty} P\left(\bigcup_{i=1}^{n} F_i\right) = \lim_{n \to \infty} P\left(\bigcup_{i=1}^{n} E_i\right) = \lim_{n \to \infty} P(E_n),$$

where the last equality follows since $\{E_n, n \geq 1\}$ is increasing and hence $\bigcup_{i=1}^{n} E_i = E_n$. This establishes the theorem for increasing sequences.

If $\{E_n, n \geq 1\}$ is decreasing, then $E_n \supseteq E_{n+1}, \forall n$, implies that $E_n^c \subseteq E_{n+1}^c, \forall n$. Therefore, the sequence $\{E_n^c, n \geq 1\}$ is increasing and

$$P(\lim_{n \to \infty} E_n) = P\left(\bigcap_{i=1}^{\infty} E_i\right) = 1 - P\left[\left(\bigcap_{i=1}^{\infty} E_i\right)^c\right] = 1 - P\left(\bigcup_{i=1}^{\infty} E_i^c\right)$$

$$= 1 - P(\lim_{n \to \infty} E_n^c) = 1 - \lim_{n \to \infty} P(E_n^c) = 1 - \lim_{n \to \infty}[1 - P(E_n)]$$

$$= 1 - 1 + \lim_{n \to \infty} P(E_n) = \lim_{n \to \infty} P(E_n). \quad \blacklozenge$$

Example 1.19   Suppose that in a population some individuals produce offspring of the same kind. The offspring of the initial population are called second generation, the offspring of the second generation are called third generation, and so on. If with probability $\exp[-(2n^2 + 7)/(6n^2)]$ the entire population by the nth generation completely dies out before producing any offspring, what is the probability that such a population survives forever?

SOLUTION: Let E_n denote the event of extinction of the entire population by the nth generation; then

$$E_1 \subseteq E_2 \subseteq E_3 \subseteq \cdots \subseteq E_n \subseteq E_{n+1} \subseteq \cdots$$

because if E_n occurs, then E_{n+1} also occurs. Hence

$$P\{\text{population survives forever}\} = 1 - P\{\text{population eventually dies out}\}$$

$$= 1 - P\left(\bigcup_{i=1}^{\infty} E_i\right) = 1 - \lim_{n\to\infty} P(E_n)$$

$$= 1 - \lim_{n\to\infty} \exp\left(-\frac{2n^2 + 7}{6n^2}\right) = 1 - e^{-1/3}. \blacklozenge$$

1.6   Probabilities 0 and 1

Events with probabilities 0 and 1 should not be misinterpreted. If E and F are events with probabilities 0 and 1, respectively, it is not correct to say that E is the sample space S and F is the empty set $\emptyset$. In fact, there are experiments in which there exist infinitely many events each with probability 1, and infinitely many events each with probability 0. An example follows.

Suppose that an experiment consists of selecting a random point from the interval $(0, 1)$. Since every point in $(0, 1)$ has a decimal representation such as

$$0.529387043219721\ldots,$$

the experiment is equivalent to picking an endless decimal from $(0, 1)$ at random (note that if a decimal terminates, all of its digits from some point on are 0). In such an experiment we want to compute the probability of selecting the point $1/3$. In other words, we want to compute the probability of choosing $0.333333\ldots$ in a random selection of an endless decimal. Let A_n be the event that the selected decimal has 3 as its first n digits; then

$$A_1 \supset A_2 \supset A_3 \supset A_4 \supset \cdots \supset A_n \supset A_{n+1} \supset \cdots,$$

since the occurrence of A_{n+1} guarantees the occurrence of A_n. Now $P(A_1) = 1/10$ because there are 10 choices $0, 1, 2, \ldots, 9$ for the first digit and we want only one of them, namely 3, to occur. $P(A_2) = 1/100$ since there are 100 choices $00, 01, \ldots, 09, 10, 11, \ldots, 19, 20, \ldots, 99$ for the first two digits and we want only one of them, 33, to occur. $P(A_3) = 1/1000$ because there are 1000 choices $000, 001, \ldots, 999$ for the first three digits and we want only one of them, 333, to occur. Continuing this argument, we have $P(A_n) = (1/10)^n$. Since $\bigcap_{n=1}^{\infty} A_n = \{1/3\}$, by Theorem 1.7,

$$P\left(\frac{1}{3} \text{ is selected}\right) = P\left(\bigcap_{n=1}^{\infty} A_n\right) = \lim_{n \to \infty} P(A_n) = \lim_{n \to \infty} \left(\frac{1}{10}\right)^n = 0.$$

Note that there is nothing special about the point $1/3$, for any other point $0.\alpha_1\alpha_2\alpha_3\alpha_4 \cdots$ from $(0, 1)$, the same argument could be used to show that the probability of its occurrence is 0 (define A_n to be the event that the first n digits of the selected decimal are $\alpha_1, \alpha_2, \ldots, \alpha_n$, respectively, and repeat the same argument). We have shown that in random selection of points from $(0, 1)$, the probability of the occurrence of any particular point is 0. Now for $t \in (0, 1)$, let $B_t = (0, 1) - \{t\}$. Then $P\{t\} = 0$ implies that

$$P(B_t) = P(\{t\}^c) = 1 - P\{t\} = 1.$$

Therefore, there are infinitely many events, B_t's, each with probability 1 and none equal to the sample space $(0, 1)$.

1.7 Random Selection of Points from Intervals

In Section 1.6, we showed that the probability of the occurrence of any particular point in a random selection of points from an interval (a, b) is 0. This implies immediately that if $[\alpha, \beta] \subseteq (a, b)$, then the events that the point falls in $[\alpha, \beta]$, (α, β), $[\alpha, \beta)$, and $(\alpha, \beta]$ are all equiprobable. Now consider the intervals $\left(a, \dfrac{a+b}{2}\right)$ and $\left(\dfrac{a+b}{2}, b\right)$, since $\dfrac{a+b}{2}$ is the midpoint of (a, b), it is reasonable to assume that

$$p_1 = p_2, \tag{1.5}$$

where p_1 is the probability that the point belongs to $\left(a, \dfrac{a+b}{2}\right)$ and p_2 is the probability that it belongs to $\left(\dfrac{a+b}{2}, b\right)$. The events that the random point belongs to $\left(a, \dfrac{a+b}{2}\right)$ and $\left(\dfrac{a+b}{2}, b\right)$ are mutually exclusive and

$$\left(a, \frac{a+b}{2}\right) \cup \left[\frac{a+b}{2}, b\right) = (a, b);$$

therefore,

$$p_1 + p_2 = 1.$$

This relation and (1.5) imply that

$$p_1 = p_2 = \frac{1}{2}.$$

Hence *the probability that a random point selected from (a, b) falls into the interval* $\left(a, \dfrac{a+b}{2}\right)$ *is 1/2. The probability that it falls into* $\left[\dfrac{a+b}{2}, b\right)$ *is also 1/2.* Note that the length of each of these intervals is 1/2 of the length of (a, b). Now consider the intervals $\left(a, \dfrac{2a+b}{3}\right]$, $\left(\dfrac{2a+b}{3}, \dfrac{a+2b}{3}\right]$, and $\left(\dfrac{a+2b}{3}, b\right)$. Since $\dfrac{2a+b}{3}$ and $\dfrac{a+2b}{3}$ are the points that divide the interval (a, b) into three subintervals with equal lengths, we can assume that

$$p_1 = p_2 = p_3, \tag{1.6}$$

where p_1, p_2, and p_3 are the probabilities that the point falls into $\left(a, \dfrac{2a+b}{3}\right]$, $\left(\dfrac{2a+b}{3}, \dfrac{a+2b}{3}\right]$, and $\left(\dfrac{a+2b}{3}, b\right)$, respectively. On the other hand, these three intervals are mutually disjoint and

$$\left(a, \frac{2a+b}{3}\right] \cup \left(\frac{2a+b}{3}, \frac{a+2b}{3}\right] \cup \left(\frac{a+2b}{3}, b\right) = (a, b).$$

Hence

$$p_1 + p_2 + p_3 = 1.$$

This relation and (1.6) imply that

$$p_1 = p_2 = p_3 = \frac{1}{3}.$$

Therefore, *the probability that a random point selected from (a, b) falls into the interval $\left(a, \dfrac{2a + b}{3}\right]$ is 1/3. The probability that it falls into $\left(\dfrac{2a + b}{3}, \dfrac{a + 2b}{3}\right]$ is 1/3, and the probability that it falls into $\left(\dfrac{a + 2b}{3}, b\right)$ is 1/3.* Note that the length of each of these intervals is 1/3 of the length of (a, b). These and other similar observations indicate that the probability of the event that a random point from (a, b) falls into a subinterval (α, β) is equal to $(\beta - \alpha)/(b - a)$.

Note that in this discussion we have assumed that subintervals of equal lengths are equiprobable. Even though two subintervals of equal lengths might differ by a finite or a countably infinite set (or even a set of measure zero), this assumption is still consistent with our intuitive understanding of choosing random points from intervals. This is because in such an experiment, the probability of the occurrence of a finite or countably infinite set (or a set of measure zero) is 0.

So far, we have based our discussion of selecting random points from intervals on our intuitive understanding of this experiment and not on a mathematical definition. Such discussions are often necessary for the creation of appropriate mathematical meanings for unclear concepts. The following definition, which is based on our intuitive analysis, gives an exact mathematical meaning to the experiment of random selection of points from intervals.

DEFINITION   *A point is said to be randomly selected from an interval (a, b) if any two subintervals of (a, b) that have the same length are equally likely to include the point. The probability associated with the event that the subinterval (α, β) contains the point is defined to be $(\beta - \alpha)/(b - a)$.*

As explained before, choosing a random number from $(0, 1)$ is equivalent to choosing all the decimal digits of the number successively. Since in practice this is impossible, choosing exact random points or numbers from $(0, 1)$ or any other interval is only a theoretical matter. Approximate random numbers, however, can be generated by computers. There are many algorithms for approximate random number generation, some of which are not accurate and do not generate good approximations. One good algorithm is given by Stephen Parker and Keith Miller in the paper "Random Number Generators: Good Ones Are Not Hard to Find," *Communications of ACM*, October 1988, Volume 31, Number 10. Simple mechanical tools can also be used to find such approximations. For example, consider a spinner mounted on a wheel of *unit circumference (radius $1/2\pi$)*. Let A be a point on the perimeter of the wheel. Each time that we flick the spinner, it stops,

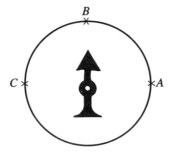

Figure 1.5   Spinner, a model to generate random numbers.

pointing toward some point B on the wheel's circumference. The length of the arc AB (directed, say, counterclockwise) is an approximate random number between 0 and 1 if the spinner does not have any "sticky" spots (see Figure 1.5).

Finally, when talking about the experiment of selecting a random point from an interval (a, b), we may think of an extremely large hypothetical box that contains infinitely many indistinguishable balls. Imagine that each ball is marked by a number from (a, b), each number of (a, b) is marked on exactly one ball, and the balls are completely mixed up, so that in a random selection of balls, any two of them have the same chance of being drawn. With this transcendental model in mind, choosing a random number from (a, b) is then equivalent to drawing a random ball from such a box and looking at its number.

EXERCISES

A

1. A bus arrives at a station every day at a random time between 1:00 P.M. and 1:30 P.M. What is the probability that a person arriving at this station at 1:00 P.M. will have to wait at least 10 minutes?

2. Past experience shows that every new book of a certain publisher captures randomly between 4 and 12% of the market. What is the probability that the next book of this publisher captures at most 6.35% of the market?

3. Which of the following statements are true? If a statement is true, prove it. If it is false, give a counterexample.
   (a) If A is an event with probability 1, then A is the sample space.
   (b) If B is an event with probability 0, then $B = \emptyset$.

4. Let A and B be two events. Show that if $P(A) = 1$ and $P(B) = 1$, then $P(AB) = 1$.

5. A point is selected at random from the interval $(0, 2000)$. What is the probability that it is an integer?

B

6. Let $A_1, A_2, \ldots, A_n$ be n events. Show that if $P(A_1) = P(A_2) = \cdots = P(A_n) = 1$, then $P(A_1 A_2 \cdots A_n) = 1$.

7. (a) Prove that $\bigcap_{n=1}^{\infty}(1/2 - 1/2n, 1/2 + 1/2n) = \{1/2\}$.
   (b) Using part (a), show that the probability of selecting $1/2$ in a random selection of a point from $(0, 1)$ is 0.

8. A point is selected at random from the interval $(0, 1)$. What is the probability that it is rational? What is the probability that it is irrational?

9. Let $\{A_1, A_2, A_3, \ldots\}$ be a sequence of events. Prove that if the series $\sum_{n=1}^{\infty} P(A_n)$ converges, then $P(\bigcap_{m=1}^{\infty} \bigcup_{n=m}^{\infty} A_n) = 0$. This is called the *Borel–Cantelli lemma*. It says that if $\sum_{n=1}^{\infty} P(A_n) < \infty$, the probability that infinitely many of the A_n's occur is 0.
   HINT: Let $B_m = \bigcup_{n=m}^{\infty} A_n$ and apply Theorem 1.7 to $\{B_m, m \geq 1\}$.

10. Show that the result of Exercise 6 is not true for an infinite number of events. That is, show that if $\{E_t : 0 < t < 1\}$ is a collection of events for which $P(E_t) = 1$, it is not necessarily true that $P\left(\bigcap_{t \in (0,1)} E_t\right) = 1$.

11. Let A be the set of rational numbers in $(0, 1)$. Since A is countable it can be written as a sequence [i.e., $A = \{r_n : n = 1, 2, 3, \ldots\}$]. Prove that for any $\varepsilon > 0$, A can be covered by a sequence of open balls whose total length is less than ε. That is, $\forall \varepsilon > 0$, there exists a sequence of open intervals (α_n, β_n) such that $r_n \in (\alpha_n, \beta_n)$ and $\sum_{n=1}^{\infty}(\beta_n - \alpha_n) < \varepsilon$. This important result explains why in a random selection of points from $(0, 1)$, the probability of choosing a rational is zero.
    HINT: Let $\alpha_n = r_n - \varepsilon/2^{n+2}$, $\beta_n = r_n + \varepsilon/2^{n+2}$.

Review Problems

1. The number of minutes it takes for a certain animal to react to a certain stimulus is a random number between 2 and 4.3. Find the probability that the reaction time of such an animal to this stimulus is no longer than 3.25 minutes.

2. Define a sample space for the experiment of choosing two distinct sets randomly from the set of all subsets of $A = \{1, 2\}$. Describe the following events: The intersection of these sets is empty; these two sets are complements of each other; one set contains more elements than the other.

3. In a certain experiment, whenever the event A occurs, the event B also occurs. Which of the following statements is true and why?
   (a) If we know that A has not occurred, we can be sure that B has not occurred as well.
   (b) If we know that B has not occurred, we can be sure that A has not occurred as well.

4. The following relations *are not* always true. In each case give an example to refute them.
   (a) $P(A \cup B) = P(A) + P(B)$.
   (b) $P(AB) = P(A)P(B)$.

5. A coin is tossed until for the first time the same result appears twice in succession. Define a sample space for this experiment.

6. The number of the current residents of a hospital is 63. Of these 37 are male and 20 are for surgery. If among those who are for surgery 12 are male, how many of the 63 patients are neither male nor for surgery?

7. Let A, B, and C be three events. Prove that

$$P(A \cup B \cup C) \leq P(A) + P(B) + P(C).$$

8. Let A, B, and C be three events. Show that

$$P(A \cup B \cup C) = P(A) + P(B) + P(C)$$

if and only if $P(AB) = P(AC) = P(BC) = 0$.

9. A bookstore receives six boxes of books per month on six random days of each month. Suppose that two of these boxes are from one publisher, two from another publisher, and the remaining two from a third publisher. Define a sample space for the possible orders in which these boxes are received in a given month by the bookstore. Describe the event that the last two boxes of books received last month are from the same publisher.

10. Suppose that in a certain town the number of people with blood type O and blood type A are approximately the same. The number of people with blood type B is 1/10 of those with blood type A and twice the number of those with blood type AB. Find the probability that the next baby born in this town has blood type AB.

11. A number is selected at random from the set of natural numbers $\{1, 2, 3, \ldots, 1000\}$. What is the probability that it is not divisible by 4, 7, or 9?

12. A number is selected at random from the set $\{1, 2, 3, \ldots, 150\}$. What is the probability that it is relatively prime to 150? See Exercise 15 of Section 1.4 for the definition of relatively prime numbers.

13. Suppose that each day the price of a stock moves up 1/8 of a point, moves down 1/8 of a point, or remains unchanged. For $i \geq 1$, let U_i and D_i be the events that the price of the stock moves up and down on the ith coming trading day, respectively. In terms of U_i's and D_i's find an expression for the event that the price of the stock market (a) remains unchanged on the ith coming trading day; (b) moves up on every day of the next n trading days; (c) remains unchanged on at least one of the next n trading days; (d) is the same as today after three trading days; (e) does not move down on any of the next n trading days.

14. A bus traveling from Baltimore to New York has a breakdown at a random location. What is the probability that it broke after passing through Philadelphia? The distances of New York and Philadelphia from Baltimore are, respectively, 199 and 96 miles.

15. The coefficient of the quadratic equation $ax^2 + bx + c = 0$ are determined by tossing a fair die three times (the first outcome is a, the second one b, and the third one c). Find the probability that the equation has no real roots.

Chapter 2

COMBINATORIAL METHODS

2.1 Introduction

In the study of many different areas of probability, such as simple games of chance, occupancy and order problems, and sampling procedures, we are usually dealing with finite sample spaces in which sample points are all equally likely. Theorem 1.3 shows that in such cases, the probability of an event A is evaluated simply by dividing the number of points of A by the total number of sample points. Therefore, some problems in probability can be solved just by counting the total number of sample points and the number of ways that an event can occur. In this chapter we study a few rules that enable us to count systematically. These rules come from an area in mathematics called *combinatorial analysis*, which deals with the methods of counting. It is a very broad field and has applications in almost every branch of applied and pure mathematics. Besides probability and statistics, it is used in information theory, coding and decoding, linear programming, transportation problems, industrial planning, scheduling production, group theory, foundations of geometry, and other fields. The history of combinatorial analysis, as a formal branch of mathematics, goes back to the time of Tartaglia in the sixteenth century. After Tartaglia, great mathematicians such as Pascal, Fermat, Chevalier Antoine de Méré (1607–1684), James Bernoulli, Gottfried Leibniz (1646–1716), and Leonhard Euler (1707–1783) made substantial contributions to this field until the twentieth century, when, because of its enormous applications, combinatorial analysis was developed very rapidly.

2.2   Counting Principle

Suppose that there are n routes from a town, A, to another town, B, and m routes from B to a third town, C. If we decide to go from A to C via B, then for each route that we choose from A to B, we have m choices from B to C. Therefore, altogether we have nm choices to go from A to C via B. This simple example motivates the following principle, which is the basis of this chapter.

THEOREM 2.1   (Counting Principle)   *If the set E contains n elements and the set F contains m elements, there are nm ways in which we can choose first an element of E and then an element of F.*

PROOF: Let $E = \{a_1, a_2, \ldots, a_n\}$ and $F = \{b_1, b_2, \ldots, b_m\}$; then the following rectangular array, which consists of nm elements, contains all possible ways in which we can choose first an element of E and then an element of F.

$$(a_1, b_1), \quad (a_1, b_2), \quad \ldots, \quad (a_1, b_m)$$
$$(a_2, b_1), \quad (a_2, b_2), \quad \ldots, \quad (a_2, b_m)$$
$$\vdots$$
$$(a_n, b_1), \quad (a_n, b_2), \quad \ldots, \quad (a_n, b_m) \quad \blacklozenge$$

Now suppose that a fourth town, D, is connected to C by l routes. If we decide to go from A to D passing through C after B, then for each pair of routes that we choose from A to C, there are l possibilities from C to D. Therefore, by the counting principle, the total number of ways we can go from A to D via B and C is the number of ways we can go from A to C through B times l, that is, nml. This motivates a generalization of the counting principle.

THEOREM 2.2   (Generalized Counting Principle)   *Let $E_1, E_2, \ldots,$ E_k be sets with $n_1, n_2, \ldots, n_k$ elements, respectively. Then there are $n_1 \times n_2 \times n_3 \times \cdots \times n_k$ ways in which we can first choose an element of E_1, then an element of E_2, then an element of $E_3, \ldots,$ and finally an element of E_k.*

In probability, this theorem is used whenever we want to compute the total number of possible outcomes when k experiments are performed. Suppose that the first experiment has n_1 possible outcomes, the second experiment has n_2 possible outcomes, $\ldots$, and the kth experiment has n_k possible outcomes. If we define E_i to be the set of all possible outcomes of the ith experiment, then the total number of possible outcomes coincides

b — # of outcomes /Trial
n — # of Trials
*b*ⁿ outcomes

with the number of ways that we can first choose an element of E_1, then an element of E_2, then an element of $E_3, \ldots$, and finally an element of E_k; that is, $n_1 \times n_2 \times \cdots \times n_k$.

Example 2.1   How many outcomes does the experiment of throwing five dice have?

SOLUTION: Let E_i, $1 \le i \le 5$, be the set of all possible outcomes of the ith die. Then $E_i = \{1, 2, 3, 4, 5, 6\}$. The number of the outcomes of the experiment of throwing five dice equals the number of ways we can first choose an element of E_1, then an element of $E_2, \ldots$, and finally, an element of E_5. Thus it is $6 \times 6 \times 6 \times 6 \times 6 = 6^5$. ◆

IMPORTANT REMARK: Consider experiments such as flipping a fair coin several times, tossing a number of fair dice, drawing a number of cards from an ordinary deck of 52 cards at random and with replacement, and drawing a number of balls from an urn at random and with replacement. In Section 3.4, when discussing the concept of *independence*, we will easily show that in such experiments all the possible outcomes are equiprobable. Until then, however, in all the problems dealing with this sort of experiments, we *assume*, without explicitly so stating in each case, that the sample points of the sample space of the experiment under consideration are all equally likely. ◆

Example 2.2   In tossing four fair dice, what is the probability of at least one 3?

SOLUTION: Let A be the event of at least one 3. Then A^c is the event of no 3 in tossing the four dice. $N(A^c)$ and N, the number of sample points of A^c and the total number of sample points, respectively, are given by $5 \times 5 \times 5 \times 5 = 5^4$ and $6 \times 6 \times 6 \times 6 = 6^4$. Therefore, $P(A^c) = N(A^c)/N = 5^4/6^4$. Hence $P(A) = 1 - P(A^c) = 1 - 625/1296 = 671/1296 \approx 0.52$. ◆

Example 2.3   Virginia wants to give her son, Brian, 14 different baseball cards within a seven-day period. If Virginia gives cards no more than once a day, in how many ways can this be done?

SOLUTION: Each of the baseball cards can be given on seven different days. Therefore, in $7 \times 7 \times \cdots \times 7 = 7^{14} \approx 6.78 \times 10^{11}$ ways Virginia can give the cards to Brian. ◆

Example 2.4   At a state university in Maryland, there is hardly enough space for students to park their cars in their own lots. Jack, a student who parks in the faculty parking lot every day, noticed that none of the last 10 tickets he got was issued on a Monday or on a Friday. Is it wise for Jack to conclude that the campus police do not patrol the faculty parking lot on Mondays and on Fridays? Assume that police give no tickets on weekends.

SOLUTION: Suppose that the answer is negative and the campus police patrol the parking lot randomly; that is, the parking lot is patrolled every day with the same probability. Let A be the event that out of 10 tickets given on random days, none is issued on a Monday or on a Friday. If $P(A)$ is very small, we can conclude that the campus police do not patrol the parking lot on these two days. Otherwise, we conclude that what happened is accidental and police patrol the parking lot randomly. To find $P(A)$, note that since each ticket has five possible days of being issued, there are 5^{10} possible ways for all tickets to have been issued. Of these in only 3^{10} ways no ticket is issued on a Monday or on a Friday. Thus $P(A) = 3^{10}/5^{10} \approx 0.006$, a rather small probability. Therefore, it is reasonable to have serious doubts that the campus police patrol the parking lot on these two days.   ♦

Example 2.5   (Standard Birthday Problem)   What is the probability that at least two students of a class of size n have the same birthday? Compute the numerical values of such probabilities for $n = 23, 30, 50$, and 60. Assume that the birth rates are constant throughout the year and that each year has 365 days.

SOLUTION: There are 365 possibilities for the birthdays of each of the n students. Therefore, the sample space has 365^n points. In $365 \times 364 \times 363 \times \cdots \times [365 - (n-1)]$ ways the birthdays of no two of the n students coincide. Hence $P(n)$, the probability that no two students have the same birthday, is

$$P(n) = \frac{365 \times 364 \times 363 \times \cdots \times [365 - (n-1)]}{365^n},$$

and therefore the desired probability is $1 - P(n)$. For $n = 23, 30, 50$, and 60 the answers are 0.507, 0.706, 0.970, and 0.995, respectively.   ♦

REMARK: In the studies of probability and statistics, birthday problems similar to Example 2.5 have been very popular since 1939, when introduced by von Mises. This is probably because when solving such problems, numerical values obtained are often surprising. Persi Diaconis and Frederick Mosteller, two Harvard professors, have mentioned that they "find the utility of birthday problems impressive as a tool for thinking about coincidences." Diaconis and Mosteller have illustrated basic statistical techniques for studying the fascinating, curious, and complicated *Theory of Coincidences* in the December 1989 issue of the *Journal of the American Statistical Association*. In their study, they have used birthday problems "as examples which make the point that in *many problems our intuitive grasp of the odds is far off."* Throughout this book, when appropriate, we will bring up some interesting versions of these problems, but now that we have cited "coincidence," let us read a few sentences from the abstract of the above-mentioned paper to get a better feeling for its meaning.

> ... Once we set aside coincidences having apparent causes, four principles account for large numbers of remaining coincidences: hidden cause; psychology, including memory and perception; multiplicity of endpoints, including the counting of "close" or nearly alike events as if they were identical; and the law of truly large numbers, which says that when enormous numbers of events and people and their interactions cumulate over time, almost any outrageous event is bound to occur. These sources account for much of the force of synchronicity. ♦

Number of Subsets of a Set

As an important application of the generalized counting principle, we now prove the following theorem, which has lots of good applications itself.

THEOREM 2.3   *A set with n elements has 2^n subsets.*

PROOF: Let $A = \{a_1, a_2, a_3, \ldots, a_n\}$ be a set with n elements. Then there is a one-to-one correspondence between the subsets of A and the sequences of 0's and 1's of length n: To a subset B of A we associate a sequence $b_1 b_2 b_3 \cdots b_n$, where $b_i = 0$ if $a_i \notin B$ and $b_i = 1$ if $a_i \in B$. For example, if $n = 3$, we associate to the empty subset of A the sequence 000, to $\{a_2, a_3\}$ the sequence 011, and to $\{a_1\}$ the sequence 100. Now by the generalized counting principle the number of sequences of 0's and 1's of length n is $2 \times 2 \times 2 \times \cdots \times 2 = 2^n$. Thus the number of subsets of A is also 2^n. ♦

Example 2.6   A restaurant has advertised that it offers over 1000 varieties of pizza. If, at the restaurant, it is possible to have on a pizza any combination of pepperoni, mushrooms, sausage, green peppers, onions, anchovies, salami, bacon, olives, and ground beef, is the restaurant's advertisement true?

SOLUTION: Any combination of the 10 ingredients that the restaurant offers can be put on a pizza. Thus the number of different types of pizza that it is possible to make is equal to the number of subsets of the set {pepperoni, mushrooms, sausage, green peppers, onions, anchovies, salami, bacon, olives, ground beef}, which is $2^{10} = 1024$. Therefore, the restaurant's advertisement is true. Note that the empty subset of the set of ingredients corresponds to a plain cheese pizza. ♦

Tree Diagrams

Tree diagrams are useful pictorial representations of the breakdown of a complex counting problem into smaller, more tractable cases. They are used in situations where the number of possible ways an experiment can be performed is finite. The following examples illustrate how tree diagrams are constructed and why they are useful. A great advantage of tree diagrams is that they identify all possible cases systematically.

Example 2.7   Bill and John keep playing chess until one of them wins two games in a row or three games altogether. In what percent of all possible cases does the game end because Bill wins three games without winning two in a row?

SOLUTION: The tree diagram of Figure 2.1 illustrates all possible cases. The total number of possible cases is equal to the number of the endpoints of the branches, which is 10. The number of cases in which Bill wins three games without winning two in a row, as seen from the figure, is one. So the answer is 10%. Note that the probability of this event is not 0.10 because not all of the branches of the tree are equiprobable.   ◆

Example 2.8   Mark has $4. He decides to bet $1 on the flip of a fair coin four times. What is the probability that (a) he breaks even; (b) he gains money?

SOLUTION: The tree diagram of the Figure 2.2 illustrates various things that can happen to Mark. The diagram has 16 endpoints, showing that the sample space has 16 elements. In six of these 16 cases, Mark breaks even and in five cases he gains money, so the desired probabilities are 6/16 and 5/16, respectively.   ◆

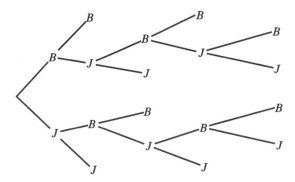

Figure 2.1   Tree diagram of Example 2.7.

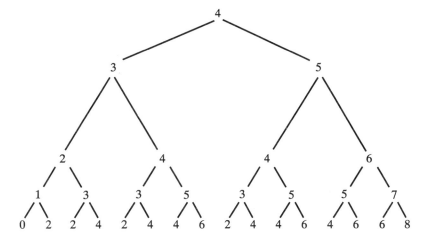

Figure 2.2   Tree diagram of Example 2.8.

EXERCISES

A

1. How many six-digit numbers are there? How many of them contain the digit 5?

2. How many different five-letter codes can be made using *a*, *b*, *c*, *d*, and *e*? How many of them start with *ab*?

3. The population of a town is 20,000. If each resident has three initials, is it true that at least two people have the same initials?

4. In how many different ways can 15 offices be painted with four different colors?

5. In flipping a fair coin 23 times, what is the probability of all heads or all tails?

6. In how many ways can we draw five cards from an ordinary deck of 52 cards (a) with replacement; (b) without replacement?

7. Two fair dice are thrown. What is the probability that the outcome is a 6 and an odd number?

8. Mr. Smith has 12 shirts, eight pairs of slacks, eight ties, and four jackets. If four shirts, three pairs of slacks, two ties, and two jackets are blue, what is the probability that an all-blue outfit is the result of a random selection?

9. A multiple-choice test has 15 questions each having four possible answers, out of which only one is correct. If the questions are answered at random, what is the probability of getting all of them right?

10. Suppose that in a state, license plates have three letters followed by three numbers in a way that no letter or number is repeated in a single plate. Determine the number of possible license plates in this state.

11. A library has 800,000 books, and the librarian wants to encode each using a code word consisting of three letters followed by two numbers. Are there enough code words to encode all these books with different code words?

12. How many $n \times m$ arrays (matrices) with entries 0 or 1 are there?

13. How many divisors does 55,125 have?
HINT: $55{,}125 = 3^2 5^3 7^2$.

14. A delicatessen has advertised that it offers over 500 varieties of sandwiches. If at this deli it is possible to have any combination of salami, turkey, bologna, corned beef, ham, and cheese on French bread with the possible additions of lettuce, tomato, and mayonnaise, is the deli's advertisement true? Assume that a sandwich necessarily has bread and at least one type of meat or cheese.

15. How many four-digit numbers can be formed using only the digits 2, 4, 6, 8, and 9? How many of these have some digit repeated?

16. In a mental health clinic there are 12 patients. A group therapist invites all these patients to join her for therapy. How many possible groups could she get?

17. Suppose that four cards are drawn from an ordinary deck of 52 cards successively, with replacement and at random. What is the probability of at least one king?

18. A campus telephone extension has four digits. How many different extensions with no repeated digits exist? Of these, (a) how many do not start with a 0; (b) how many do not have 01 as the first two digits?

19. There are N types of drugs in a market to reduce acid indigestion. A random sample of n drugs is taken with replacement. What is the probability that brand A is included?

20. Jenny, a probability student, having seen Example 2.5 and its solution, becomes convinced that it is a nearly even bet that someone among the next 22 people she meets randomly will have the same birthday as she does. What is the fallacy in Jenny's thinking? What is the minimum number of people that Jenny must meet before the chances are better than even that someone shares her birthday?

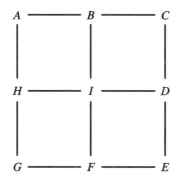

Figure 2.3 Islands and connecting bridges of Exercise 21.

21. A salesperson covers islands $A, B, \ldots, I$. These islands are connected by the bridges shown in the Figure 2.3. While at an island, the salesperson takes one of the possible bridges at random and goes to another one. She does her business in this new island and then takes a bridge at random to go to the next one. She continues this until she reaches an island for the second time on a day. She stays there overnight and then continues her trips the day after. If she starts her trips from island I tomorrow, in what percent of all possible trips will she end up staying overnight again at island I?

22. A drunk man is at an intersections O and does not really know which way he is going. At the end of each block he either continues forward with probability 1/2 or he turns back with probability 1/2. Draw a tree diagram to find the probability that after walking four blocks (a) he is back at intersection O; (b) he is only one block away from intersection O.

B

= 1 - Prob contains no 5's.

23. An integer is selected at random from the set $\{1, 2, \ldots, 1,000,000\}$. What is the probability that it contains the digit 5?

24. How many divisors does a natural number N have?
HINT: N can be written as $p_1^{n_1} p_2^{n_2} \cdots p_k^{n_k}$, where $p_1, p_2, \ldots, p_k$ are distinct primes.

25. In tossing four fair dice, what is the probability of at most one 3?

26. A delicatessen has advertised that it offers over 3000 varieties of sandwiches. If at this deli it is possible to have any combination of salami, turkey, bologna, corned beef, and ham with or without Swiss and/or American cheese on French, white, or whole wheat bread and possible additions of lettuce, tomato, and mayonnaise, is the deli's

advertisement true? Assume that a sandwich necessarily has bread and at least one type of meat or cheese.

27. One of the five elevators of the Evans Hall at the University of California at Berkeley leaves the basement with eight passengers and stops at all of the remaining 11 floors. If it is equally likely that a passenger gets off at any of these 11 floors, what is the probability that no two of these eight passengers will get off at the same floor?

28. The elevator of a four-floor building leaves the first floor with six passengers and stops at all of the remaining three floors. If it is equally likely that a passenger gets off at any of these three floors, what is the probability that at each stop of the elevator at least one passenger departs?

29. A number is selected randomly from the set

$$\{0000, 0001, 0002, 0003, \ldots, 9998, 9999\}.$$

What is the probability that the sum of the first two digits of the number selected is equal to the sum of its last two digits?

30. What is the probability that a random r-digit number ($r \geq 3$) contains at least one 0, at least one 1, and at least one 2?

2.3   Permutations

To count the number of outcomes of an experiment or the number of possible ways an event can occur, it is often useful to look for special patterns. Sometimes patterns help us develop techniques for counting. Two simple cases in which patterns enable us to count easily are permutations and combinations. We study these two patterns in this section and the next.

An ordered arrangement of r objects from a set A containing n objects ($0 < r \leq n$) is called an *r-element permutation of A* or a *permutation of the elements of A taken r at a time.* For example, if $A = \{a, b, c, d\}$, then ab is a two-element permutation of A, acd is a three-element permutation of A, and $adcb$ is a four-element permutation of A. The order in which objects are arranged is important. For example, ab and ba are considered different two-element permutations, abc and cba are distinct three-element permutations, and $abcd$ and $cbad$ are different four-element permutations.

Let $_nP_r$ denote the number of permutations of a set A containing n elements taken r at a time ($1 \leq r \leq n$). $_nP_r$ is computed using the generalized counting principle as follows: Since A has n elements, the number of choices for the first object in the r-element permutation is n. For the second object, the number of choices is the remaining $n - 1$ elements of

A. For the third one, the number of choices is the remaining $n - 2, \ldots$, and finally, for the rth object the number of choices is $n - (r - 1) = n - r + 1$. Hence

$$_nP_r = n(n - 1)(n - 2) \cdots (n - r + 1). \tag{2.1}$$

An n-element permutation of a set with n objects is simply called a *permutation*. $_nP_n$, the number of permutations of a set containing n elements, is evaluated from (2.1) by putting $r = n$.

$$_nP_n = n(n - 1)(n - 2) \cdots (n - n + 1) = n!. \tag{2.2}$$

The formula $n!$ for the number of permutations of a set of n objects has been well known for a long time. Although it first appeared in the works of Persian and Arab mathematicians in the twelfth century, there are indications that the mathematicians of India were aware of this rule a few hundred years before Christ. However, the surprise notation "!" used for factorial was introduced by Christian Kramp in 1808. He chose this symbol perhaps because $n!$ gets *surprisingly* large even for small numbers. For example, we have that $19! \approx 1.216451 \times 10^{17}$, a number which according to Karl Smith[†] is greater than 10 times "the total of all copies of all words ever printed (including everything from the Sunday newspaper, novels, textbooks, back to the Gutenberg Bible)."

There is a popular alternative for relation (2.1). It is obtained by multiplying both sides of (2.1) by $(n - r)! = (n - r)(n - r - 1) \cdots 3 \cdot 2 \cdot 1$. We get

$$_nP_r \cdot (n - r)!$$
$$= [n(n - 1)(n - 2) \cdots (n - r + 1)][(n - r)(n - r - 1) \cdots 3 \cdot 2 \cdot 1].$$

This gives $_nP_r \cdot (n - r)! = n!$. Therefore,

$$_nP_r = \frac{n!}{(n - r)!}. \tag{2.3}$$

Note that for $r = n$, this relation implies that $_nP_n = n!/0!$. But by (2.2), $_nP_n = n!$. Therefore, for $r = n$, to make (2.3) consistent with (2.2), we define $0! = 1$.

[†] Karl J. Smith, *The Nature of Mathematics*, 6th ed., Brooks/Cole, Pacific Grove, Calif., 1991, p. 32.

Example 2.9   Three people, Brown, Smith, and Jones, must be scheduled for job interviews. In how many different orders can this be done?

SOLUTION: The number of different orders is equal to the number of permutations of the set {Brown, Smith, Jones}. So there are $3! = 6$ possible orders for the interviews. ◆

Example 2.10   Suppose that two anthropology, four computer science, three statistics, three biology, and five music books are put on a bookshelf with a random arrangement. What is the probability that the books of the same subject are together?

SOLUTION: Let A be the event that all the books of the same subject are together. Then $P(A) = N(A)/N$, where $N(A)$ is the number of arrangements in which the books dealing with the same subject are together and N is the total number of possible arrangements. Since there are 17 books and each of their arrangements is a permutation of the set of these books, $N = 17!$. To calculate $N(A)$, note that there are $2! \times 4! \times 3! \times 3! \times 5!$ arrangements in which anthropology books are first, computer science books are next, then are statistics books, after that biology, and finally, music. Also, there are the same number of arrangements for each possible ordering of the subjects. Since the subjects can be ordered in $5!$ ways, $N(A) = 5! \times 2! \times 4! \times 3! \times 3! \times 5!$. Hence

$$P(A) = \frac{5! \times 2! \times 4! \times 3! \times 3! \times 5!}{17!} \approx 6.996 \times 10^{-8}. \quad ◆$$

Example 2.11   If five boys and five girls sit in a row in a random order, what is the probability that no two children of the same sex sit together?

SOLUTION: There are $10!$ ways for 10 persons to sit in a row. In order that no two of the same sex sit together, boys must occupy positions 1, 3, 5, 7, 9 and girls positions 2, 4, 6, 8, 10, or vice versa. In each case there are $5! \times 5!$ possibilities. Therefore, the desired probability is equal to

$$\frac{2 \times 5! \times 5!}{10!} \approx 0.008. \quad ◆$$

We showed that the number of permutations of a set of n objects is $n!$. This formula is valid only if all of the objects of the set are distinguishable from each other. Otherwise, the number of permutations is different. For example, there are $8!$ permutations of the eight letters $STANFORD$ because all of these letters are distinguishable from each other. But the number of permutations of the 8 letters $BERKELEY$ is less than $8!$ since the second, the fifth, and the seventh letters in *BERKELEY* are indistinguishable. If in any permutation of these letters we change the positions of these three indistinguishable *E*'s with each other, no new permutations would be generated. Let us calculate the number of permutations of the letters *BERKELEY*. Suppose that there are x such permutations and consider any particular one of them, say *BYERELEK*. If we label the *E*'s: $BYE_1RE_2LE_3K$ so that all of the letters are distinguishable, then by arranging *E*'s among themselves we get 3! new permutations, namely,

$$BYE_1RE_2LE_3K \quad BYE_2RE_3LE_1K$$
$$BYE_1RE_3LE_2K \quad BYE_3RE_1LE_2K$$
$$BYE_2RE_1LE_3K \quad BYE_3RE_2LE_1K$$

which are otherwise all the same. Therefore, if all the letters were different, then for each one of the x permutations we would have 3! times as many. That is, the total number of permutations would have been $x \times 3!$. But since eight different letters generate exactly 8! permutations, we must have $x \times 3! = 8!$. This gives $x = 8!/3!$. We have shown that *the number of distinguishable permutations of the letters BERKELEY is* 8!/3!. This sort of reasoning will lead us to the following general theorem.

THEOREM 2.4   *The number of distinguishable permutations of n objects of k different types, where n_1 are alike, n_2 are alike, ..., n_k are alike and $n = n_1 + n_2 + \cdots + n_k$ is*

$$\frac{n!}{n_1! \times n_2! \times \cdots \times n_k!}.$$

Example 2.12   How many different 10-letter codes can be made using three a's, four b's, and three c's?

SOLUTION: By Theorem 2.4, the number of such codes is $10!/(3! \times 4! \times 3!) = 4200$. ♦

Example 2.13   In how many ways can we paint 11 offices so that four of them will be painted green, three yellow, two white, and the remaining two pink?

SOLUTION: Let "*ggypgwpygwy*" represent the situation in which the first office is painted green, the second office is painted green, the third one yellow, and so on, with similar representations for other cases. Then the answer is equal to the number of distinguishable permutations of "*ggggyyywwpp*," which by Theorem 2.4 is $11!/(4! \times 3! \times 2! \times 2!) = 69,300$. ♦

Example 2.14   A fair coin is flipped 10 times. What is the probability of exactly three heads?

SOLUTION: The set of all sequences of H(heads) and T(tails) of length 10 forms the sample space and contains 2^{10} elements. Of all these, those with three H and seven T are desirable. But the number of distinguishable sequences with three H's and seven T's is equal to $10!/(3! \times 7!)$. Therefore, the probability of exactly three heads is $\left(\dfrac{10!}{3! \times 7!} \right) /2^{10} \approx 0.12$.   ♦

EXERCISES

A

1. How many permutations of the set $\{a, b, c, d, e\}$ begin with a and end with c?

2. How many different messages can be sent by five dashes and three dots?

3. Robert has eight guests, two of whom are Jim and John. If the guests will arrive in a random order, what is the probability that John will not arrive right after Jim?

4. Let A be the set of all sequences of 0's, 1's, and 2's of length 12.
   (a) How many elements are there in A?
   (b) How many elements of A have exactly six 0's and 6 ones?
   (c) How many elements of A have exactly three 0's, four 1's, and five 2's?

5. Professor Haste is somewhat familiar with six languages. To translate texts from one language into another directly, how many one-way dictionaries does he need?

6. In an exhibition, 20 cars of the same style that are distinguishable only by their colors are to be parked in a row, all facing a certain window. If four of the cars are blue, three are black, five are yellow, and eight are white, how many choices are there?

7. At various yard sales, a woman has acquired five forks, of which no two are alike. The same applies to her four knives and seven spoons. In how many different ways can three place settings be chosen if each place setting consists of exactly one fork, one knife, and one spoon? Assume that the arrangment of the place settings on the table is unimportant.

8. In a conference, Dr. Richman's lecture is related to Dr. Chollet's and should not precede it. If there are six more speakers, how many schedules could be arranged?

WARNING: Dr. Richman's lecture is not necessarily scheduled right after Dr. Chollet's lecture.

9. A dancing contest has 11 competitors, of whom three are Americans, two are Mexicans, three are Russians, and three are Italians. If the contest result lists only the nationality of the dancers, how many outcomes are possible?

10. Six fair dice are tossed. What is the probability that at least two of them show the same face?

11. Find the number of distinguishable permutations of the letters MISSISSIPPI.

12. A fair die is tossed eight times. What is the probability of exactly two 3's, three 1's, and three 6's?

13. In drawing nine cards with replacement from an ordinary deck of 52 cards, what is the probability of three aces of spades, three queens of hearts, and three kings of clubs?

14. At a party, n men and m women put their drinks on a table and go out on the floor to dance. When they return, none of them recognizes his or her drink, so everyone takes a drink at random. What is the probability that each man selects his own drink?

15. There are 20 chairs in a room numbered 1 through 20. If eight girls and 12 boys sit on these chairs at random, what is the probability that the thirteenth chair is occupied by a boy?

16. There are 12 students in a class. What is the probability that their birthdays fall in 12 different months? Assume that all months have the same probability of including the birthday of a randomly selected person.

17. If we put five math, six biology, eight history, and three literature books on a bookshelf at random, what is the probability that all the math books are together?

18. One of the five elevators of Evans Hall at the University of California at Berkeley starts with seven passengers and stops at nine floors. Assuming that it is equally likely that a passenger gets off at any of these nine floors, find the probability that at least two of these passengers will get off at the same floor.

19. Five boys and five girls sit in a row at random. What is the probability that the boys are together and the girls are together?

20. If n balls are randomly placed into n cells, what is the probability that each cell will be occupied?

21. A town has six parks. On a Saturday, six classmates, who are unaware of each other's decision, choose a park at random and go there at the same time. What is the probability that at least two of them go to the same park? Convince yourself that this exercise is the same as Exercise 10, only expressed in a different context.

22. A club of 136 members is in the process of choosing a president, a vice president, a secretary, and a treasurer. If two of the members are not on speaking terms and do not serve together, in how many ways can these four people be chosen?

B

23. Let S and T be finite sets with n and m elements, respectively.
   (a) How many functions $f : S \to T$ can be defined?
   (b) If $m \geq n$, how many injective (one-to-one) function $f : S \to T$ can be defined?
   (c) If $m = n$, how many surjective (onto) functions $f : S \to T$ can be defined?

24. A fair die is tossed eight times. What is the probability of exactly two 3's, exactly three 1's, and exactly two 6's?

25. Suppose that 20 sticks are broken each into one long and one short part. By pairing them randomly, the 40 parts are then used to make 20 new sticks. What is the probability that long parts are all paired with short ones? What is the probability that the new sticks are exactly the same as the old ones?

26. At a party, 15 married couples are seated at random at a round table. What is the probability that all men are sitting next to their wives? Suppose that of these married couples, five husbands and their wives are older than 50 and the remaining husbands and wives are all younger than 50. What is the probability that all men over 50 are sitting next to their wives? Note that when people are sitting around a round table, only their seats relative to each other matters. The exact position of a person is not important.

27. A box contains five blue and eight red balls. Jim and Jack start drawing balls from the box, respectively, one at a time, at random, and without replacement until a blue ball is drawn. What is the probability that Jack draws the blue ball?

2.4   Combinations

In many combinatorial problems, unlike permutations, the order in which elements are arranged is immaterial. For example, suppose that in a contest the 10 semifinalists are determined and we want to count the number of possible ways that three contestants enter the finals. If we argue that there are $10 \times 9 \times 8$ such possibilities, we are wrong since the contestants cannot be ordered. If A, B, and C are three of the semifinalists, then ABC, BCA, ACB, BAC, CAB, and CBA are all the same event and have the same meaning: "A, B, and C are the finalists." Combinations are introduced to handle such problems.

DEFINITION   *An unordered arrangement of r objects from a set A containing n objects $(r \leq n)$ is called an r-element combination of A or a combination of the elements of A taken r at a time.*

Therefore, two combinations are different only if they differ in composition. Let x be the number of r-element combinations of a set A of n objects. If all the permutations of each r-element combination are found, then all the r-element permutations of A are found. Since for each r-element combination of A there are $r!$ permutations and the total number of r-element permutations is $_nP_r$, we have

$$x \cdot r! = {_nP_r}.$$

Hence $x \cdot r! = n!/(n-r)!$, so $x = n!/[(n-r)!\,r!]$. Therefore, we have shown that *the number of r-element combinations of n objects is given by*

$$_nC_r = \frac{n!}{(n-r)!\,r!}.$$

Historically, a formula equivalent to $n!/[(n-r)!\,r!]$ turned up in the works of the Indian mathematician Bhaskara II (1114–1185) in the middle of the twelfth century. Therefore, the rule for calculation of the number of r-element combinations of n objects has been known for a long time.

NOTATION: By the symbol $\binom{n}{r}$ (read: n choose r) we mean the number of all r-element combinations of n objects. Therefore, for $r \leq n$,

$$\binom{n}{r} = \frac{n!}{r!\,(n-r)!}.$$

Observe that $\binom{n}{0} = \binom{n}{n} = 1$ and $\binom{n}{1} = \binom{n}{n-1} = n$. Also for any $0 \leq r \leq n$,

$$\binom{n}{r} = \binom{n}{n-r}$$

and

$$\binom{n+1}{r} = \binom{n}{r} + \binom{n}{r-1}. \tag{2.4}$$

These relations can be proved algebraically or verified combinatorially. Let us prove (2.4) by a combinatorial argument. Consider a set of $n+1$ objects, $\{a_1, a_2, \ldots, a_n, a_{n+1}\}$. There are $\binom{n+1}{r}$ r-element combinations of this set. Now we separate these r-element combinations into two disjoint classes: one class consisting of all r-element combinations of $\{a_1, a_2, \ldots, a_n\}$ and another consisting of all $(r-1)$-element combinations of $\{a_1, a_2, \ldots, a_n\}$ attached to a_{n+1}. The latter class contains $\binom{n}{r-1}$ elements and the former contains $\binom{n}{r}$ elements, showing that (2.4) is valid.

Example 2.15   In how many ways can two math and three biology books be selected from eight math and six biology books?

SOLUTION: There are $\binom{8}{2}$ possible ways to select two math books and $\binom{6}{3}$ possible ways to select three biology books. Therefore, by the counting principle,

$$\binom{8}{2} \times \binom{6}{3} = \frac{8!}{6!\,2!} \times \frac{6!}{3!\,3!} = 560$$

is the total number of ways in which two math and three biology books can be selected.   ◆

Example 2.16   A random sample of 45 instructors from different state universities were selected randomly and asked whether they are happy with their teaching loads.   The responses of 32 of them were negative.   If Drs. Smith, Brown, and Jones were among those questioned, what is the probability that all three of them gave negative responses?

SOLUTION: There are $\binom{45}{32}$ different possible groups with negative responses. If three of them are Drs. Smith, Brown, and Jones, the other 29 are from the remaining 42 faculty members questioned. Hence the desired probability is

$$\frac{\binom{42}{29}}{\binom{45}{32}} \approx 0.35. \quad \blacklozenge$$

Example 2.17 In a small town, 11 of the 25 schoolteachers are against abortion, eight are for abortion, and the rest are indifferent. A random sample of five schoolteachers is selected for an interview. What is the probability that (a) all of them are for abortion; (b) all of them have the same opinion?

SOLUTION: (a) There are $\binom{25}{5}$ different ways to select random samples of size 5 out of 25 teachers. Of these, only $\binom{8}{5}$ are all for abortion. Hence the desired probability is

$$\frac{\binom{8}{5}}{\binom{25}{5}} \approx 0.0011.$$

(b) By an argument similar to part (a), the desired probability equals

$$\frac{\binom{11}{5} + \binom{8}{5} + \binom{6}{5}}{\binom{25}{5}} \approx 0.0099. \quad \blacklozenge$$

Example 2.18 In Maryland's lottery, players pick 6 different integers between 1 and 49, order of selection being irrelevant. The lottery commission then selects six of these as the *winning numbers*. A player wins the grand prize if all six numbers that he or she has selected match the winning numbers. He or she wins the second prize if exactly five and the third prize if exactly four of the six numbers chosen match with the winning ones. Find the probability that a certain choice of a bettor wins the grand, the second, and the third prizes, respectively.

SOLUTION: The probability of winning the grand prize is

$$\frac{1}{\binom{49}{6}} = \frac{1}{13,983,816}.$$

The probability of winning the second prize is

$$\frac{\binom{6}{5}\binom{43}{1}}{\binom{49}{6}} = \frac{258}{13,983,816} \approx \frac{1}{54,200},$$

and the probability of winning the third prize is

$$\frac{\binom{6}{4}\binom{43}{2}}{\binom{49}{6}} = \frac{13,545}{13,983,816} \approx \frac{1}{1032}. \quad \blacklozenge$$

Example 2.19   From an ordinary deck of 52 cards, seven cards are drawn at random and without replacement. What is the probability that at least one of them is a king?

SOLUTION: In $\binom{52}{7}$ ways seven cards can be selected from an ordinary deck of 52 cards. In $\binom{48}{7}$ of these, none of the cards selected is a king. Therefore, the desired probability is

$$P(\text{at least one king}) = 1 - P(\text{no kings})$$

$$= 1 - \frac{\binom{48}{7}}{\binom{52}{7}} = 0.4496. \quad \blacklozenge$$

WARNING: A common mistake is to calculate this and similar probabilities as follows: To make sure that there is at least one king among the seven cards drawn, we will first choose a king; there are $\binom{4}{1}$ possibilities. Then we choose the remaining six cards from the remaining 51 cards; there are $\binom{51}{6}$ possibilities for this. Thus the answer is

$$\frac{\binom{4}{1}\binom{51}{6}}{\binom{52}{7}} = 0.5385.$$

This solution is wrong because it counts some of the possible outcomes several times. For example, the hand K_H, 5_C, 6_D, 7_H, K_D, J_C, and 9_S is counted twice. Once when K_H is selected as the first card from the kings and 5_C, 6_D, 7_H, K_D, J_C, and 9_S from the remaining 51, and once when K_D is selected as the first card from the kings and K_H, 5_C, 6_D, 7_H, J_C, and 9_S from the remaining 51 cards. ◆

Example 2.20   What is the probability that a poker hand is a full house? A poker hand consists of five randomly selected cards from an ordinary deck of 52 cards. It is a full house if three cards are of one denomination and two cards are of another denomination: for example, three queens and two 4's.

SOLUTION: The number of different poker hands is $\binom{52}{5}$. To count the number of full houses, let us call a hand of type (Q,4), if it has three queens and two 4's, with similar representations for other types of full houses. Observe that (Q,4) and (4,Q) are different full houses, and types such as (Q,Q) and (K,K) do not exist. Hence there are 13×12 different types of full houses. Since for every particular type, say (4,Q), there are $\binom{4}{3}$ ways to select three 4's and $\binom{4}{2}$ ways to select two Q's, the desired probability equals

$$\frac{13 \times 12 \times \binom{4}{3} \times \binom{4}{2}}{\binom{52}{5}} \approx 0.0014. \quad ◆$$

Example 2.21   An absentminded professor wrote n letters and sealed them in envelopes before writing the addresses on the envelopes. Then he wrote the n addresses on the envelopes at random. What is the probability that at least one letter was addressed correctly?

SOLUTION: The total number of ways that one can write n addresses on n envelopes is $n!$; thus the sample space contains $n!$ points. Now we calculate the number of outcomes in which at least one envelope is addressed correctly. To do this, let E_i be the event that the ith letter is addressed correctly; then

$E_1 \cup E_2 \cup \cdots \cup E_n$ is the event that at least one letter is addressed correctly. To calculate $P(E_1 \cup E_2 \cup \cdots \cup E_n)$, we use the inclusion–exclusion principle. To do so we must calculate the probabilities of all possible intersections of the events from $E_1, \ldots, E_n$, add the probabilities that are obtained by intersecting an odd number of the events, and subtract all the probabilities that are obtained by intersecting an even number of the events. Therefore, we need to know the number of elements of E_i's, $E_i \cap E_j$'s, $E_i \cap E_j \cap E_k$'s, and so on. Now E_i contains $(n-1)!$ points, because when the ith letter is addressed correctly, there are $(n-1)!$ ways to address the remaining $n-1$ envelopes.   So $P(E_i) = \dfrac{(n-1)!}{n!}$.   Similarly, $E_i \cap E_j$ contains $(n-2)!$ points, because if the ith and jth envelopes are addressed correctly, the remaining $n-2$ envelopes can be addressed in $(n-2)!$ ways. Thus $P(E_i \cap E_j) = \dfrac{(n-2)!}{n!}$.   Similarly, $P(E_i \cap E_j \cap E_k) = \dfrac{(n-3)!}{n!}$, and so on. Now in computing $P(E_1 \cup E_2 \cup \cdots \cup E_n)$, there are n terms of the form $P(E_i)$, $\dbinom{n}{2}$ terms of the form $P(E_i \cap E_j)$, $\dbinom{n}{3}$ terms of the form $P(E_i \cap E_j \cap E_k)$, and so on. Hence

$$P(E_1 \cup E_2 \cup \cdots \cup E_n) = n\frac{(n-1)!}{n!} - \binom{n}{2}\frac{(n-2)!}{n!} + \cdots$$

$$+ (-1)^{n-2}\binom{n}{n-1}\frac{[n-(n-1)]!}{n!}$$

$$+ (-1)^{n-1}\binom{n}{n}\frac{1}{n!}.$$

This expression simplifies to

$$P(E_1 \cup E_2 \cup \cdots \cup E_n) = 1 - \frac{1}{2!} + \frac{1}{3!} - \frac{1}{4!} + \cdots + \frac{(-1)^{n-1}}{n!}. \quad \blacklozenge$$

REMARK: Note that since $e^x = \displaystyle\sum_{n=0}^{\infty}(x^n/n!)$, if $n \to \infty$, then

$$P\left(\bigcup_{i=1}^{\infty} E_i\right) = 1 - \frac{1}{2!} + \frac{1}{3!} - \frac{1}{4!} + \cdots + \frac{(-1)^{n-1}}{n!} + \cdots$$

$$= 1 - \left(1 - 1 + \frac{1}{2!} - \frac{1}{3!} + \frac{1}{4!} - \cdots + \frac{(-1)^n}{n!} + \cdots\right)$$

$$= 1 - \sum_{n=0}^{\infty}\frac{(-1)^n}{n!} = 1 - \frac{1}{e} \approx 0.632.$$

Hence even if the number of the envelopes is very large, there is still a very good chance for at least one envelope to be addressed correctly. ◆

One of the most important applications of combinatorics is that the formula for the r-element combinations of n objects enables us to find an algebraic expansion for $(x + y)^n$.

THEOREM 2.5   (Binomial Expansion)   *For any integer $n \geq 0$,*

$$(x + y)^n = \sum_{i=0}^{n} \binom{n}{i} x^{n-i} y^i.$$

PROOF: By looking at some special cases, such as

$$(x + y)^2 = (x + y)(x + y) = x^2 + xy + yx + y^2$$

and

$$(x + y)^3 = (x + y)(x + y)(x + y)$$
$$= x^3 + x^2 y + yx^2 + xy^2 + yx^2 + xy^2 + y^2 x + y^3,$$

it should become clear that when we carry out the multiplication

$$(x + y)^n = \underbrace{(x + y)(x + y) \cdots (x + y)}_{n \text{ times}},$$

we obtain only terms of the form $x^{n-i} y^i$, $0 \leq i \leq n$. Therefore, all we have to do is to find out how many times the term $x^{n-i} y^i$ appears, $0 \leq i \leq n$. This is easily seen to be $\binom{n}{n-i} = \binom{n}{i}$ because $x^{n-i} y^i$ emerges only whenever the x's of $n - i$ of the n factors of $(x + y)$ are multiplied by the y's of the remaining i factors of $(x + y)$. Hence

$$(x + y)^n = \binom{n}{0} x^n + \binom{n}{1} x^{n-1} y + \binom{n}{2} x^{n-2} y^2 + \cdots$$
$$+ \binom{n}{n-1} xy^{n-1} + \binom{n}{n} y^n. \quad ◆$$

Example 2.22   What is the coefficient of $x^2 y^3$ in the expansion of $(2x + 3y)^5$?

SOLUTION: Let $u = 2x$ and $v = 3y$; then $(2x + 3y)^5 = (u + v)^5$. The coefficient of u^2v^3 in the expansion of $(u + v)^5$ is $\binom{5}{3}$ and $u^2v^3 = (2^2 \cdot 3^3)x^2y^3$; therefore, the coefficient of x^2y^3 in the expansion of $(2x + 3y)^5$ is $\binom{5}{3}(2^2 \cdot 3^3) = 1080$. ◆

Example 2.23   Evaluate the sum

$$\binom{n}{0} + \binom{n}{1} + \binom{n}{2} + \binom{n}{3} + \cdots + \binom{n}{n}.$$

SOLUTION: A set containing n elements has $\binom{n}{i}$, $0 \le i \le n$, subsets with i elements. So the given expression is the total number of the subsets of a set of n elements and therefore it equals 2^n. A second way to see this is to note that by the binomial expansion, the given expression equals

$$\sum_{i=0}^{n} \binom{n}{i}1^{n-i}1^i = (1 + 1)^n = 2^n. ◆$$

Example 2.24   Evaluate the sum

$$\binom{n}{1} + 2\binom{n}{2} + 3\binom{n}{3} + \cdots + n\binom{n}{n}.$$

SOLUTION:

$$i\binom{n}{i} = i \cdot n!/[i! \, (n - i)!] = n \cdot (n - 1)!/[(i - 1)! \, (n - i)!]$$

$$= n\binom{n-1}{i-1}.$$

So

$$\binom{n}{1} + 2\binom{n}{2} + 3\binom{n}{3} + \cdots + n\binom{n}{n}$$

$$= n\left[\binom{n-1}{0} + \binom{n-1}{1} + \binom{n-1}{2} + \cdots + \binom{n-1}{n-1}\right] = n \cdot 2^{n-1}$$

by Example 2.23. ◆

Example 2.25   Prove that

$$\binom{2n}{n} = \sum_{i=0}^{n} \binom{n}{i}^2.$$

SOLUTION: We show this using a combinatorial argument. For an analytic proof, see Exercise 43. Let A= $\{a_1, a_2, \ldots, a_n\}$ and B= $\{b_1, b_2, \ldots, b_n\}$ be two disjoint sets. The number of subsets of $A \cup B$ with n elements is $\binom{2n}{n}$. On the other hand, any subset of $A \cup B$ with n elements is the union of a subset of A with i elements and a subset of B with $n - i$ elements for some $0 \leq i \leq n$. Since for each i there are $\binom{n}{i}\binom{n}{n-i}$ such subsets, we have that the total number of subsets of $A \cup B$ with n elements is $\sum_{i=0}^{n} \binom{n}{i}\binom{n}{n-i}$. But since $\binom{n}{n-i} = \binom{n}{i}$, we have identity.   ♦

Example 2.26 Suppose that we want to distribute n distinguishable balls into k distinguishable cells so that n_1 balls are distributed into the first cell, n_2 balls into the second cell, $\ldots$, n_k balls into the kth cell, where $n_1 + n_2 + n_3 + \cdots + n_k = n$. To count the number of ways that this is possible, note that we have $\binom{n}{n_1}$ choices to distribute n_1 balls into the first cell; for each choice of n_1 balls in the first cell, we then have $\binom{n - n_1}{n_2}$ choices to distribute n_2 balls into the second cell; for each choice of n_1 balls in the first cell and n_2 balls in the second cell, we have $\binom{n - n_1 - n_2}{n_3}$ choices to distribute n_3 balls into the third cell; and so on. Hence by the generalized counting principle the total number of such distributions is

$$\binom{n}{n_1}\binom{n - n_1}{n_2}\binom{n - n_1 - n_2}{n_3} \cdots \binom{n - n_1 - n_2 - \cdots - n_{k-1}}{n_k}$$

$$= \frac{n!}{(n - n_1)!\, n_1!} \times \frac{(n - n_1)!}{(n - n_1 - n_2)!\, n_2!} \times \frac{(n - n_1 - n_2)!}{(n - n_1 - n_2 - n_3)!\, n_3!} \times \cdots$$

$$\times \frac{(n - n_1 - n_2 - n_3 - \cdots - n_{k-1})!}{(n - n_1 - n_2 - n_3 - \cdots - n_{k-1} - n_k)!\, n_k!} = \frac{n!}{n_1!\, n_2!\, n_3! \cdots n_k!}$$

because $n_1 + n_2 + n_3 + \cdots + n_k = n$ and $(n - n_1 - n_2 - \cdots - n_k)! = 0! = 1$.   ♦

As an application of Example 2.26, we state a generalization of the binomial expansion (Theorem 2.5) and leave its proof as an exercise (see Exercise 26).

THEOREM 2.6   **(Multinomial Expansion)**   *In the expansion of*

$$(x_1 + x_2 + \cdots + x_k)^n,$$

the coefficient of the term $x_1^{n_1} x_2^{n_2} x_3^{n_3} \cdots x_k^{n_k}, n_1 + n_2 + \cdots + n_k = n$, *is given by*

$$\frac{n!}{n_1! \, n_2! \, \cdots \, n_k!}.$$

Therefore,

$$(x_1 + x_2 + \cdots + x_k)^n = \sum_{n_1 + n_2 + \cdots + n_k = n} \frac{n!}{n_1! \, n_2! \, \cdots \, n_k!} \, x_1^{n_1} x_2^{n_2} x_3^{n_3} \cdots x_k^{n_k}.$$

Note that the sum is taken over all nonnegative integers $n_1, n_2, \ldots, n_k$ *such that* $n_1 + n_2 + \cdots + n_k = n$.

EXERCISES

A

1. Jim has 20 friends. If he decides to invite six of them to his birthday party, how many choices does he have?

2. Each of the 50 U.S. states has two senators. In how many ways may a majority be achieved in the U.S. Senate? Ignore the possibility of absence or abstention. Assume that all senators are present and voting.

3. A panel consists of 20 men and 25 women. How many choices do we have for a jury of six men and six women from this panel?

4. From an ordinary deck of 52 cards, five are drawn randomly. What is the probability of exactly three face cards?

5. A random sample of n elements is taken from a population of size N without replacement. What is the probability that a fixed element of the population is included? Simplify your answer.

6. Judy puts one piece of fruit in her child's lunch bag everyday. If she has three oranges and two apples for the next five days, in how many ways can she do this?

7. Ann puts at most one piece of fruit in her child's lunch bag everyday. If she has only three oranges and two apples for the next eight lunches of her child, in how many ways can she do this?

8. Lily has 20 friends. Among them are Kevin and Gerry, who are husband and wife. Lily wants to invite six of her friends to her birthday party. If neither Kevin nor Gerry will go to a party without the other, how many choices does Lily have?

9. In front of Jeff's office there is a parking lot with 13 parking spots in a row. When a car arrives at this lot, it is parked randomly at one of the empty spots. Jeff parks his car in the only empty spot that is left. Then he goes to his office. On his return he finds that there are seven empty spots. If he has not parked his car at either end of the parking area, what is the probability that both of the parking spaces that are next to Jeff's car are empty?

10. Find the coefficient of x^9 in the expansion of $(2 + x)^{12}$.

11. Find the coefficient of $x^3 y^4$ in the expansion of $(2x - 4y)^7$.

12. A team consisting of three boys and four girls must be formed from a group of nine boys and eight girls. If two of the girls are feuding and refuse to play on the same team, how many possibilities do we have?

13. A fair coin is tossed 10 times. What is the probability of (a) five heads; (b) at least five heads?

14. If five numbers are selected at random from the set $\{1, 2, 3, \ldots, 20\}$, what is the probability that their minimum is larger than 5?

15. From a faculty of six professors, six associate professors, 10 assistant professors, and 12 instructors, a committee of size 6 is formed randomly. What is the probability that (a) there are exactly two professors on the committee; (b) all committee members are of the same rank?

16. A lake contains 200 trout, 50 of them are caught randomly, tagged, and returned. If again we catch 50 trout at random, what is the probability of getting exactly five tagged trout?

17. Find the values of $\displaystyle\sum_{i=0}^{n} 2^i \binom{n}{i}$ and $\displaystyle\sum_{i=0}^{n} x^i \binom{n}{i}$.

18. A fair die is tossed six times. What is the probability of exactly two 6's?

19. Suppose that 12 married couples take part in a contest. If 12 persons each win a prize, what is the probability that from every couple one of them is a winner? Assume that all of the $\binom{24}{12}$ possible sets of winners are equally probable.

20. Poker hands are classified into the following 10 disjoint categories in increasing order of likelihood. Calculate the probability of the occurrence of each class separately. Recall that a poker hand consists of five cards selected randomly from an ordinary deck of 52 cards.

- *Royal flush:* The 10, jack, queen, king, and ace of the same suit.
- *Straight flush:* All cards in the same suit, with consecutive denominations except for the royal flush.
- *Four of a kind:* Four cards of one denomination and one card of a second denomination: for example, four 8's and a jack.
- *Full house:* Three cards of one denomination and two cards of a second denomination: for example, three 4's and two queens.
- *Flush:* Five cards all in one suit but not a straight or royal flush.
- *Straight:* Cards of distinct consecutive denominations, not all in one suit: for example, 3 of hearts, 4 of hearts, 5 of spades, 6 of hearts, and 7 of clubs.
- *Three of a kind:* Three cards of one denomination, a fourth card of a second denomination, and a fifth card of a third denomination.
- *Two pairs:* Two cards from one denomination, another two from a second denomination, and the fifth card from a third denomination.
- *One pair:* Two cards from one denomination, with the third, fourth, and fifth cards from a second, third, and fourth denomination, respectively: for example, two 8's, a king, a 5, and a 4.
- *None of the above.*

21. A history professor who teaches three sections of the same course every semester decides to make several tests and use them for the next 10 years (20 semesters) as final exams. The professor has two policies: (1) not to give the same test to more than one class in a semester, and (2) not to repeat the same combination of three tests on any two semesters. Determine the minimum number of different tests that the professor should make.

22. Using Theorem 2.6, expand $(x + y + z)^2$.

23. What is the coefficient of $x^2 y^3 z^2$ in the expansion of $(2x - y + 3z)^7$?

24. What is the coefficient of $x^3 y^7$ in the expansion of $(2x - y + 3)^{13}$?

25. An ordinary deck of 52 cards is dealt among four players, 13 each, at random. What is the probability that each player gets 13 cards of the same suit?

26. Using induction, binomial expansion, and the identity

$$\binom{n}{n_1} \frac{(n - n_1)!}{n_2! \, n_3! \, \cdots \, n_k!} = \frac{n!}{n_1! \, n_2! \, \cdots \, n_k!},$$

prove the formula of multinomial expansion.

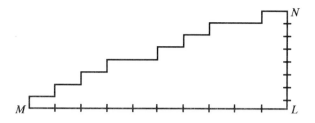

Figure 2.4 Staircase of Exercise 27.

27. A staircase is to be constructed between M and N (see Figure 2.4). The distances from M to L, and from L to N, are 5 and 2 meters, respectively. If the height of a step is 25 centimeters and its width can be any integer multiple of 50 centimeters, how many different choices do we have?

28. Each of the 50 U.S. states has two senators. What is the probability that in a random committee of 50 senators (a) Maryland is represented; (b) all states are represented?

B

29. Prove the binomial expansion formula by induction.
HINT: Use the identity $\binom{n}{k-1} + \binom{n}{k} = \binom{n+1}{k}$.

30. A class has 30 students. What is the probability that there are six months each containing the birthdays of two students and six months each containing the birthdays of three students? Assume that all months have the same probability of including the birthday of a randomly selected person.

31. In a closet there are 10 pairs of shoes. If six shoes are selected at random, what is the probability of (a) no complete pairs; (b) exactly one complete pair; (c) exactly two complete pairs; (d) exactly three complete pairs?

32. An ordinary deck of 52 cards is divided into two equal sets randomly. What is the probability that each set contains exactly 13 red cards?

33. A train consists of n cars. Each of m passengers, $m > n$, will choose a car at random to ride in. What is the probability that (a) there will be at least one passenger in each car; (b) exactly r ($r < n$) cars remain unoccupied?

34. Suppose that n indistinguishable balls are placed at random into n distinguishable cells. What is the probability that exactly one cell remains empty?

35. Prove that

$$\binom{n}{0} - \binom{n}{1} + \binom{n}{2} - \cdots + (-1)^k \binom{n}{k} + \cdots + (-1)^n \binom{n}{n} = 0.$$

36. Show that

$$\binom{n}{0} + \binom{n+1}{1} + \cdots + \binom{n+r}{r} = \binom{n+r+1}{r}.$$

HINT: $\binom{n}{r} = \binom{n+1}{r} - \binom{n}{r-1}$.

37. By a combinatorial argument, prove that for $r \le n$ and $r \le m$,

$$\binom{n+m}{r} = \binom{m}{0}\binom{n}{r} + \binom{m}{1}\binom{n}{r-1} + \cdots + \binom{m}{r}\binom{n}{0}.$$

38. Evaluate the following sum

$$\binom{n}{0} + \frac{1}{2}\binom{n}{1} + \frac{1}{3}\binom{n}{2} + \cdots + \frac{1}{n+1}\binom{n}{n}.$$

39. Suppose that five points are selected at random from the interval $(0, 1)$. What is the probability that exactly two of them are between 0 and $1/4$? HINT: For any point there are four equally likely possibilities: to fall into $(0, 1/4)$, $[1/4, 1/2)$, $[1/2, 3/4)$, and $[3/4, 1)$.

40. A lake has N trout, and t of them are caught at random, tagged, and returned. We catch n trout at a later time randomly and observe that m of them are tagged.
   (a) Find P_N, the probability of what we observed happen.
   (b) To estimate the number of trout in the lake, statisticians find the value of N that maximizes P_N. Such a value is called the *maximum likelihood estimator of N*. Show that the maximum of P_N is $[nt/m]$, where by $[nt/m]$ we mean the greatest integer less than or equal to nt/m. That is, prove that the maximum likelihood estimator of the number of trout in the lake is $[nt/m]$.
   HINT: Investigate for what values of N P_N is increasing and for what values it is decreasing.

41. In how many ways can 10 different photographs be placed in six different envelopes, no envelope being left empty?
   HINT: An easy way to do this problem is to use the following version of the inclusion–exclusion principle: Let $A_1, A_2, \ldots, A_n$ be n subsets of

a finite set Ω with N elements. Let $N(A_i)$ be the number of elements of A_i, $1 \le i \le n$, and $N(A_i^c) = N - N(A_i)$. Let S_k be the sum of the elements of all those intersections of $A_1, A_2, \ldots, A_n$ that are formed of exactly k sets. That is,

$$S_1 = N(A_1) + N(A_2) + \cdots + N(A_n),$$
$$S_2 = N(A_1 A_2) + N(A_1 A_3) + \cdots + N(A_{n-1} A_n),$$

and so on. Then

$$N(A_1^c A_2^c \cdots A_n^c) = N - S_1 + S_2 - S_3 + \cdots + (-1)^n S_n.$$

Now to solve the problem, let N be the number of ways that 10 different photographs can be placed in six different envelopes, allowing for the possibility of empty envelopes. Let A_i be the set of all situations in which envelope i is empty. Then the desired quantity is $N(A_1^c A_2^c \cdots A_6^c)$.

42. We are given n $(n > 5)$ points in space no three of which lie on the same straight line. Let Ω be the family of planes defined by any three of these points. Suppose that the points are situated in a way that no four of them are coplanar and no two planes of Ω are parallel. From the set of the lines of the intersections of the planes of Ω, a line is selected at random. What is the probability that it passes through none of the n points?

    HINT: For $i = 0, 1, 2$, let A_i be the set of all lines of the intersections that are determined by planes having i of the given n points in common. If $|A_i|$ denotes the number of elements of A_i, the answer is $|A_0|/(|A_0| + |A_1| + |A_2|)$.

43. Using the binomial theorem, calculate the coefficient of x^n in the expansion of $(1 + x)^{2n} = (1 + x)^n (1 + x)^n$ to prove that

$$\binom{2n}{n} = \sum_{i=0}^{n} \binom{n}{i}^2.$$

    For a combinatorial proof of this relation, see Example 2.25.

44. An absentminded professor wrote n letters and sealed them in envelopes without writing the addresses on them. Then he wrote the n addresses on the envelopes at random. What is the probability that exactly k of the envelopes were addressed correctly?

    HINT: Consider a particular set of k letters. Let M be the total number of ways that only these k letters can be addressed correctly. The desired probability is the quantity $\binom{n}{k} M/n!$; using Example 2.21, argue that M satisfies $\sum_{i=2}^{n-k} (-1)^i / i! = M/(n-k)!$.

45. A fair coin is tossed n times. Calculate the probability of no successive heads.

HINT: Let x_i be the number of sequences of H's and T's of length i with no successive H's. Show that x_i satisfies $x_i = x_{i-1} + x_{i-2}$, $i \geq 2$, where $x_0 = 1$ and $x_1 = 2$. The answer is $x_n/2^n$. Note that $\{x_i\}_{i=1}^{\infty}$ is a Fibonacci-type sequence.

46. What is the probability that the birthdays of at least two students of a class of size n are at most k days apart? Assume that the birth rates are constants throughout the year and that each year has 365 days.

HINT: A solution is given by Abramson and Moser in the October 1970 issue of the *American Mathematical Monthly*.

2.5 Stirling's Formula

To estimate $n!$ for large values of n, the following formula, which is very important in analytic probability theory, is used. It was discovered in 1730 by James Stirling (1692–1770) and had appeared in his book *Methodus Differentialis* (Bowyer, London, 1730). Also, independently, a slightly less general version of the formula was discovered by De Moivre, who used it to prove the central limit theorem.

THEOREM 2.7   (Stirling's Formula)

$$n! \sim \sqrt{2\pi n}\, n^n e^{-n},$$

where the sign $\sim$ means

$$\lim_{n \to \infty} \frac{n!}{\sqrt{2\pi n}\, n^n e^{-n}} = 1.$$

Stirling's formula, which usually gives excellent approximations in numerical computations, is also often used to prove theoretical problems as well. One should note that although the ratio $R(n) = n!/(\sqrt{2\pi n}\, n^n e^{-n})$ becomes 1 at ∞, it is still close to 1 even for very small values of n. The following table shows this fact.

n	$n!$	$\sqrt{2\pi n}\; n^n e^{-n}$	$R(n)$
1	1	0.922	1.084
2	2	1.919	1.042
5	120	118.019	1.017
8	40,320	39,902.396	1.010
10	3,628,800	3,598,695.618	1.008
12	479,001,600	475,687,486.474	1.007

We should be careful that even though $\lim_{n\to\infty} R(n)$ is 1, the difference between $n!$ and $\sqrt{2\pi n}\; n^n e^{-n}$ increases as n gets larger. In fact, it goes to ∞ as $n \to \infty$.

Example 2.27   Approximate the value of $[2^n (n!)^2]/(2n)!$ for large n.

SOLUTION: By Stirling's formula, we have

$$\frac{2^n (n!)^2}{(2n)!} \sim \frac{2^n 2\pi n (n^n)^2 e^{-2n}}{\sqrt{4\pi n}\,(2n)^{2n} e^{-2n}} = \frac{\sqrt{\pi n}}{2^n}. \quad \blacklozenge$$

EXERCISE

1. Use Stirling's formula to approximate $\binom{2n}{n} 1/2^{2n}$ and $[(2n)!]^3/$ $[(4n)!\,(n!)^2]$ for large n.

Review Problems

1. Albert goes to the grocery store to buy fruit. There are seven different varieties of fruit and Albert is determined to buy no more than one of any variety. How many different orders can he place?

2. Virginia has 1 one-dollar bill, 1 two-dollar bill, 1 five-dollar bill, 1 ten-dollar bill, and 1 twenty-dollar bill. She decides to give some money to her son Brian without asking for change. How many choices does she have?

3. If four fair dice are tossed, what is the probability that they show four different faces?

4. From the 10 points that are placed on a circumference, two are selected randomly. What is the probability that they are adjacent?

5. A father buys nine different toys for his four children. In how many ways can he give one child three toys and the remaining three children two toys each?

6. A window dresser has decided to display five different dresses in a circular arrangement. How many choices does she have?

7. Judy has three sets of classics in literature, each set having four volumes. In how many ways can she put them in a bookshelf so that books of each set are not separated?

8. Suppose that 30 lawn mowers, of which seven have defects, are sold to a hardware store. If the inspector of the store inspects six of the lawn mowers randomly, what is the probability that he finds at least one defective lawn mower?

9. In how many ways can 23 identical refrigerators be allocated among four stores so that one store gets eight refrigerators, another four, a third store five, and the last one six refrigerators?

10. In how many arrangements of the letters *BERKELEY* are all three *E*'s adjacent?

11. Bill and John play in a backgammon tournament. A player is the winner if he wins three games in a row or four games altogether. In what percent of all possible cases does the tournament end because John wins four games without winning three in a row?

12. In a small town, both of the accidents that occurred during the week of June 8, 1988, were on Friday the 13th. Is this a good excuse for a superstitious person to argue that Friday the 13th's are inauspicious?

13. How many eight-digit numbers without two identical successive digits are there?

14. A *palindrome* is a sequence of characters that reads the same forward and backward. For example, *rotator, madam, Hannah*, the German name *Otto*, and an Indian language, *Malayalam*, are palindromes. So are the following expressions: *"Put up," "Madam I'm Adam," "Was it a cat I saw?"* and these two sentences in Latin concerning St. Martin, Bishop of Tours: *"Signa te, signa; temere me tangis et angis. Roma tibi subito motibus ibit amor."* (Cross, cross yourself; you annoy and vex me needlessly. Through my exertions, Rome, your desire, will soon be near.) Determine the number of palindromes of length 11 that can

be made with (a) no letters repeating more than twice; (b) one letter repeating three times and no other letter more than twice (the chemical term *detartrated* is one such palindrome); (c) one letter repeating three times, two each repeating two times, and one repeating four times.

15. By mistake, a student who is running a computer program enters with negative signs two of the six positive numbers and with positive signs two of the four negative numbers. If at some stage the program chooses three distinct numbers from these 10 at random and multiplies them, what is the probability that at that stage no mistake will occur?

16. Suppose that four women and two men enter a restaurant and sit at random around a table that has four chairs on one side and another four on the other side. What is the probability that the men are not sitting on one side?

17. Cyrus and 27 other students are taking a course in probability this semester. If their professor chooses eight students at random and with replacement to ask them eight different questions, what is the probability that one of them is Cyrus?

18. If five Americans, five Italians, and five Mexicans sit randomly at a round table, what is the probability that the persons of the same nationality sit together?

19. The chairperson of the industry–academic partnership of a town invites all 12 members of the board and their spouses to his house for a Christmas party. If a board member may attend without his spouse, but not vice versa, how many different groups can the chairperson get?

20. In a bridge game, each of the four players gets 13 random cards. What is the probability that in a bridge game every player has an ace?

21. From a faculty of six professors, six associate professors, 10 assistant professors, and 12 instructors, a committee of size 6 is formed randomly. What is the probability that there is at least one person from each rank on the committee?
HINT: : Be careful, the answer is not

$$\frac{\binom{6}{1}\binom{6}{1}\binom{10}{1}\binom{12}{1}\binom{30}{2}}{\binom{34}{6}} = 1.397.$$

To find the correct answer, use the inclusion–exclusion principle explained in Section 1.4.

22. An urn contains 15 white and 15 black balls. Suppose that 15 persons each draw two balls blindfolded from the urn without replacement. What is the probability that each of them draws one white ball and one black ball?

23. In a lottery the tickets are numbered 1 through N. A person purchases n $(1 \le n \le N)$ tickets at random. What is the probability that the ticket numbers are consecutive? (This is a special case of a problem posed by Euler in 1763.)

24. An ordinary deck of 52 cards is dealt at random among A, B, C, and D, 13 each. What is the probability that (a) A and B together get two aces; (b) A gets all the face cards; (c) A gets five hearts and B gets the remaining eight hearts?

25. To test if a computer program works properly, we run it with 12 different data sets using four computers, each running three data sets. If the data sets are distributed randomly among different computers, how many possibilities are there?

26. If two fair dice are tossed six times, what is the probability that the sixth sum obtained is not a repetition?

Chapter 3

CONDITIONAL
PROBABILITY
AND INDEPENDENCE

3.1   Conditional Probability

To introduce the notion of conditional probability, let us first examine the following question: Suppose that all of the freshmen of an engineering college took calculus and discrete math last semester. Suppose that 70% of the students passed calculus, 55% passed discrete math, and 45% passed both. If a randomly selected freshman is found to have passed calculus last semester, what is the probability that he or she also passed discrete math last semester? To answer this question, let A and B be the events that the randomly selected freshman passed discrete math and calculus last semester, respectively. Note that the quantity we are asked to find is not $P(A)$, which is 0.55; it would have been if we were not aware that B has occurred. Knowing that B has occurred changes the chances of the occurrence of A. To find the desired probability, denoted by the symbol $P(A \mid B)$ [read: probability of A given B] and called *the conditional probability of A given B*, let n be the number of all the freshmen in the engineering college. Then $(0.7)n$ is the number of freshmen who passed calculus and $(0.45)n$ is the number of those who passed both calculus and discrete math. Therefore, of the $(0.7)n$ freshmen who passed calculus, $(0.45)n$ of them passed discrete math as well. It is given that the randomly selected student is one of the $(0.7)n$ who passed calculus; we also want to find the probability that he or she is one of the $(0.45)n$ who passed discrete math. This is obviously equal to $(0.45)n/(0.7)n = 0.45/0.7$. Hence

$$P(A \mid B) = \frac{0.45}{0.7}.$$

But 0.45 is $P(AB)$ and 0.7 is $P(B)$. Therefore, this example suggests that

$$P(A \mid B) = \frac{P(AB)}{P(B)}. \tag{3.1}$$

As another example, suppose that a bus arrives at a station every day at a random time between 1:00 P.M. and 1:30 P.M. Suppose that at 1:10 the bus has not arrived and we want to calculate the probability that it will arrive in not more than 10 minutes from now. Let A be the event that the bus will arrive between 1:10 and 1:20 and B be the event that it will arrive between 1:10 and 1:30. Then the desired probability is

$$P(A \mid B) = \frac{20 - 10}{30 - 10} = \frac{\dfrac{20 - 10}{30 - 0}}{\dfrac{30 - 10}{30 - 0}} = \frac{P(AB)}{P(B)}. \tag{3.2}$$

The relation $P(A \mid B) = P(AB)/P(A)$ discovered in (3.1) and (3.2) can also be verified for other types of conditional probability problems. It is compatible with our intuition and is a logical generalization of the existing relation between the conditional relative frequency of the event A with respect to the condition B and the ratio of the relative frequencies of AB and the relative frequency of B. For these reasons it is taken as a definition.

DEFINITION   *If $P(B) > 0$, the conditional probability of A given B, denoted by $P(A \mid B)$, is*

$$P(A \mid B) = \frac{P(AB)}{P(B)}. \tag{3.3}$$

If $P(B) = 0$, formula (3.3) makes no sense, so the conditional probability is defined only for $P(B) > 0$. This formula may be expressed in words by saying that the conditional probability of A given that B has occurred is the ratio of the probability of joint occurrence of A and B and the probability of B. It is important to note that (3.3) is neither an axiom nor a theorem. It is a definition. As explained above, this definition is fully justified and is not made arbitrarily. Indeed, historically, it was used implicitly even before it was formally introduced by De Moivre in the book *The Doctrine of Chance*.

Example 3.1 In a certain region of Russia, the probability that a person lives at least 80 years is 0.75 and the probability that he or she lives at least 90 years is 0.63. What is the probability that a randomly selected 80-year-old person from this region will survive to become 90?

SOLUTION: Let B and A be the events that the person selected survives to become 80 and 90 years old, respectively. We are interested in $P(A \mid B)$. By definition,

$$P(A \mid B) = \frac{P(AB)}{P(B)} = \frac{P(A)}{P(B)} = \frac{0.63}{0.75} = 0.84. \quad \blacklozenge$$

Example 3.2 From the set of all families with two children, a *family* is selected at random and is found to have a girl. What is the probability that the other child of the family is a girl? Assume that in a two-child family all sex distributions are equally probable.

SOLUTION: Let B and A be the events that the family has a girl and the family has two girls, respectively. We are interested in $P(A \mid B)$. Now in a family with two children there are four equally likely possibilities: (boy, boy), (girl, girl), (boy, girl), (girl, boy), where by, say, (girl, boy), we mean that the older child is a girl and the younger is a boy. Thus $P(B) = 3/4$, $P(AB) = 1/4$. Hence

$$P(A \mid B) = \frac{P(AB)}{P(B)} = \frac{1/4}{3/4} = \frac{1}{3}. \quad \blacklozenge$$

Example 3.3 From the set of all families with two children, a *child* is selected at random and is found to be a girl. What is the probability that the second child of this girl's family is also a girl? Assume that in a two-child family all sex distributions are equally probable.

SOLUTION: Let B be the event that the child selected at random is a girl, and let A be the event that the second child of her family is also a girl. We want to calculate $P(A \mid B)$. Now the set of all possibilities is as follows: The child is a girl with a sister, the child is a girl with a brother, the child is a boy with a sister, and the child is a boy with a brother. Thus $P(B) = 2/4$ and $P(AB) = 1/4$. Hence

$$P(A \mid B) = \frac{P(AB)}{P(B)} = \frac{1/4}{2/4} = \frac{1}{2}. \quad \blacklozenge$$

NOTE: There is a major difference between Examples 3.2 and 3.3. In Example 3.2 a family is selected and found to have a girl. Therefore, we have three equally likely possibilities: (girl, girl), (boy, girl), and (girl, boy), of which only one is desirable: (girl, girl). So the required probability is 1/3. In Example 3.3, since rather than a family a child is selected, families with (girl, girl), (boy, girl), and (girl, boy) are not equally likely to be chosen. In fact, the probability that a family with two girls is selected equals twice the probability that a family with one girl is selected. That is,

$$P(\text{girl, girl}) = 2P(\text{girl, boy}) = 2P(\text{boy, girl}).$$

This and the fact that the sum of the probabilities of (girl, girl), (girl, boy), (boy, girl) is 1 give

$$P(\text{girl, girl}) + \frac{1}{2}P(\text{girl, girl}) + \frac{1}{2}P(\text{girl, girl}) = 1,$$

which implies that

$$P(\text{girl, girl}) = \frac{1}{2}. \quad \blacklozenge$$

Example 3.4 An English class consists of 10 Koreans, five Italians, and 15 Hispanics. A paper belonging to one of the students of this class is found. If the name on the paper is not Korean, what is the probability that it is Italian? Assume that names completely identify ethnic groups.

SOLUTION: Let A be the event that the name on the paper is Italian, and let B be the event that it is not Korean. To calculate $P(A \mid B)$, note that $AB = A$, hence $P(AB) = 5/30$. Therefore,

$$P(A \mid B) = \frac{P(AB)}{P(B)} = \frac{5/30}{20/30} = \frac{1}{4}. \quad \blacklozenge$$

Example 3.5 We draw eight cards at random from an ordinary deck of 52 cards. Given that three of them are spades, what is the probability that the remaining five are also spades?

SOLUTION: Let B be the event that at least three of them are spades, and let A denote the event that all of the eight cards selected are spades. Then $P(A \mid B)$ is the desired quantity and is calculated from $P(A \mid B) = P(AB)/P(B)$ as follows: $P(AB) = \binom{13}{8} \Big/ \binom{52}{8}$ since AB is the event that all eight cards selected are spades. $P(B) = \sum_{x=3}^{8} \left[\binom{13}{x}\binom{39}{8-x} \right] \Big/ \binom{52}{8}$ because at least three of the cards selected are spades if exactly three of them are spades, or exactly four of them are spades, ..., or all eight of them are spades. Hence

$$P(A \mid B) = \frac{P(AB)}{P(B)} = \frac{\dfrac{\binom{13}{8}}{\binom{52}{8}}}{\displaystyle\sum_{x=3}^{8} \dfrac{\binom{13}{x}\binom{39}{8-x}}{\binom{52}{8}}} \approx 5.44 \times 10^{-6}. \quad \blacklozenge$$

An important feature of conditional probabilities (stated in the following theorem) is that they satisfy the same axioms that ordinary probabilities satisfy. This enables us to use the theorems that are true for probabilities for conditional probabilities as well.

THEOREM 3.1   *Let S be the sample space of an experiment, and let B be an event of S with $P(B) > 0$. Then*

(a) $P(A \mid B) \geq 0$ *for any event A of S.*
(b) $P(S \mid B) = 1.$
(c) *If $A_1, A_2, \ldots$ is a sequence of mutually exclusive events, then*

$$P\left(\bigcup_{i=1}^{\infty} A_i \mid B\right) = \sum_{i=1}^{\infty} P(A_i \mid B).$$

The proof of this theorem is left as an exercise.

Reduction of Sample Space

Let B be an event of a sample space S with $P(B) > 0$. For a subset A of B, define $Q(A) = P(A \mid B)$. Then Q is a function from the set of subsets of B to $[0, 1]$. Clearly, $Q(A) \geq 0$, $Q(B) = P(B \mid B) = 1$, and by Theorem 3.1, if $A_1, A_2, \ldots$ is a sequence of mutually exclusive subsets of B, then

$$Q\left(\bigcup_{i=1}^{\infty} A_i\right) = P\left(\bigcup_{i=1}^{\infty} A_i \mid B\right) = \sum_{i=1}^{\infty} P(A_i \mid B) = \sum_{i=1}^{\infty} Q(A_i).$$

Thus Q satisfies the axioms that probabilities do and hence it is a probability function. Note that while P is defined for all subsets of S, the probability function Q is defined only for subsets of B. Therefore, for Q, the sample space is reduced from S to B. This reduction of sample space is sometimes very helpful in calculation of conditional probabilities. Suppose that we are interested in $P(E \mid B)$, where $E \subseteq S$. One way to calculate this quantity is to reduce S to B and find $Q(EB)$. It is usually easier to compute the unconditional probability $Q(EB)$ rather than the conditional probability $P(E \mid B)$. Examples follow.

Example 3.6   A child mixes 10 good and three dead batteries. To find the dead batteries, his father tests them one by one and without replacement. If the first four batteries tested are all good, what is the probability that the fifth one is dead?

SOLUTION: Using the information that the first four batteries tested are all good, we rephrase the problem in the reduced sample space: six good and three dead batteries are mixed. A battery is selected at random, what is the probability that it is dead? The solution to this trivial question is 3/9 = 1/3. Note that without reducing the sample space, the solution of this problem is not so easy.   ◆

Example 3.7   A farmer decides to test four fertilizers for his soybean fields. He buys 32 bags of fertilizers, eight bags from each kind, and tries them randomly on 32 plots, eight plots from each of fields A, B, C, and D, one bag per plot. If from type I fertilizer one bag is tried on field A and 3 on field B, what is the probability that two bags of type I fertilizer are tried on field D?

SOLUTION: Using the information that one and three bags of type I fertilizer are tried on fields A and B, respectively, we rephrase the problem in the reduced sample space: 16 bags of fertilizers, of which four are type I, are tried randomly, eight on field C and eight on field D. What is the probability that two bags of type I fertilizer are tried on field D? The solution to this easier unconditional version of the problem is

$$\frac{\binom{4}{2}\binom{12}{6}}{\binom{16}{8}} \approx 0.43. \quad ◆$$

Law of Multiplication

The relation $P(A \mid B) = P(AB)/P(B)$ is also useful for calculating $P(AB)$. If we multiply both sides of this relation by $P(B)$ [note that $P(B) > 0$], we get

$$P(AB) = P(B)P(A \mid B), \tag{3.4}$$

which means that the probability of the joint occurrence of A and B is the product of the probability of B and the conditional probability of A given that B has occurred. If $P(A) > 0$, then by letting $A = B$ and $B = A$ in (3.4), we obtain

$$P(BA) = P(A)P(B \mid A).$$

Since $P(BA) = P(AB)$, this relation gives

$$P(AB) = P(A)P(B \mid A). \tag{3.5}$$

Thus, to calculate $P(AB)$, depending on which of the quantities $P(A \mid B)$ and $P(B \mid A)$ is known, we may use (3.4) or (3.5), respectively. The following example clarifies the usefulness of these relations.

Example 3.8   Suppose that five good fuses and two defective ones have been mixed up. To find the defective fuses, we test them one by one, at random, and without replacement. What is the probability that we are lucky and find both of the defective fuses in the first two tests?

SOLUTION: Let D_1 and D_2 be the events of finding a defective fuse in the first and second tests, respectively. We are interested in $P(D_1 D_2)$. Using (3.5), we get

$$P(D_1 D_2) = P(D_1)P(D_2 \mid D_1) = \frac{2}{7} \times \frac{1}{6} = \frac{1}{21}. \quad \blacklozenge$$

Relation (3.5) can be generalized for calculating the probability of the joint occurrence of several events. For example, if $P(AB) > 0$, then

$$P(ABC) = P(A)P(B \mid A)P(C \mid AB). \tag{3.6}$$

To see this, note that $P(AB) > 0$ implies that $P(A) > 0$; therefore,

$$P(A)P(B \mid A)P(C \mid AB) = P(A)\frac{P(AB)}{P(A)}\frac{P(ABC)}{P(AB)} = P(ABC).$$

The following theorem will generalize (3.5) and (3.6) to n events and can be shown in the same way as (3.6) is shown.

THEOREM 3.2 *If $P(A_1A_2A_3 \cdots A_{n-1}) > 0$, then*

$$P(A_1A_2A_3 \cdots A_{n-1}A_n)$$
$$= P(A_1)P(A_2 \mid A_1)P(A_3 \mid A_1A_2) \cdots P(A_n \mid A_1A_2A_3 \cdots A_{n-1}).$$

Example 3.9 Suppose that five good and two defective fuses have been mixed up. To find the defective ones, we test them one by one, at random, and without replacement. What is the probability that we find both of the defective fuses in exactly three tests?

SOLUTION: Let D_1, D_2, and D_3 be the events that the first, second, and third fuses tested are defective, respectively. Let G_1, G_2, and G_3 be the events that the first, second, and third fuses tested are good, respectively. We are interested in the probability of the event $G_1D_2D_3 \cup D_1G_2D_3$. We have

$$P(G_1D_2D_3 \cup D_1G_2D_3) = P(G_1D_2D_3) + P(D_1G_2D_3)$$
$$= P(G_1)P(D_2 \mid G_1)P(D_3 \mid G_1D_2)$$
$$+ P(D_1)P(G_2 \mid D_1)P(D_3 \mid D_1G_2)$$
$$= \frac{5}{7} \times \frac{2}{6} \times \frac{1}{5} + \frac{2}{7} \times \frac{5}{6} \times \frac{1}{5} \approx 0.095. \quad \blacklozenge$$

EXERCISES

A

1. Suppose that 15% of the population of a country are unemployed women and 25% are unemployed. What percent of the unemployed are women?

2. Suppose that 41% of Americans have blood type A and 4% have blood type AB. If in the blood of a randomly selected American soldier the A antigen is found, what is the probability that his blood type is A? The A antigen is found only in blood types A and AB.

3. In a technical college all students are required to take calculus and physics. Statistics shows that 32% of the students of this college get A's in calculus and 20% of them get A's in both calculus and physics. Gino, a randomly selected student of this college, has passed calculus with an A. What is the probability that he got an A in physics?

4. Suppose that two fair dice have been tossed and the total of their top faces is found to be divisible by 5. What is the probability that both of them have landed 5?

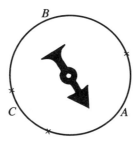

Figure 3.1   Spinner of Exercise 7.

5. A bus arrives at a station every day at a random time between 1:00 P.M. and 1:30 P.M. A person arrives at this station at 1 and waits for the bus. If at 1:15 the bus has not yet arrived, what is the probability that the person should have to wait at least an additional 5 minutes?

6. Prove that $P(A \mid B) > P(A)$ if and only if $P(B \mid A) > P(B)$. In probability, if for two events A and B, $P(A \mid B) > P(A)$, we say that A and B are *positively correlated*. If $P(A \mid B) < P(A)$, A and B are said to be *negatively correlated*.

7. A spinner is mounted on a wheel of unit circumference (radius $1/2\pi$). Arcs A, B, and C of lengths $1/3$, $1/2$, and $1/6$ are marked off on the wheel's perimeter (see Figure 3.1). The spinner is flicked and we know that it is not pointing toward C. What is the probability that it points toward A?

8. In throwing two fair dice, what is the probability of a sum of 5 if they land on different numbers?

9. In a small lake, it is estimated that there are approximately 105 fish, of which 40 are trout and 65 are carp. A fisherman caught eight fish; what is the probability that exactly two of them are trout if we know that at least three of them are not?

10. From 100 cards numbered $00, 01, \ldots, 99$, one card is drawn. Suppose that α and β are the sum and the product, respectively, of the digits of the card selected. Calculate $P\{\alpha = i \mid \beta = 0\}$, $i = 0, 1, 2, 3, \ldots, 18$.

11. From families with three children, a family is selected at random and found to have a boy. What is the probability that the boy has (a) an older brother and a younger sister; (b) an older brother; (c) a brother and a sister? Assume that in a three-child family all gender distributions have equal probabilities.

12. Prove that if $P(A) = a$ and $P(B) = b$, then $P(A \mid B) \geq (a + b - 1)/b$.

13. Prove Theorem 3.1.

14. Prove that if $P(E \mid F) \geq P(G \mid F)$ and $P(E \mid F^c) \geq P(G \mid F^c)$, then $P(E) \geq P(G)$.

15. The theaters of a town are showing seven comedies and nine dramas. Marlon has seen five of them so far. If the first three movies he has seen are dramas, what is the probability that the last two are comedies? Assume that Marlon chooses the shows at random and sees each movie at most once.

16. Suppose that 28 crayons, of which four are red, are divided randomly among Jack, Marty, Sharon, and Martha (seven each). If Sharon has exactly one red crayon, what is the probability that Marty has the remaining three?

17. If eight defective and 12 nondefective items are inspected one by one, at random, and without replacement, what is the probability that (a) the first four items inspected are defective; (b) from the first three items at least two are defective?

18. There are five boys and six girls in a class. For an oral exam, their teacher calls them one by one and randomly. (a) What is the probability that the boys and the girls alternate? (b) What is the probability that the boys are called first? Compare the answers to parts (a) and (b).

B

19. A number is selected at random from the set $\{1, 2, \ldots, 10{,}000\}$ and observed to be odd. What is the probability that, it is (a) divisible by three; (b) divisible by neither 3 nor 5?

20. A retired person chooses randomly one of the six parks of his town everyday and goes there for hiking. We are told that he was seen in one of these parks, Oregon Ridge, once during the last 10 days. What is the probability that during this period he has hiked in this park two or more times?

21. An urn contains five white and three red chips. Each time we draw a chip, we look at its color. If it is red, we replace it along with two new red chips, and if it is white, we replace it along with three new white chips. What is the probability that in successive drawing of chips the colors of the first four chips alternate?

22. A big urn contains 1000 red chips numbered 1 through 1000 and 1750 blue chips numbered 1 through 1750. A chip is removed at random and its number is found to be divisible by 3. What is the probability that its number is also divisible by 5?

23. There are three types of animals in a laboratory: 15 type I, 13 type II, and 12 type III. Animals of type I react to a particular stimulus in 5 seconds, animals of types II and III react to the same stimulus in 4.5 and 6.2 seconds, respectively. A psychologist selects 10 of these animals at random and finds that exactly four of them react to this stimulus in 6.2 seconds. What is the probability that at least two of them react to the same stimulus in 4.5 seconds?

24. In an international school, 60 students, of whom 15 are Korean, 20 are French, eight are Greek, and the rest are Chinese, are divided randomly into four classes of 15 each. If there are a total of eight French and six Korean students in classes A and B, what is the probability that class C has four of the remaining 12 French and three of the remaining nine Korean students?

25. From an ordinary deck of 52 cards, cards are drawn one by one, at random and without replacement. What is the probability that the fourth heart is drawn on the tenth draw?
    HINT: Let F denote the event that in the first nine draws there are exactly three hearts and E be the event that the tenth draw is a heart. Use $P(FE) = P(F)P(E \mid F)$.

26. In a series of games, the winning number of the nth game, $n = 1, 2, 3, \ldots$, is a number selected at random from the set of integers $\{1, 2, \ldots, n+2\}$. Don bets on 1 in each game and says that he quits as soon as he wins. What is the probability that he has to play indefinitely?
    HINT: Let A_n be the event that Don loses the first n games. To calculate the desired probability, $P(\cap_{i=1}^{\infty} A_i)$, use Theorem 1.7.

3.2   Law of Total Probability

Sometimes it is not possible to calculate the probability of the occurrence of an event A directly, but it is possible to find $P(A \mid B)$ and $P(A \mid B^c)$ for some event B. In such cases, the following theorem, which is conceptually rich and has enormous applications, is used. It states that $P(A)$ is the weighted average of the probability of A given that B has occurred and given that it has not occurred.

THEOREM 3.3   (Law of Total Probability)   *Let B be an event with $P(B) > 0$ and $P(B^c) > 0$. Then for any event A,*

$$P(A) = P(A \mid B)P(B) + P(A \mid B^c)P(B^c).$$

PROOF: Let S be the sample space of the experiment. Clearly, $AB \cup AB^c = A$. Since $(AB)(AB^c) = \emptyset$, we have

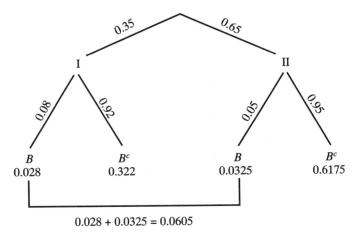

Figure 3.2   Tree diagram of Example 3.10.

$$P(A) = P(AB \cup AB^c) = P(AB) + P(AB^c). \qquad (3.7)$$

Now $P(B) > 0$ and $P(B^c) > 0$. These imply that $P(AB) = P(A \mid B)P(B)$ and $P(AB^c) = P(A \mid B^c)P(B^c)$. Putting these in (3.7), we have the theorem. ◆

Example 3.10   An insurance company rents 35% of the cars for its customers from agency I and 65% from agency II. If 8% of the cars of agency I and 5% of the cars of agency II break down during the rental periods, what is the probability that a car rented by this insurance company breaks down?

SOLUTION: Let A be the event that a car rented by this insurance company breaks down. Let I and II be the events that it is rented from agencies I and II, respectively. Then by the law of total probability:

$$P(A) = P(A \mid \text{I})P(\text{I}) + P(A \mid \text{II})P(\text{II})$$
$$= (0.08)(0.35) + (0.05)(0.65) = 0.0605.$$

Tree diagrams facilitate solutions to this kind of problem.   Let B and B^c stand for breakdown and not breakdown during the rental period, respectively.   Then as Figure 3.2 shows, to find the probability that a car breaks down, all we need to do is to compute, by multiplication,   the probability of each path that leads to a point B and then add them up.   So as seen from the tree, the probability that the car breaks down is

$(0.35)(0.08) + (0.65)(0.05) = 0.06$ and the probability that it does not break down is $(0.35)(0.92) + (0.65)(0.95) = 0.94$.   ◆

Example 3.11   (Polya's Urn Model)   An urn contains w white and b blue chips. A chip is drawn at random and then is returned to the urn with $c > 0$ chips of the same color. Prove that if $n = 2, 3, 4, \ldots$, such experiments are made, then at each draw the probability of a white chip is still $w/(w+b)$ and the probability of a blue chip is $b/(w+b)$. This model was first introduced in preliminary studies of "contagious diseases" and "accident proneness" in actuarial mathematics.

PROOF: For $1 \le i \le n$, let W_i be the event that the ith draw is white and B_i be the event that it is blue. To show that $P(W_n) = w/(w+b)$ and $P(B_n) = b/(w+b)$ for all n, we use induction. Clearly, $P(W_1) = w/(w+b)$ and $P(B_1) = b/(w+b)$. Assume that $P(W_{k-1}) = w/(w+b)$ and $P(B_{k-1}) = b/(w+b)$. To show that $P(W_k) = w/(w+b)$ and $P(B_k) = b/(w+b)$, suppose that after the $(k-2)$nd experiment there are w_1 white and b_1 blue chips in the urn. Then $P(W_{k-1}) = w_1/(w_1 + b_1)$ and $P(B_{k-1}) = b_1/(w_1 + b_1)$; thus

$$\frac{w_1}{w_1 + b_1} = \frac{w}{w+b}, \quad \frac{b_1}{w_1 + b_1} = \frac{b}{w+b}.$$

These give

$$P(W_k) = P(W_k \mid W_{k-1})P(W_{k-1}) + P(W_k \mid B_{k-1})P(B_{k-1})$$
$$= \frac{w_1 + c}{w_1 + b_1 + c} \cdot \frac{w_1}{w_1 + b_1} + \frac{w_1}{w_1 + b_1 + c} \cdot \frac{b_1}{w_1 + b_1}$$
$$= \frac{w_1}{w_1 + b_1} = \frac{w}{w+b}$$

and $P(B_k) = 1 - P(W_k) = 1 - w/(w+b) = b/(w+b)$.   ◆

Example 3.12   (Gambler's Ruin Problem)   Two gamblers play the game of "heads or tails," in which each time a fair coin lands heads player A wins \$1 from B and each time it lands tails player B wins \$1 from A. Suppose that player A initially has a dollars and player B has b dollars. If they continue to play this game successively, what is the probability that (a) A will be ruined; (b) the game goes forever with nobody winning?

SOLUTION: (a) Let E be the event that A will be ruined if he or she starts with i dollars, and let $p_i = P(E)$. Our aim is to calculate p_a. To do so, we define F to be the event that A wins the first game. Then

$$P(E) = P(E \mid F)P(F) + P(E \mid F^c)P(F^c).$$

In this formula, $P(E \mid F)$ is the probability that A will be ruined given that he wins the first game, so $P(E \mid F)$ is the probability that A will be ruined if his capital is $i + 1$, that is $P(E \mid F) = p_{i+1}$. Similarly, $P(E \mid F^c) = p_{i-1}$. Hence

$$p_i = p_{i+1} \cdot \frac{1}{2} + p_{i-1} \cdot \frac{1}{2}. \tag{3.8}$$

Now $p_0 = 1$ because if A starts with 0 dollars, he or she is already ruined. Also, if the capital of A reaches $a + b$, then B is ruined, thus $p_{a+b} = 0$. Therefore, we have to solve the system of recursive equations (3.8) subject to the boundary conditions $p_0 = 1$ and $p_{a+b} = 0$. To do so, note that (3.8) implies that

$$p_{i+1} - p_i = p_i - p_{i-1}.$$

Hence, letting $p_1 - p_0 = \alpha$, we get

$$p_i - p_{i-1} = p_{i-1} - p_{i-2} = p_{i-2} - p_{i-3} = \cdots = p_2 - p_1 = p_1 - p_0 = \alpha.$$

Thus

$$p_1 = p_0 + \alpha$$
$$p_2 = p_1 + \alpha = p_0 + \alpha + \alpha = p_0 + 2\alpha$$
$$p_3 = p_2 + \alpha = p_0 + 2\alpha + \alpha = p_0 + 3\alpha$$
$$\vdots$$
$$p_i = p_0 + i\alpha$$
$$\vdots$$

Now $p_0 = 1$ gives $p_i = 1 + i\alpha$. But $p_{a+b} = 0$, thus $0 = 1 + (a + b)\alpha$. This gives $\alpha = -1/(a + b)$; therefore

$$p_i = 1 - \frac{i}{a+b} = \frac{a+b-i}{a+b}.$$

In particular, $p_a = b/(a + b)$. That is, the probability that A will be ruined is $b/(a + b)$.

(b) The same method can be used with obvious modifications to calculate q_i, the probability that B is ruined if he or she starts with i dollars. The result is

$$q_i = \frac{a+b-i}{a+b}.$$

Since B starts with b dollars, he or she will be ruined with probability $q_b = a/(a+b)$. Thus the probability that the game goes on forever with nobody winning is $1 - (q_b + p_a)$. But $1 - (q_b + p_a) = 1 - a/(a+b) - b/(a+b) = 0$. Therefore, if this game is played successively, eventually either A is ruined or B is ruined. ◆

NOTE: If the game is not fair and on each play gambler A wins \$1 from B with probability p, $0 < p < 1$, $p \neq 1/2$, and loses \$1 to B with probability $q = 1 - p$, relation (3.8) becomes

$$p_i = pp_{i+1} + qp_{i-1},$$

but $p_i = (p+q)p_i$, so $(p+q)p_i = pp_{i+1} + qp_{i-1}$. This gives

$$q(p_i - p_{i-1}) = p(p_{i+1} - p_i)$$

which reduces to

$$p_{i+1} - p_i = \frac{q}{p}(p_i - p_{i-1}).$$

Using this and following the same line of arguments lead us to

$$p_i = \left[\left(\frac{q}{p}\right)^{i-1} + \left(\frac{q}{p}\right)^{i-2} + \cdots + \left(\frac{q}{p}\right) + 1\right]\alpha + p_0,$$

where $\alpha = p_1 - p_0$. Using $p_0 = 1$, we obtain

$$p_i = \frac{(q/p)^i - 1}{(q/p) - 1}\alpha + 1.$$

After finding α from $p_{a+b} = 0$ and substituting in this equation, we get

$$p_i = \frac{(q/p)^i - (q/p)^{a+b}}{1 - (q/p)^{a+b}}.$$

In particular,

$$p_a = \frac{1 - (p/q)^b}{1 - (p/q)^{a+b}},$$

and similarly,

$$q_b = \frac{1 - (q/p)^a}{1 - (q/p)^{a+b}}.$$

In this case also, $p_a + q_b = 1$, meaning that eventually either A or B will be ruined and the game does not go on forever. For comparison, suppose that A and B both start with \$10. If they play a fair game, the probability that A will be ruined is 1/2 and the probability that B will be ruined is also 1/2. If they play an unfair game with $p = 3/4$, $q = 1/4$, then p_{10}, the probability that A will be ruined, is almost 0.00002. ◆

To generalize Theorem 3.3, we will now make a definition.

DEFINITION *Let $\{B_1, B_2, \ldots, B_n\}$ be a set of nonempty subsets of the sample space S of an experiment. If the events $B_1, B_2, \ldots, B_n$ are mutually exclusive and $\bigcup_{i=1}^{n} B_i = S$, the set $\{B_1, B_2, \ldots, B_n\}$ is called a partition of S.*

For example, if S of Figure 3.3 is the sample space of an experiment, then B_1, B_2, B_3, B_4, and B_5 of the same figure form a partition of S. As another example, consider the experiment of drawing a card from an ordinary deck of 52 cards. Let B_1, B_2, B_3, and B_4 denote the events that the card is a spade, a club, a diamond, and a heart, respectively. Then $\{B_1, B_2, B_3, B_4\}$ is a partition of the sample space of this experiment. If A_i, $1 \le i \le 10$, denotes the event that the value of the card drawn is i and A_{11}, A_{12}, and A_{13} are the events of jack, queen, and king, respectively, then $\{A_1, A_2, \ldots, A_{13}\}$ is another partition of the same sample space. *Note that for an experiment*

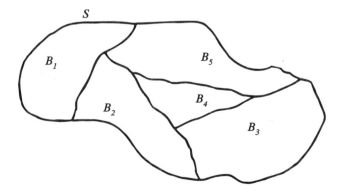

Figure 3.3 Partition of the given sample space S.

with sample space S, for any event A that is nonempty, the set $\{A, A^c\}$ is a partition.

As observed, Theorem 3.3 is used whenever it is not possible to calculate $P(A)$ directly, but it is possible to find $P(A \mid B)$ and $P(A \mid B^c)$ for some event B. In many situations, it is neither possible to find $P(A)$ directly, nor possible to find a single event B that enables us to use

$$P(A) = P(A \mid B)P(B) + P(A \mid B^c)P(B^c).$$

In such situations Theorem 3.4, which is a generalization of Theorem 3.3, might be applicable.

THEOREM 3.4   *If $\{B_1, B_2, \ldots, B_n\}$ is a partition of the sample space of an experiment and $P(B_i) > 0$ for $i = 1, 2, \ldots, n$, then for any event A of S,*

$$P(A) = P(A \mid B_1)P(B_1) + P(A \mid B_2)P(B_2) + \cdots + P(A \mid B_n)P(B_n)$$
$$= \sum_{i=1}^{n} P(A \mid B_i)P(B_i).$$

PROOF: Since $B_1, B_2, \ldots, B_n$ are mutually exclusive, $B_i B_j = \emptyset$ for $i \neq j$. Thus $(AB_i)(AB_j) = \emptyset$ for $i \neq j$. Hence $\{AB_1, AB_2, \ldots, AB_n\}$ is a set of mutually exclusive events. Now

$$S = B_1 \cup B_2 \cup \cdots \cup B_n$$

gives

$$A = AS = AB_1 \cup AB_2 \cup \cdots \cup AB_n,$$

therefore,

$$P(A) = P(AB_1) + P(AB_2) + \cdots + P(AB_n).$$

But $P(AB_i) = P(A \mid B_i)P(B_i)$ for $i = 1, 2, \ldots, n$, so

$$P(A) = P(A \mid B_1)P(B_1) + P(A \mid B_2)P(B_2) + \cdots + P(A \mid B_n)P(B_n). \quad \blacklozenge$$

When using this theorem, one should be very careful to choose B_1, B_2, $B_3, \ldots, B_n$ so that they form a partition of the sample space.

Example 3.13   Suppose that 80% of the seniors, 70% of the juniors, 50% of the sophomores, and 30% of the freshmen of a college use the library

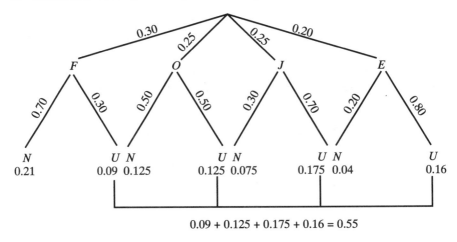

$$0.09 + 0.125 + 0.175 + 0.16 = 0.55$$

Figure 3.4   Tree diagram of Example 3.13.

of their campus frequently. If 30% of all students are freshmen, 25% are sophomores, 25% are juniors, and 20% are seniors, what percent of all students use the library frequently?

SOLUTION: Let A be the event that a randomly selected student is using the library frequently. Let F, O, J, and E be the events that he or she is a freshman, sophomore, junior, or senior, respectively. Then $\{F, O, J, E\}$ is a partition of the sample space. Thus

$$P(A) = P(A \mid F)P(F) + P(A \mid O)P(O)$$
$$+ P(A \mid J)P(J) + P(A \mid E)P(E)$$
$$= (0.30)(0.30) + (0.50)(0.25) + (0.70)(0.25) + (0.80)(0.20) = 0.55.$$

Therefore, 55% of these students use the campus library frequently. The same calculation can be carried out easily from the tree diagram of Figure 3.4, in which U means they use the library frequently and N means that they do not. ♦

Example 3.14   An urn contains 10 white and 12 red chips. Two chips are drawn at random and without looking at their colors are discarded. What is the probability that a third chip drawn is red?

SOLUTION: For $i \geq 1$, let R_i be the event that the ith chip drawn is red and W_i be the event that it is white. Intuitively, it should be clear that the two discarded chips provide no information, so $P(R_3) = 12/22$, the same as if it were the first chip drawn from the urn. To prove this mathematically, note that $\{R_2W_1, W_2R_1, R_2R_1, W_2W_1\}$ is a partition of the sample space; therefore,

$$P(R_3) = P(R_3 \mid R_2W_1)P(R_2W_1) + P(R_3 \mid W_2R_1)P(W_2R_1)$$
$$+ P(R_3 \mid R_2R_1)P(R_2R_1) + P(R_3 \mid W_2W_1)P(W_2W_1). \quad (3.9)$$

Now

$$P(R_2W_1) = P(R_2 \mid W_1)P(W_1) = \frac{12}{21} \times \frac{10}{22} = \frac{20}{77},$$

$$P(W_2R_1) = P(W_2 \mid R_1)P(R_1) = \frac{10}{21} \times \frac{12}{22} = \frac{20}{77},$$

$$P(R_2R_1) = P(R_2 \mid R_1)P(R_1) = \frac{11}{21} \times \frac{12}{22} = \frac{22}{77},$$

and

$$P(W_2W_1) = P(W_2 \mid W_1)P(W_1) = \frac{9}{21} \times \frac{10}{22} = \frac{15}{77}.$$

Substituting these values in (3.9), we get

$$P(R_3) = \frac{11}{20} \times \frac{20}{77} + \frac{11}{20} \times \frac{20}{77} + \frac{10}{20} \times \frac{22}{77} + \frac{12}{20} \times \frac{15}{77} = \frac{12}{22}. \quad \blacklozenge$$

EXERCISES

A

1. If 5% of men and 0.25% of women are color blind, what is the probability that a randomly selected person is color blind?

2. Suppose that 40% of the students of a campus are girls. If 20% of the girls and 16% of the boys of this campus are A students, what percent of all of them are A students?

3. Jim has three cars of different models: A, B, and C. The probabilities that models A, B, and C use over 3 gallons of gasoline from Jim's house to his work are 0.25, 0.32, and 0.53, respectively. On a certain day, all three of Jim's cars have 3 gallons of gasoline each. Jim chooses one of his cars at random and without paying attention to the amount of gasoline in the car drives it toward his office. What is the probability that he makes it?

4. One of the cards of an ordinary deck of 52 cards is lost. What is the probability that a random card drawn from this deck is a spade?

5. Two cards from an ordinary deck of 52 cards are missing. What is the probability that a random card drawn from this deck is a spade?

6. Of the patients of a hospital, 20% of those with, and 35% of those without, myocardial infarction have had strokes. If 40% of the patients have had myocardial infarction, what percent of the patients have had strokes?

7. Suppose that 37% of a community are at least 45 years old. If 80% of the time a person who is 45 or older tells the truth and 65% of the time a person below 45 tells the truth, what is the probability that a randomly selected person answers a question truthfully?

8. A person has six guns. The probability of hitting a target when properly aimed and fired are 0.6, 0.5, 0.7, 0.9, 0.7, and 0.8, respectively. What is the probability of hitting a target if a gun is selected at random, aimed properly, and fired?

9. A factory produces its entire output by three machines. Machines I, II, and III produce 50%, 30%, and 20% of the output, where 4%, 2%, and 4% of their outputs are defective, respectively. What fraction of the total output is defective?

10. Suppose that five coins, of which exactly three are gold, are distributed among five persons, one each, at random, and one by one. Are the chances of getting a gold coin equal for all participants? Why or why not?

11. A child gets lost in the Epcot Center in Florida. The father of the child believes that the probability of his being lost in the east wing of the center is 0.75 and in the west wing is 0.25. The security department sends an officer to the east and an officer to the west to look for the child. If the probability that a security officer who is looking in the correct wing finds the child is 0.4, find the probability that the child is found.

12. Suppose that there exist N families on the earth and that the maximum number of children a family has is c. Let α_j $(0 \le j \le c$, $\sum_{j=0}^{c} \alpha_j = 1)$ be the fraction of families with j children. Find the fraction of all children in the world who are the kth born of their families $(k = 1, 2, \ldots, c)$.

13. Let B be an event of a sample space S with $P(B) > 0$. For a subset A of S, define $Q(A) + P(A \mid B)$. By Theorem 3.1 we know that Q is a probability function. For E and F, events of $S[P(FB) > 0]$, show that $Q(E \mid F) = P(E \mid FB)$.

B

14. Suppose that 40% of the students on a campus who are married to students on the same campus are female. Moreover, suppose that 30% of those who are married but not to students at this campus are also female. If one-third of the married students on this campus are married to other students on this campus, what is the probability that a randomly selected married student from this campus is a woman?

HINT: Let M, C, and F denote the events that the random student is married, is married to a student on the same campus, and is female. For any event A, let $Q(A) = P(A \mid M)$. Then by Theorem 3.1, Q satisfies the same axioms that probabilities satisfy. Applying Theorem 3.3 to Q and using the result of Exercise 13, we obtain

$$P(F \mid M) = P(F \mid MC)P(C \mid M) + P(F \mid MC^c)P(C^c \mid M).$$

15. Suppose that 10 good and three dead batteries are mixed up. Jack tests them one by one, at random, and without replacement. But before testing the fifth battery he realizes that he does not remember whether the first one tested is good or is dead. All he remembers is that the last three that were tested are all good. What is the probability that the first one is also good?

16. A box contains 18 tennis balls, of which eight are new. Suppose that three balls are selected randomly, played with, and after play are returned to the box. If another three balls are selected for a second play, what is the probability that they are all new?

17. From families with three children, a *child* is selected at random and found to be a girl. What is the probability that she has an older sister? Assume that in a three-child family all sex distributions are equally probable.

HINT: Let G be the event that the randomly selected child is girl, A be the event that she has an older sister, and O, M, and Y be the events that she is the oldest, the middle, and the youngest child of the family, respectively. For any subset B of the sample space let $Q(B) = P(B \mid G)$, then apply Theorem 3.3 to Q. (See also Exercises 13 and 14.)

18. Suppose that three numbers are selected one by one, at random, and without replacement from the set of numbers $\{1, 2, 3, \ldots, n\}$. What is the probability that the third number falls between the first two if the first number is smaller than the second?

19. (*Shrewd Prisoner's Dilemma*)   Because of a prisoner's constant supplication, the king grants him this favor: He is given $2N$ balls, which differ from each other only in that half of them are green and half are red. The king instructs the prisoner to divide up the balls between two identical urns. One of the urns will then be selected at random and the prisoner will be asked to choose a ball at random from the urn chosen. If the ball turns out to be green, the prisoner will be freed. How should he distribute the balls in the urn to maximize his chances of freedom?

HINT: Let g be the number of green balls and r be the number of red balls in the first urn. The corresponding numbers in the second urn are $N - g$ and $N - r$. The probability that a green ball is drawn is $f(g, r)$:

$$f(g, r) = \frac{1}{2} \left(\frac{g}{g + r} + \frac{N - g}{2N - g - r} \right).$$

Find the maximum of this function of two variables (r and g). Note that the maximum need not occur at an interior point of the domain.

3.3   Bayes' Formula

To introduce Bayes' formula, let us first examine the following problem. In a bolt factory 30, 50, and 20% of the production is manufactured by machines I, II, and III, respectively. If 4, 5, and 3% of the outputs of these respective machines is defective, what is the probability that a randomly selected bolt that is found to be defective is manufactured by machine III? To solve this problem, let A be the event that a random bolt is defective and B_3 be the event that it is manufactured by machine III. We are asked to find $P(B_3 \mid A)$. Now

$$P(B_3 \mid A) = \frac{P(B_3 A)}{P(A)}, \tag{3.10}$$

so we need to know the quantities $P(B_3 A)$ and $P(A)$. But neither of these is given. To find $P(B_3 A)$, note that since $P(A \mid B_3)$ and $P(B_3)$ are known we can use the relation

$$P(B_3 A) = P(A \mid B_3) P(B_3). \tag{3.11}$$

To calculate $P(A)$, we must use the law of total probability. Let B_1 and B_2 be the events that the bolt is manufactured by machines I and II, respectively. Then $\{B_1, B_2, B_3\}$ is a partition of the sample space, hence

$$P(A) = P(A \mid B_1) P(B_1) + P(A \mid B_2) P(B_2) + P(A \mid B_3) P(B_3). \tag{3.12}$$

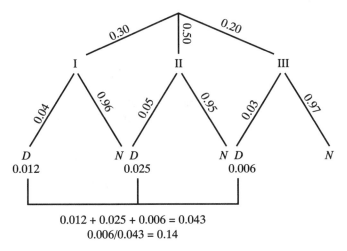

$$0.012 + 0.025 + 0.006 = 0.043$$
$$0.006/0.043 = 0.14$$

Figure 3.5 Tree diagram for relation (3.14).

Substituting (3.11) and (3.12) in (3.10), we arrive at the celebrated *Bayes' formula*:

$$P(B_3 \mid A) = \frac{P(B_3 A)}{P(A)}$$

$$= \frac{P(A \mid B_3)P(B_3)}{P(A \mid B_1)P(B_1) + P(A \mid B_2)P(B_2) + P(A \mid B_3)P(B_3)} \qquad (3.13)$$

$$= \frac{(0.03)(0.20)}{(0.04)(0.30) + (0.05)(0.50) + (0.03)(0.20)} \approx 0.14. \qquad (3.14)$$

Relation (3.13) is a particular case of Bayes' formula (Theorem 3.5). We will now explain how a tree diagram is used to write relation (3.14). Figure 3.5, in which D stands for "defective" and N for "not defective," is a tree diagram for this problem. To find the desired probability all we need do is find (by multiplication) the probability of the required path, the path from III to D, and divide it by the sum of the probabilities of the paths that lead to D's.

In general, modifying the argument from which (3.13) was deduced, we arrive at the following celebrated theorem.

THEOREM 3.5 (Bayes' Theorem) *Let* $\{B_1, B_2, \ldots, B_n\}$ *be a partition of the sample space* S *of an experiment. If for* $i = 1, 2, \ldots, n$, $P(B_i) > 0$, *then for any event* A *of* S *with* $P(A) > 0$,

$P(B_k \mid A)$

$$= \frac{P(A \mid B_k)P(B_k)}{P(A \mid B_1)P(B_1) + P(A \mid B_2)P(B_2) + \cdots + P(A \mid B_n)P(B_n)}.$$

In statistical applications of Bayes' theorem, $B_1, B_2, \ldots, B_n$ are called *hypotheses*, $P(B_i)$ is called the *prior* probability of B_i, and the conditional probability $P(B_i \mid A)$ is called the *posterior* probability of B_i after the occurrence of A.

Note that for any event B of S, the set $\{B, B^c\}$ is a partition of S. Thus by Theorem 3.5, if $P(B) > 0$ and $P(B^c) > 0$, then for any event A of S with $P(A) > 0$,

$$P(B \mid A) = \frac{P(A \mid B)P(B)}{P(A \mid B)P(B) + P(A \mid B^c)P(B^c)}$$

and

$$P(B^c \mid A) = \frac{P(A \mid B^c)P(B^c)}{P(A \mid B)P(B) + P(A \mid B^c)P(B^c)}.$$

These are the simplest forms of Bayes' formula. They are used whenever the quantities $P(A \mid B)$, $P(A \mid B^c)$, and $P(B)$ are given or can be calculated.

Theorem 3.5, in its present form, is due to Laplace, who named it after Thomas Bayes (1701–1761). Bayes, a prominent English philosopher and an ordained minister, did a comprehensive study of the calculation of $P(B \mid A)$ in terms of $P(A \mid B)$. His work was continued by such great mathematicians as Laplace and Gauss.

We will now present several examples concerning Bayes' theorem. To emphasize the convenience of tree diagrams, in Example 3.18 we will use a tree diagram as well.

Example 3.15   A judge is 65% sure that a suspect has committed a crime. During the course of the trial a witness convinces the judge that there is an 85% chance that the criminal is left-handed. If 23% of the population is left-handed and the suspect is also left-handed, with this new information, how certain should the judge be of the guilt of the suspect?

SOLUTION: Let G and I be the events that the suspect is guilty and innocent, respectively. Let A be the event that the suspect is left-handed. Since $\{G, I\}$ is a partition of sample space, we can use Bayes' formula to calculate $P(G \mid A)$, the probability that the suspect has committed the crime in view of the new evidence.

$$P(G \mid A) = \frac{P(A \mid G)P(G)}{P(A \mid G)P(G) + P(A \mid I)P(I)}$$

$$= \frac{(0.85)(0.65)}{(0.85)(0.65) + (0.23)(0.35)} \approx 0.87. \quad \blacklozenge$$

Example 3.16 On the basis of reconnaissance reports, Colonel Smith decides that the probabilities of the enemy attack against the left is 0.20, against the center is 0.50, and against the right is 0.30. A flurry of enemy radio traffic occurs in preparation for the attack. Since deception is normal as a prelude to battle, Colonel Brown, having intercepted the radio traffic, tells General Quick that if the enemy wanted to attack on the left, the probability is 0.20 that he would have sent this particular radio traffic. He tells the general that the corresponding probabilities for an attack on the center or the right are 0.70 and 0.10, respectively. How should General Quick use these two equally reliable staff members' views to get the best probability profile for the forthcoming attack?

SOLUTION: Let A be the event that the attack would be against the left, B be the event that the attack would be against the center, and C be the event that the attack would be against the right. Let Δ be the event that this particular flurry of radio traffic occurs. The general should take the opinion of Colonel Smith as prior probabilities for A, B, and C. That is, $P(A) = 0.20$, $P(B) = 0.50$, and $P(C) = 0.30$. Then he should calculate $P(A \mid \Delta)$, $P(B \mid \Delta)$, and $P(C \mid \Delta)$ based on Colonel Brown's view, using Bayes' theorem, as follows.

$$P(A \mid \Delta) = \frac{P(\Delta \mid A)P(A)}{P(\Delta \mid A)P(A) + P(\Delta \mid B)P(B) + P(\Delta \mid C)P(C)}$$

$$= \frac{(0.2)(0.2)}{(0.2)(0.2) + (0.7)(0.5) + (0.1)(0.3)} = \frac{(0.2)(0.2)}{0.42} \approx 0.095.$$

Similarly,

$$P(B \mid \Delta) = \frac{(0.7)(0.5)}{0.42} \approx 0.83 \text{ and } P(C \mid \Delta) = \frac{(0.1)(0.3)}{0.42} \approx 0.071. \quad \blacklozenge$$

Example 3.17 (Jailer's Paradox) The jailer of a prison in which Abe, Bill, and Tim are held is the only one who knows which of these three prisoners is condemned to death and which two will be freed. Abe has written a letter to his fiancée and wants to give it to either Bill or Tim, whoever goes free, to deliver (note that since only one will die, either Bill or Tim will be freed). So Abe asks the jailer to tell him which of the two will be freed. The jailer refuses to give that information to Abe, explaining that if he does, the probability of Abe dying increases from 1/3 to 1/2. As Zweifel notes in the June 1986 issue of *Mathematics Magazine*, page 156, "this seems intuitively suspect, since the jailer is providing no new information to Abe, so why should his probability of dying change?" Abe already knows that one of the other two prisoners will go free, and obviously, knowing the name of the one who will be freed does not change the probability of his dying. To explain this paradox, we will use Bayes' theorem to prove that even if the jailer tells the name of one of the other two who will go free, the probability of Abe dying is still 1/3. Let A, B, T, and J be the events that "Abe dies," "Bill dies," "Tim dies," and "the jailer says that Tim goes free," respectively. Then

$$P(A \mid J) = \frac{P(J \mid A)P(A)}{P(J \mid A)P(A) + P(J \mid B)P(B) + P(J \mid T)P(T)}$$

$$= \frac{\dfrac{1}{2} \times \dfrac{1}{3}}{\dfrac{1}{2} \times \dfrac{1}{3} + 1 \times \dfrac{1}{3} + 0 \times \dfrac{1}{3}} = \frac{1}{3}.$$

Similarly, if D is the event that "the jailer says that Bill goes free," then $P(A \mid D) = 1/3$. In this problem, the controversy hinges on a misunderstanding of what information the jailer is giving to Abe. Zweifel in his explanation of this paradox adds:

> Aside from the formal application of Bayes' theorem, one would like to understand this "paradox" from an intuitive point of view. The crucial point which can perhaps make the situation clear is the discrepancy between $P(J \mid A)$ and $P(J \mid B)$ noted above. Bridge players call this the "Principle of Restricted Choice." The probability of a restricted choice is obviously greater than that of a free choice, and a common error made by those who attempt to solve such problems intuitively is to overlook this point. In the case of the jailer's paradox, if the jailer says "Tim will go free" this is twice as likely to occur when Bill is scheduled to die (restricted choice; jailer *must* say "Tim") as when Abe is scheduled to die (free choice; jailer could say either "Tim" or "Bill").

The jailer's paradox and its solution have been around for a long time. Despite this, when the same problem in a different context came up in the "Ask Marilyn" column of *Parade Magazine* on September 9, 1990 (see Exercise 10) and Marilyn Vos Savant[†] gave the correct answer to the

---

[†] Ms. Vos Savant is the writer of a reader-correspondence column and is listed in the *Guinness Book of Records* Hall of Fame for "highest IQ."

problem in the December 2, 1990 issue, she was savagely taken to task by three mathematicians. By the time the February 17, 1991 issue of *Parade Magazine* was published, Vos Savant had received 2000 letters on the problem, of which 92% of general respondents and 65% of university respondents were against her answer. This story reminds us of the famous statement of De Moivre in the dedication to *The Doctrine of Chance*:

> Some of the Problems about Chance having a great appearance of Simplicity, the Mind is easily drawn into a belief, that their Solution may be attained by the meer Strength of natural good Sense; which generally proving otherwise and the Mistakes occasioned thereby being not unfrequent, 'tis presumed that a Book of this Kind, which teaches to distinguish Truth from what seems so nearly to resemble it, will be looked upon as a help to good Reasoning. ♦

Example 3.18 A box contains seven red and 13 blue balls. Two balls are selected at random and are discarded without their colors being seen. If a third ball is drawn randomly and observed to be red, what is the probability that both of the discarded balls were blue?

SOLUTION: Let BB, BR, and RR be the events that the discarded balls are blue and blue, blue and red, and red and red, respectively. Also, let R be the event that the third ball drawn is red. Since $\{BB, BR, RR\}$ is a partition of sample space, Bayes' formula can be used to calculate $P(BB \mid R)$.

$$P(BB \mid R) = \frac{P(R \mid BB)P(BB)}{P(R \mid BB)P(BB) + P(R \mid BR)P(BR) + P(R \mid RR)P(RR)}.$$

Now

$$P(BB) = \frac{13}{20} \times \frac{12}{19} = \frac{39}{95},$$

$$P(RR) = \frac{7}{20} \times \frac{6}{19} = \frac{21}{190},$$

and

$$P(BR) = \frac{13}{20} \times \frac{7}{19} + \frac{7}{20} \times \frac{13}{19} = \frac{91}{190},$$

where the last equation follows since BR is the union of two disjoint events: namely, the first ball discarded was blue, the second was red, and vice versa. Thus

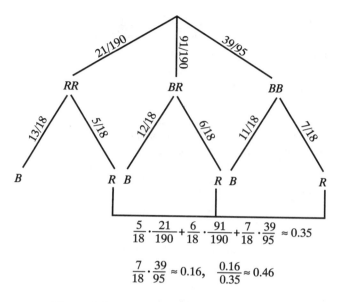

$$\frac{5}{18} \cdot \frac{21}{190} + \frac{6}{18} \cdot \frac{91}{190} + \frac{7}{18} \cdot \frac{39}{95} \approx 0.35$$

$$\frac{7}{18} \cdot \frac{39}{95} \approx 0.16, \quad \frac{0.16}{0.35} \approx 0.46$$

Figure 3.6   Tree diagram for Example 3.18.

$$P(BB \mid R) = \frac{\dfrac{7}{18} \times \dfrac{39}{95}}{\dfrac{7}{18} \times \dfrac{39}{95} + \dfrac{6}{18} \times \dfrac{91}{190} + \dfrac{5}{18} \times \dfrac{21}{190}} \approx 0.46.$$

This can be found easily from the tree diagram of Figure 3.6 as well. Alternatively, reducing sample space, given that the third ball was red, there are 13 blue and six red balls which could have been discarded. Thus

$$P(BB \mid R) = \frac{13}{19} \times \frac{12}{18} \approx 0.46. \quad \blacklozenge$$

EXERCISES

A

1. In transmitting the signals dot and dash, a communication system changes 1/4 of the dots to dashes and 1/3 of the dashes to dots. If 40% of the signals transmitted are dots and 60% are dashes, what is the probability that a dot received was actually a transmitted dot?

2. On a multiple-choice exam with four choices for each question, a student either knows the answer to a question or marks it at random.

If the probability that he or she knows the answers is 2/3, what is the probability that an answer that was marked correctly was not marked randomly?

3. Suppose that 5% of the men and 2% of the women working for a corporation make over $120,000 a year. If 30% of the employees of the corporation are women, what percent of those who make over $120,000 a year are women?

4. A stack of cards consists of six red and five blue cards. A second stack of cards consists of nine red cards. A stack is selected at random and three of its cards are drawn. If all of them are red, what is the probability that the first stack was selected?

5. A certain cancer is found in 1 person in 5000. If a person does have the disease, in 92% of the cases the diagnostic procedure will show that he or she actually has it. If a person does not have the disease, the diagnostic procedure in 1 out of 500 cases gives a false positive result. Determine the probability that a person with a positive test result has the cancer.

6. Urns I, II, and III contain three pennies and four dimes, two pennies and five dimes, three pennies and one dime, respectively. One coin is selected at random from each urn. If two of the three coins are dimes, what is the probability that the coin selected from urn I is a dime?

7. In a study it was discovered that 25% of the paintings of a certain gallery are not original. A collector in 15% of the cases makes a mistake in judging if a painting is authentic or a copy. If she buys a piece thinking that it is original, what is the probability that it is not?

8. There are three identical cards that differ only in color. Both sides of one are black, both sides of the second one are red, and one side of the third card is black and its other side is red. These cards are mixed up and one of them is selected at random. If the upper side of this card is red, what is the probability that its other side is black?

9. With probability of 1/6 there are i defective fuses among 1000 fuses ($i = 0, 1, 2, 3, 4, 5$). If among 100 fuses selected at random, none was defective, what is the probability of no defective fuses at all?

10. On a TV game show, there are three curtains. Behind two curtains there is nothing, but behind the third there is a prize that the player might win. The probability that the prize is behind a given curtain is 1/3. The game begins with the contestant guessing a curtain. The host of the show (master of ceremonies), who *knows* behind which curtain the prize is, will then pull back a curtain other than the choice of the player and reveal that the prize is not behind that curtain. So the host will not

pull back the curtain selected by the player, neither will he pull back the one with the prize, if different from the player's choice. At this point, the master of ceremonies gives the player the opportunity to change her choice of curtain and select the other one. The question is whether the player should change her choice. That is, has the probability of the prize being behind the curtain of the contestant's choice changed from 1/3 to 1/2 or it is still 1/3? If it is still 1/3, the contestant should definitely change her choice. Otherwise, there is no point in doing so.

B

11. There are two stables on a farm, one that houses 20 horses and 13 mules, the other with 25 horses and eight mules. Without any pattern, animals occasionally leave their stables and then return to their stables. Suppose that during a period when all the animals are in their stables, a horse comes out of a stable and then goes back. What is the probability that the next animal coming out of the same stable will also be a horse?

12. An urn contains five red and three blue chips. Suppose that four of these chips are selected at random and transferred to a second urn, which was originally empty. If a random chip from this second urn is blue, what is the probability that two red and two blue chips were transferred from the first urn to the second urn?

13. The advantage of a certain blood test is that 90% of the time it is positive for patients having a certain disease. Its disadvantage is that 25% of the time it is also positive in healthy people. In a certain location 30% of the people have the disease and anybody with a positive blood test is given a drug that cures the disease. If 20% of the time the drug produces a characteristic rash, what is the probability that a person from this location who has the rash had the disease in the first place?

3.4   Independence

Let A and B be two events of a sample space S and assume that $P(A) > 0$ and $P(B) > 0$. We have seen that, in general, the conditional probability of A given B is not equal to the probability of A. However, if it is, that is, if $P(A \mid B) = P(A)$, we say that A is *independent* of B. This means that if A is independent of B, knowledge regarding the occurrence of B does not change the chance of the occurrence of A. The relation $P(A \mid B) = P(A)$ is equivalent to the relations $P(AB)/P(B) = P(A)$, $P(AB) = P(A)P(B)$, $P(BA)/P(A) = P(B)$, and $P(B \mid A) = P(B)$. The equivalence of the first and last of these relations implies that if A is independent of B, then B is independent of A. In other words, if knowledge regarding the occurrence of B does not change the chance of occurrence of A, then knowledge regarding the occurrence of A does not change the chance of occurrence of B. Hence independence is a symmetric relation on the set of all events of a sample space. As a result of this property, instead of making the definitions "A is independent of B" and "B is independent of A", we simply define the concept of the "independence of A and B." To do so, we take $P(AB) = P(A)P(B)$ as the definition. We do this because a symmetrical definition relating A and B does not readily follow from either of the other relations given [i.e., $P(A \mid B) = P(A)$ or $P(B \mid A) = P(B)$]. Moreover, these relations require either that $P(B) > 0$ or $P(A) > 0$, whereas our definition does not.

DEFINITION   *Two events A and B are called independent if*

$$P(AB) = P(A)P(B).$$

If two events are not independent, they are called dependent. If A and B are independent, we say that $\{A, B\}$ is an independent set of events.

Note that in this definition we did not require $P(A)$ or $P(B)$ to be strictly positive. Hence by this definition *any* event A with $P(A) = 0$ or 1 is independent of *every* event B (see Exercise 8).

Example 3.19   In the experiment of tossing a fair coin twice, let A and B be the events of getting heads on the first and second tosses, respectively. Intuitively, it is clear that A and B are independent. To prove this mathematically, note that $P(A) = 1/2$ and $P(B) = 1/2$. But since the sample space of this experiment consists of the four equally probable events: HH, HT, TH, and TT, we have $P(AB) = P(\text{HH}) = 1/4$. Hence $P(AB) = P(A)P(B)$ is valid, implying the independence of A and B.   ♦

Example 3.20   In the experiment of drawing a card from an ordinary deck of 52 cards, let A and B be the events of getting a heart and an ace, respectively. Whether or not A and B are independent cannot be answered easily on the basis of intuition alone. However, using the defining formula, $P(AB) = P(A)P(B)$, this can be answered at once since $P(AB) = 1/52$, $P(A) = 1/4$, $P(B) = 1/13$, and $1/52 = 1/4 \times 1/13$. Hence A and B are independent events.   ♦

Example 3.21   An urn contains five red and seven blue balls. Suppose that two balls are selected at random and with replacement. Let A and B be the events that the first and the second balls are red, respectively.

Then using the counting principle we get $P(AB) = (5 \times 5)/(12 \times 12)$. Now $P(AB) = P(A)P(B)$ since $P(A) = 5/12$ and $P(B) = 5/12$. Thus A and B are independent. If we do the same experiment without replacement, then $P(B \mid A) = 4/11$ while

$$P(B) = P(B \mid A)P(A) + P(B \mid A^c)P(A^c)$$

$$= \frac{4}{11} \times \frac{5}{12} + \frac{5}{11} \times \frac{7}{12} = \frac{5}{12},$$

as expected. Thus $P(B \mid A) \neq P(B)$, implying that A and B are dependent. ◆

Example 3.22　In the experiment of selecting a random number from the set of natural numbers $\{1, 2, 3, \ldots, 100\}$, let A, B, and C denote the events that they are divisible by 2, 3, and 5, respectively. Clearly, $P(A) = 1/2$, $P(B) = 33/100$, $P(C) = 1/5$, $P(AB) = 16/100$, and $P(AC) = 1/10$. Hence A and B are dependent while A and C are independent. Note that if the random number is selected from $\{1, 2, 3, \ldots, 300\}$, then each of $\{A, B\}$, $\{A, C\}$, and $\{B, C\}$ is an independent set of events. This is because 300 is divisible by 2, 3, and 5 but 100 is not divisible by 3. ◆

Example 3.23　A spinner is mounted on a wheel. Arcs A, B, and C, of equal length, are marked off on the wheel's perimeter (see Figure 3.7). In a game of chance, the spinner is flicked, and depending on whether it stops on A, B, or C, the player wins 1, 2, or 3 points, respectively. Suppose that a player plays this game twice. Let E denote the event that he wins 1 point in the first game and any number of points in the second. Let F be the event that he wins a total of 3 points in both games, and G be the event that he wins a total of 4 points in both games. The sample space of this experiment has $3 \times 3 = 9$ elements, $E = \{(1, 1), (1, 2), (1, 3)\}$, $F = \{(1, 2), (2, 1)\}$, and $G = \{(1, 3), (2, 2), (3, 1)\}$. Therefore, $P(E) = 1/3$, $P(G) = 1/3$,

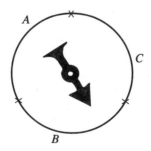

Figure 3.7　Spinner of Example 3.23.

$P(F) = 2/9$, $P(FE) = 1/9$, and $P(GE) = 1/9$. These show that E and G are independent, whereas E and F are not. To justify these intuitively, note that if we are interested in a total of 3 points, then getting 1 point in the first game is good luck, since obtaining 3 points in the first game makes it impossible to win a sum of 3. However, if we are interested in a sum of 4 points, it does not matter what we win in the first game. We have the same chance of obtaining a sum of 4 if the first game results in any of the numbers 1, 2, or 3. ◆

We now prove that if A and B are independent events, so are A and B^c.

THEOREM 3.6 *If A and B are independent, then A and B^c are independent as well.*

PROOF: Since $AB \subseteq A$, by Theorem 1.5:

$$P(AB^c) = P(A - AB) = P(A) - P(AB)$$
$$= P(A) - P(A)P(B) = P(A)[1 - P(B)] = P(A)P(B^c). \quad ◆$$

COROLLARY *If A and B are independent, then A^c and B^c are independent as well.*

PROOF: A and B are independent, so by Theorem 3.6, the events A and B^c are independent. Now using the same theorem again we have that A^c and B^c are independent. ◆

Thus, if A and B are independent, knowledge about the occurrence or nonoccurrence of A does not change the chances of the occurrence or nonoccurrence of B, and vice versa.

IMPORTANT REMARK: *If A and B are mutually exclusive events and $P(A) > 0$, $P(B) > 0$, then they are dependent. This is because if we are given that one has occurred, the chance of the occurrence of the other one is zero.* That is, the occurrence of one of them precludes the occurrence of the other. For example, let A be the event that the next president of the United States is a Democrat and B be the event that he or she is a Republican. Then A and B are mutually exclusive, hence they are dependent. If A occurs, that is, if the next president is a Democrat, the probability that B occurs, that is, he or she is a Republican is zero, and vice versa. ◆

The following example shows that if A is independent of B and if A is independent of C, then A is not necessarily independent of BC or of $B \cup C$.

Example 3.24   A person arrives at his office every day at a random time between 8:00 A.M. and 9:00 A.M. Let A be the event that he arrives at his office tomorrow between 8:15 A.M. and 8:45 A.M. Let B be the event that he arrives between 8:30 A.M. and 9:00 A.M., and let C be the event that he arrives either between 8:15 A.M. and 8:30 A.M. or between 8:45 A.M. and 9:00 A.M. Then AB, AC, BC, and $B \cup C$ are the events that the person arrives at his office between 8:30 and 8:45, 8:15 and 8:30, 8:45 and 9:00, and 8:15 and 9:00, respectively. Thus $P(A) = P(B) = P(C) = 1/2$ and $P(AB) = P(AC) = 1/4$. So $P(AB) = P(A)P(B)$ and $P(AC) = P(A)P(C)$; that is, A is independent of B and it is independent of C. However, since BC and A are mutually exclusive, they are dependent. Also, $P(A \mid B \cup C) = 2/3 \neq P(A)$. Thus $B \cup C$ and A are dependent as well.   ♦

Example 3.25   (Hardy–Weinberg Law in Genetics)   In diploid organisms, which we all are, each hereditary character of each individual is carried by a pair of genes. A gene is either recessive a or dominant A, so that the possible pairs of genes are AA, Aa (same as aA), and aa. Let $r = P(AA)$, $s = P(Aa)$, and $t = P(aa)$, $r + s + t = 1$. Let E be the event that the father transmits an A. Then

$$P(E) = P(E \mid \text{father is } AA)P(\text{father is } AA)$$

$$+ P(E \mid \text{father is } Aa)P(\text{father is } Aa)$$

$$+ P(E \mid \text{father is } aa)P(\text{father is } aa)$$

$$= 1 \cdot r + \frac{1}{2} \cdot s + 0 \cdot t = r + \frac{s}{2}.$$

Similarly, the probability that the father transmits a is $t + s/2$. The probabilities that the mother transmits A and a are also $r + s/2$ and $t + s/2$, respectively. Now suppose that during the process of reproduction, all matings are random, so that the contributions of the parents to the genetic heritage of their offspring are independent of each other. Then the probabilities that an offspring selected at random is AA, Aa, and aa are $r_1 = (r + s/2)^2$, $s_1 = 2(r + s/2)(t + s/2)$, and $t_1 = (t + s/2)^2$, respectively. These formulas will easily give us the probabilities that a randomly selected offspring from the third generation are AA, Aa, and aa: $r_2 = (r_1 + s_1/2)^2$, $s_2 = 2(r_1 + s_1/2)(t_1 + s_1/2)$, and $t_2 = (t_1 + s_1/2)^2$. Some algebra shows that $r_2 = r_1$, $t_2 = t_1$ and $s_2 = s_1$. For example,

$$r_2 = \left(r_1 + \frac{s_1}{2}\right)^2 = \left[\left(r + \frac{s}{2}\right)^2 + \left(r + \frac{s}{2}\right)\left(t + \frac{s}{2}\right)\right]^2$$

$$= \left[\left(r + \frac{s}{2}\right)\left(r + \frac{s}{2} + t + \frac{s}{2}\right)\right]^2$$

$$= \left[\left(r + \frac{s}{2}\right)(r + t + s)\right]^2 = \left(r + \frac{s}{2}\right)^2 = r_1.$$

Therefore, the proportions of AA, Aa, and aa in the third generation are the same as in the second generation. This implies the well-known Hardy–Weinberg law: *The proportions of AA, Aa, and aa in all future generations remain the same as those in the second generation.* Note that this result is true because of random mating and might not be valid if the randomness condition is lacking. For example, it would not be true on genes that have effects on heights in populations in which short and tall marry within their own groups. ◆

We now extend the concept of independence to three events. A, B, and C are called *independent* if knowledge about the occurrence of any of them or the joint occurrence of any two of them does not change the chances of the occurrence of the remaining events. That is, A, B, and C are independent if $\{A, B\}$, $\{A, C\}$, $\{B, C\}$, $\{A, BC\}$, $\{B, AC\}$, and $\{C, AB\}$ are all independent sets of events. Hence A, B, and C are independent if

$$P(AB) = P(A)P(B),$$
$$P(AC) = P(A)P(C),$$
$$P(BC) = P(B)P(C),$$
$$P(A(BC)) = P(A)P(BC),$$
$$P(B(AC)) = P(B)P(AC),$$
$$P(C(AB)) = P(C)P(AB).$$

Now note that these relations can be reduced since the first three and the relation $P(ABC) = P(A)P(B)P(C)$ imply the last three relations. Hence the definition of the independence of three events can be shortened as follows.

DEFINITION   *The events A, B, and C are called independent if*

$$P(AB) = P(A)P(B),$$
$$P(AC) = P(A)P(C),$$
$$P(BC) = P(B)P(C),$$
$$P(ABC) = P(A)P(B)P(C).$$

If A, B, and C are independent events, we say that $\{A, B, C\}$ is an independent set of events.

The following example demonstrates that $P(ABC) = P(A)P(B)P(C)$, in general, does not imply that $\{A, B, C\}$ is a set of independent events.

Example 3.26 Let an experiment consist of throwing a die twice. Let A be the event that in the second throw the die lands 1, 2, or 5, B be the event that in the second throw it lands 4, 5 or 6, and C be the event that the sum of the two outcomes is 9. Then $P(A) = P(B) = 1/2$, $P(C) = 1/9$, and

$$P(AB) = \frac{1}{6} \neq \frac{1}{4} = P(A)P(B),$$

$$P(AC) = \frac{1}{36} \neq \frac{1}{18} = P(A)P(C),$$

$$P(BC) = \frac{1}{12} \neq \frac{1}{18} = P(B)P(C),$$

while

$$P(ABC) = \frac{1}{36} = P(A)P(B)P(C).$$

Thus the validity of $P(ABC) = P(A)P(B)P(C)$ is not sufficient for the independence of A, B, and C. $\blacklozenge$

If A, B, and C are three events and the occurrence of any of them does not change the chances of the occurrence of the remaining two, we say that A, B, and C are *pairwise independent.* Thus $\{A, B, C\}$ forms a set of pairwise independent events if $P(AB) = P(A)P(B)$, $P(AC) = P(A)P(C)$, and $P(BC) = P(B)P(C)$. The difference between pairwise independent events and independent events is that in the former, knowledge about the joint occurrence of any two of them may change the chances of the occurrence of the remaining one, while in the latter it would not. The following example illuminates the difference between pairwise independence and independence.

Example 3.27 A regular tetrahedron is a body that has four faces and if it is tossed, the probability that it lands on any face is 1/4. Suppose that one face of a regular tetrahedron has three colors: red, green, and blue. The other three faces each have only one color: red, blue, and green, respectively. We throw the tetrahedron once and let R, G, and B be the events that the face on which it lands contains red, green, and blue, respectively. Then $P(R \mid G) = 1/2 = P(R)$, $P(R \mid B) = 1/2 = P(R)$, and $P(B \mid G) = 1/2 = P(B)$. Thus the events R, B, and G are pairwise independent. However, R, B, and G are not independent events since $P(R \mid GB) = 1 \neq P(R)$. ♦

The independence of more than three events may be defined in a similar manner. A set of n events $A_1, A_2, \ldots, A_n$ is said to be *independent* if knowledge about the occurrence of any of them or the joint occurrence of any number of them does not change the chances of the occurrence of the remaining events. If we write this definition in terms of formulas, we get many equations. Similar to the case of three events, if we reduce the number of these equations to the minimum number that can be used to have all of them satisfied, we reach the following definition.

DEFINITION   *The set of events* $\{A_1, A_2, \ldots, A_n\}$ *is called independent if for every subset* $\{A_{i_1}, A_{i_2}, \ldots, A_{i_k}\}$, $k \geq 2$, *of* $\{A_1, A_2, \ldots, A_n\}$,

$$P(A_{i_1} A_{i_2} \cdots A_{i_k}) = P(A_{i_1}) P(A_{i_2}) \cdots P(A_{i_k}). \tag{3.15}$$

This definition is not in fact limited to finite sets and is extended to infinite sets of events (countable or uncountable) in the obvious way. For example, the sequence of events $\{A_i\}_{i=1}^{\infty}$ is called independent if for any of its subsets $\{A_{i_1}, A_{i_2}, \ldots, A_{i_k}\}$, $k \geq 2$, (3.15) is valid.

By definition, events $A_1, A_2, \ldots, A_n$ are independent if for all combinations $1 \leq i < j < k < \cdots \leq n$, the relations

$$P(A_i A_j) = P(A_i) P(A_j)$$

$$P(A_i A_j A_k) = P(A_i) P(A_j) P(A_k)$$

$$\vdots$$

$$P(A_1 A_2 \cdots A_n) = P(A_1) P(A_2) \cdots P(A_n)$$

are valid. Now we see that the first line stands for $\binom{n}{2}$ equations, the second line stands for $\binom{n}{3}$ equations, $\ldots$, and the last line stands for $\binom{n}{n}$ equations. Therefore, $A_1, A_2, \ldots, A_n$ are independent if all of the above $\binom{n}{2} + \binom{n}{3} + \cdots + \binom{n}{n}$ relations are satisfied. Note that by the binomial theorem (see Example 2.23),

$$\binom{n}{2} + \binom{n}{3} + \cdots + \binom{n}{n} = (1+1)^n - \binom{n}{1} - \binom{n}{0} = 2^n - n - 1.$$

Thus the number of these equations is $2^n - n - 1$. Although these equations seem to be cumbersome to check, it usually turns out that they are obvious and checking is not necessary.

Example 3.28   We draw cards, one at a time, at random, and successively from an ordinary deck of 52 cards with replacement. What is the probability that an ace appears before a face card?

SOLUTION: There are two different techniques that may be used to solve this type of problem. We will explain both.
    Technique 1: Let E be the event of an ace before a face card. Let A, F, and B be the events of ace, face card, and neither in the first experiment, respectively. then by the law of total probability.

$$P(E) = P(E \mid A)P(A) + P(E \mid F)P(F) + P(E \mid B)P(B).$$

Thus

$$P(E) = 1 \times \frac{4}{52} + 0 \times \frac{12}{52} + P(E \mid B) \times \frac{36}{52}. \qquad (3.16)$$

Now note that since the outcomes of successive experiments are all independent of each other, when the second experiment begins, the whole process probabilistically starts all over again. Therefore, if in the first experiment neither a face card nor an ace are drawn, the probability of E before doing the first experiment and after it would be the same; that is, $P(E \mid B) = P(E)$. Thus Equation (3.16) gives

$$P(E) = \frac{4}{52} + P(E) \times \frac{36}{52}.$$

Solving this equation for $P(E)$, we obtain $P(E) = 1/4$, a quantity expected because the number of face cards is three times the number of aces (see also Exercise 29).
    Technique 2: Let A_n be the event that no face card or ace appears on the first $(n - 1)$ drawings and the nth draw is an ace. Then the event of "an ace before a face card" is $\bigcup_{n=1}^{\infty} A_n$. Now $\{A_n, n \geq 1\}$ forms a sequence of mutually exclusive events because if $n \neq m$, simultaneous occurrence of A_n and A_m is the impossible event that an ace appears for the first time in the nth and mth draws. Hence

$$P\left(\bigcup_{n=1}^{\infty} A_n\right) = \sum_{n=1}^{\infty} P(A_n).$$

To compute $P(A_n)$, note that $P(\text{an ace on any draw}) = 1/13$ and $P(\text{no face card and no ace in any trial}) = 9/13$. By the independence of trials we obtain

$$P(A_n) = \left(\frac{9}{13}\right)^{n-1}\left(\frac{1}{13}\right).$$

Therefore,

$$P\left(\bigcup_{n=1}^{\infty} A_n\right) = \sum_{n=1}^{\infty}\left(\frac{9}{13}\right)^{n-1}\left(\frac{1}{13}\right) = \frac{1}{13}\sum_{n=1}^{\infty}\left(\frac{9}{13}\right)^{n-1}$$

$$= \frac{1}{13}\cdot\frac{1}{1-9/13} = \frac{1}{4}.$$

Here $\sum$ is calculated from the geometric series theorem: For $a \neq 0$, $|r| < 1$, the geometric series $\sum_{n=m}^{\infty} ar^n$ converges to $ar^m/(1-r)$. ◆

Example 3.29   Figure 3.8 shows an electric circuit in which each of the switches located at 1, 2, 3, and 4 is independently closed or open with probabilities p and $1-p$, respectively. If a signal is fed in at the input, what is the probability that it is transmitted to the output?

SOLUTION: Let E_i be the event that the switch at location i is closed, $1 \leq i \leq 4$. A signal fed in at the input will be transmitted to the output if at least one of the events $E_1 E_2$ and $E_3 E_4$ occurs. Hence the desired probability is

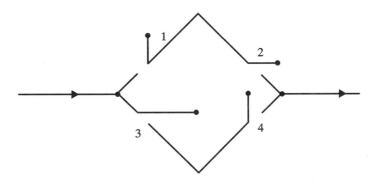

Figure 3.8   Electric circuit of Example 3.29.

$$P(E_1E_2 \cup E_3E_4) = P(E_1E_2) + P(E_3E_4) - P(E_1E_2E_3E_4)$$
$$= p^2 + p^2 - p^4 = p^2(2 - p^2). \quad \blacklozenge$$

Example 3.30   Adam tosses a fair coin $n + 1$ times, Andrew tosses the same coin n times. What is the probability that Adam gets more heads than Andrew?

SOLUTION: Let H_1 and H_2 be the number of heads obtained by Adam and Andrew, respectively. Also let T_1 and T_2 be the number of tails obtained by Adam and Andrew, respectively. Since the coin is fair,

$$P(H_1 > H_2) = P(T_1 > T_2).$$

But

$$P(T_1 > T_2) = P(n + 1 - H_1 > n - H_2) = P(H_1 \le H_2).$$

Therefore, $P(H_1 > H_2) = P(H_1 \le H_2)$. So

$$P(H_1 > H_2) + P(H_1 \le H_2) = 1$$

implies that

$$P(H_1 > H_2) = P(H_1 \le H_2) = \frac{1}{2}.$$

Note that a combinatorial solution to this problem is neither elegant nor easy to handle:

$$P(H_1 > H_2) = \sum_{i=0}^{n} P(H_1 > H_2 \mid H_2 = i)P(H_2 = i)$$

$$= \sum_{i=0}^{n} \sum_{j=i+1}^{n+1} P(H_1 = j)P(H_2 = i)$$

$$= \sum_{i=0}^{n} \sum_{j=i+1}^{n+1} \frac{\dfrac{(n+1)!}{j!\,(n+1-j)!}}{2^{n+1}} \; \frac{\dfrac{n!}{i!\,(n-i)!}}{2^n}$$

$$= \frac{1}{2^{2n+1}} \sum_{i=0}^{n} \sum_{j=i+1}^{n+1} \binom{n+1}{j}\binom{n}{i}.$$

However, comparing these two solutions, we obtain the following interesting identity.

$$\sum_{i=0}^{n} \sum_{j=i+1}^{n+1} \binom{n+1}{j}\binom{n}{i} = 2^{2n}. \quad \blacklozenge$$

EXERCISES

A

1. Jean Le Rond d'Alembert (1717–1783), a great French mathematician, believed that in successive flips of a fair coin, after a long run of heads, a tail is more likely. Do you agree with d'Alembert on this? Explain.

2. A fair die is rolled twice. Let A denote the event that the sum of the outcomes is odd and B denote the event that it lands 2 on the first toss. Are A and B independent? Why or why not?

3. An urn has three red and five blue balls. Suppose that eight balls are selected at random and with replacement. What is the probability that the first three are red and the rest are blue balls?

4. Suppose that two points are selected at random and independently from the interval $(0, 1)$. What is the probability that the first one is less than 3/4 and the second one is greater than 1/4 ?

5. According to a recent mortality table, the probability that a 35-year-old U.S. citizen will live to age 65 is 0.725. What is the probability that John and Jim, two 35-year-old Americans who are not relatives, both live to age 65? What is the probability that neither man lives to that age?

6. *Chevalier de Méré's Paradox*[†]: In the seventeenth century in France there were two popular games, one to obtain at least one 6 in four throws of a fair die and the other to bet on at least one double 6 in 24 throws of two fair dice. French nobleman and mathematician Chevalier de Méré had argued that the probabilities of a 6 in one throw of a fair die and a double 6 in one throw of two fair dice are 1/6 and 1/36, respectively. Therefore, the probability of at least one 6 in four throws of a fair die and at least one double 6 in 24 throws of two fair dice are $4 \times 1/6 = 2/3$ and $24 \times 1/36 = 2/3$, respectively. However, even then experience had convinced gamblers that the probability of winning the first game was higher than winning the second game, a contradiction.

[†] Some people consider this problem to be the birth of modern probability theory.

Explain why Chevalier de Méré's argument is false and calculate the correct probabilities of winning in these two games.

7. Find an example in which $P(AB) < P(A)P(B)$.

8. (a) Show that if $P(A) = 1$, then $P(AB) = P(B)$.
   (b) Prove that *any* event A with $P(A) = 0$ or $P(A) = 1$ is independent of *every* event B.

9. Show that if an event A is independent of itself, then $P(A) = 0$ or 1.

10. Show that if A and B are independent and $A \subseteq B$, then either $P(A) = 0$ or $P(B) = 1$.

11. Suppose that 55% of the customers of a shoestore buy black shoes. Find the probability that at least one of the next six customers who purchase a pair of shoes from this store buy black shoes. Assume that these customers decide independently.

12. Three missiles are fired at a target and hit it independently with probabilities 0.7, 0.8, and 0.9, respectively. What is the probability that the target is hit?

13. In a tire factory, the quality control inspector examines a randomly chosen sample of 15 tires. When more than one defective tire is found, the line is halted, the existing tires are recycled, and production is restarted. The purpose of this is to ensure that the rate of defects is no higher than 6%. Occasionally, by a stroke of bad luck, even if the defect rate is, in fact, no higher than 6%, more than one defective tire occurs among the 15 chosen. Determine the fraction of the time that this happens.

14. In a community of M men and W women, m men and w women smoke ($m \leq M$, $w \leq W$). If a person is selected at random and A and B are the events that the person is a man and smokes, respectively, under what conditions are A and B independent?

15. Prove that if A, B, and C are independent, then A and $B \cup C$ are independent. Also show that $A - B$ and C are independent.

16. A fair die is rolled six times. If on the ith roll, $1 \leq i \leq 6$, the outcome is i, we say that a match has occurred. What is the probability that at least one match occurs?

17. There are n cards in a box numbered 1 through n. We draw cards successively and at random with replacement. If the ith draw is the card numbered i, we say that a match has occurred. What is the probability of at least one match in n trials? What happens if n increases without bound?

18. In a certain county, 15% of patients suffering heart attacks are younger than 40, 20% are between 40 and 50, 30% are between 50 and 60, and 35% are above 60. On a certain day, 10 unrelated patients suffering heart attacks are transferred to a county hospital. If among them there is at least one patient younger than 40, what is the probability that there are two or more patients younger than 40?

19. If the events A and B are independent and the events B and C are independent, is it true that the events A and C are also independent? Why or why not?

20. From the set of all families with three children a *family* is selected at random. Let A be the event that "the family has children of both sexes" and B be the event that "there is at most one girl in the family." Are A and B independent? Answer the same question for families with two children and families with four children. Assume that for any family size all sex distributions have equal probabilities.

21. An event occurs at least once in four independent trials with probability 0.59. What is the probability of its occurrence in one trial?

22. Let $\{A_1, A_2, \ldots, A_n\}$ be an independent set of events and $P(A_i) = p_i$, $1 \le i \le n$.
  (a) What is the probability that at least one of the events $A_1, A_2, \ldots, A_n$ occurs?
  (b) What is the probability that none of the events $A_1, A_2, \ldots, A_n$ occurs?

23. Figure 3.9 shows an electric circuit in which each of the switches located at 1, 2, 3, 4, 5, and 6 is independently closed or open with probabilities p and $1 - p$, respectively. If a signal is fed in at the input, what is the probability that it is transmitted to the output?

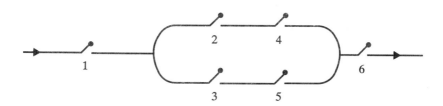

Figure 3.9 Electric circuit of Exercise 23.

B

24. An urn contains two red and four white balls. Balls are drawn from the urn successively, at random and with replacement. What is the probability that exactly three whites occur in the first five trials?

25. A fair coin is tossed n times. Show that the events "at least two heads" and "one or two tails" are independent if $n = 3$ but dependent if $n = 4$.

26. A fair coin is flipped indefinitely. What is the probability of (a) at least one head in the first n flips; (b) exactly k heads in the first n flips; (c) getting heads in all of the flips indefinitely?

27. Suppose that an airplane passenger whose itinerary requires a change of airplanes in Ankara, Turkey, has a 4% chance of losing each piece of his or her luggage independently. Suppose that the probability of losing each piece of luggage in this way is 5% at Da Vinci airport in Rome, 5% at Kennedy airport in New York, and 4% at O'Hare airport in Chicago. Dr. May travels from Bombay to San Francisco with a suitcase. He changes airplanes in Ankara, Rome, New York, and Chicago.
    (a) What is the probability that his suitcase does not reach his destination with him?
    (b) If the suitcase does not reach his destination with him, what is the probability that it was lost at Da Vinci airport in Rome?

28. An experiment consists of first tossing a fair coin and then drawing a card randomly from an ordinary deck of 52 cards with replacement. If we perform this experiment successively, what is the probability of obtaining heads on the coin before an ace from the cards?
    HINT: See Example 3.28.

29. Let S be the sample space of a repeatable experiment. Let A and B be mutually exclusive events of S. Prove that in independent trials of this experiment, the event A occurs before the event B with probability $P(A)/[P(A) + P(B)]$.
    HINT: See Example 3.28; this can be done the same way.

30. In the experiment of rolling two fair dice successively, what is the probability that a sum of 5 appears before a sum of 7?
    HINT: See Exercise 29.

31. An urn contains nine red and one blue balls. A second urn contains one red and five blue balls. One ball is removed from each urn at random and without replacement and all of the remaining balls are put into a third urn. If we draw two balls randomly from the third urn, what is the probability that one of them is red and the other one is blue?

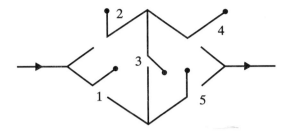

Figure 3.10   Electric circuit of Exercise 33.

32. Suppose that $n \geq 2$ missiles are fired at a target and hit it independently. If the probability that the ith missile hits it is p_i, $i = 1, 2, \ldots, n$, find the probability that at least two missiles hit the target.

33. Figure 3.10 shows an electric circuit in which each of the switches located at 1, 2, 3, 4, and 5 is independently closed or open with probabilities p and $1 - p$, respectively. If a signal is fed in at the input, what is the probability that it is transmitted to the output?

34. From a population of people with unrelated birthdays, 30 people are selected at random. What is the probability that exactly four people of this group have the same birthday and that all the others have different birthdays (exactly 27 birthdays altogether)? Assume that the birth rates are constant throughout the first 365 days of a year but that on the 366th day it is one-fourth that of the other days.

35. In a contest, contestants A, B, and C are each asked, in turn, a general scientific question. If a contestant gives a wrong answer to a question, he goes out of the game. The remaining two will continue to compete until one of them goes out. The last person remaining is the winner. Suppose that a contestant knows the answer to a question independent of the other contestants with probability p. Let CAB represent the event that C goes out first, A next, and B wins, with similar representations for other cases. Calculate and compare the probabilities of ABC, BCA, and CAB.

36. Hemophilia is a hereditary disease. If a mother has it, then with probability 1/2, any of her sons independently will inherit it. Otherwise, none of the sons becomes hemophiliac. Julie is the mother of two sons and from her family's medical history it is known that with the probability 1/4, she is hemophiliac. What is the probability that (a) her first son is hemophiliac; (b) her second son is hemophiliac; (c) none of her sons are hemophiliac?
HINT: Let H, H_1, and H_2 denote the events that the mother, the first son, and the second son are hemophiliac, respectively. Note that H_1 and H_2

are *conditionally independent* given H. That is, if we are given that the mother is hemophiliac, knowledge about one son being hemophiliac does not change the chance of the other son being hemophiliac. However, H_1 and H_2 are not independent. This is because if we know that one son is hemophiliac, the mother is hemophiliac and therefore with probability 1/2 the other son is also hemophiliac.

37. (*Laplace's Law of Succession*)   Suppose that $n+1$ urns are numbered 0 through n and the ith urn contains i red and $n-i$ white balls, $0 \le i \le n$. An urn is selected at random and then the balls it contains are removed one by one, at random, and with replacement. If the first m balls are all red, what is the probability that the $(m+1)$st ball is also red?

HINT: Let U_i be the event that the ith urn is selected, R_m be the event that the first m balls drawn are all red, and R be the event that the $(m+1)$st ball drawn is red. Note that R and R_m are conditionally independent given U_i; that is, given that the ith urn is selected, R and R_m are independent. Hence

$$P(R \mid R_m U_i) = P(R \mid U_i) = \frac{i}{n}.$$

To find $P(R \mid R_m)$, use

$$P(R \mid R_m) = \sum_{i=0}^{n} P(R \mid R_m U_i) P(U_i \mid R_m),$$

where $P(U_i \mid R_m)$ is obtained from Bayes' theorem. The final answer is

$$P(R \mid R_m) = \frac{\sum_{i=0}^{n} \left(\frac{i}{n}\right)^{m+1}}{\sum_{k=0}^{n} \left(\frac{k}{n}\right)^{m}}.$$

Note that if we consider the function $f(x) = x^{m+1}$ on the interval $[0,1]$, the right Riemann sum of this function on the partition $0 = x_0 < x_1 < x_2 < \cdots < x_n = 1$, $x_i = i/n$ is

$$\sum_{i=1}^{n} f(x_i)(x_i - x_{i-1}) = \sum_{i=1}^{n} x_i^{m+1} \frac{1}{n} = \frac{1}{n} \sum_{i=0}^{n} \left(\frac{i}{n}\right)^{m+1}.$$

Therefore, for large n, $(1/n) \sum_{i=0}^{n} (i/n)^{m+1}$ is approximately equal to $\int_0^1 x^{m+1} dx = 1/(m+2)$. Similarly, $(1/n) \sum_{k=0}^{n} (k/n)^m$ is approximately $\int_0^1 x^m dx = 1/(m+1)$. Hence

$$P(R \mid R_m) = \frac{\dfrac{1}{n}\displaystyle\sum_{i=0}^{n}\left(\dfrac{i}{n}\right)^{m+1}}{\dfrac{1}{n}\displaystyle\sum_{k=0}^{n}\left(\dfrac{k}{n}\right)^{m}} \approx \frac{1/(m+2)}{1/(m+1)} = \frac{m+1}{m+2}.$$

Laplace designed and solved this problem for philosophical reasons. He used it to argue that the sun will rise tomorrow with probability $(m+1)/(m+2)$ if we know that it has risen in the m preceding days. Therefore, according to this, as time passes by, the probability that the sun rises again becomes higher and the event of the sun rising becomes more and more certain day after day. It is clear that to argue this way, we should accept the problematic assumption that the phenomenon of the sun rising is "random" and the model of this problem is identical to its random behavior.

Review Problems

1. Two persons arrive at a train station independent of each other at random times between 1:00 P.M. and 1:30 P.M. What is the probability that one will arrive between 1:00 and 1:12, and the other between 1:17 and 1:30?

2. A polygraph operator detects innocent suspects as being guilty 3% of the time. If during a crime investigation six innocent suspects are examined by the operator, what is the probability that at least one of them is detected as guilty?

3. In statistical surveys where people are selected randomly and are asked questions, experience has shown that only 48% of those under 25 years of age, 67% between 25 and 50, and 89% above 50 will respond. A social scientist is about to send a questionnaire to a group of randomly selected people. If 30% of the population are younger than 25 and 17% are older than 50, what percent will answer her questionnaire?

4. Diseases D_1, D_2, and D_3 cause symptom A with probabilities 0.5, 0.7, and 0.8, respectively. If 5% of a population have disease D_1, 2% have disease D_2, and 3.5% have disease D_3, what percent of the population have symptom A? Assume that the only possible causes of symptom A are D_1, D_2, and D_3 and that no one carries more than one of these three diseases.

5. Professor Stern has three cars. The probability that on a given day car 1 is operative is 0.95, that car 2 is operative is 0.97, and that car 3 is

operative is 0.85. If Professor Stern's cars operate independently, find the probability that on the next Thanksgiving day (a) all three of his cars are operative; (b) at least one of his cars is operative; (c) at most two of his cars are operative; (d) none of his cars are operative.

6. A bus traveling from Baltimore to New York has a breakdown at a random location. If the bus was seen running at Wilmington, what is the probability that it broke after passing through Philadelphia? The distances of New York, Philadelphia, and Wilmington from Baltimore are, respectively, 199, 96, and 67 miles.

7. There are only three roads for the prisoners of a prison to escape from: roads A, B, and C. The records of the prison show that of the prisoners who try to escape, 30% use road A, 50% use road B, and 20% use road C. These records also show that 80% of those who try to escape from A, 75% of those who try to escape from B, and 92% of those who try to escape from C are arrested. What is the probability that a prisoner who succeeded in escaping used road C?

8. From an ordinary deck of 52 cards, 10 cards are drawn at random. If exactly four of them are hearts, what is the probability of at least one spade among them?

9. A fair die is thrown twice. If the second outcome is 6, what is the probability that the first one is 6 as well?

10. Suppose that 10 dice are thrown and we are told that among them at least one has landed 6. What is the probability that there are two or more sixes?

11. Urns I and II contain three pennies and four dimes, and two pennies and five dimes, respectively. One coin is selected at random from each urn. If exactly one of them is a dime, what is the probability that the coin selected from urn I is the dime?

12. An experiment consists of first tossing an unbiased coin and then rolling a fair die. If we perform this experiment successively, what is the probability of obtaining heads on the coin before one or two on the die?

13. Six fair dice are tossed independently. Find the probability that the number of 1's minus the number of 2's will be 3.

14. Urn I contains 25 white and 20 black balls. Urn II contains 15 white and 10 black balls. An urn is selected at random and one of its balls is drawn randomly and observed to be black and returned to the same urn. If a second ball is drawn at random from this urn, what is the probability that it is black?

15. An urn contains nine red and one blue balls. A second urn contains one red and five blue balls. One ball is removed from each urn at random

and without replacement and all of the remaining balls are put into a third urn. What is the probability that a ball drawn randomly from the third urn is blue?

16. A child gets lost at the Epcot Center in Florida. The father of the child believes that the probability of his being lost in the east wing of the center is 0.75 and in the west wing is 0.25. The security department sends three officers to the east and two to the west to look for the child. Suppose that an officer who is looking in the correct wing (east or west) finds the child, independent of the other officers, with probability 0.4. Find the probability that the child is found.

17. Solve the following problem, asked of Marilyn Vos Savant in the "Ask Marilyn" column of *Parade Magazine*, August 9, 1992.

> Three of us couples are going to Lava Hot Springs next weekend. We're staying two nights, and we've rented two studios, because each holds a maximum of only four people. One couple will get their own studio on Friday, a different couple on Saturday, and one couple will be out of luck. We'll draw straws to see which are the two lucky couples. I told my wife we should just draw once, and the loser would be the couple out of luck both nights. I figure we'll have a two-out-of-three ($66\frac{2}{3}$%) chance of winning one of the nights to ourselves. But she contends that we should draw straws *twice*—first on Friday and then, for the remaining two couples only, on Saturday—reasoning that a one-in-three ($33\frac{1}{3}$%) chance for Friday and a one-in-two (50%) chance for Saturday will give us better odds. Which way should we go?

18. A student at a certain university will pass the oral Ph.D. qualifying examination if at least two of the three examiners pass her or him. Past experience shows that (a) 15% of the students who take the qualifying exam are not prepared, and (b) each examiner will independently pass 85% of the prepared and 20% of the unprepared students. Kevin took his Ph.D. qualifying exam with Professor Smith, Brown, and Rose. What is the probability that Professor Rose has passed Kevin if we know that neither Professor Brown nor Professor Smith has passed him? Let S, B, and R be the respective events that Professor Smith, Brown, and Rose have passed Kevin. Are these three events independent? Are they conditionally independent given that Kevin is prepared? (See Exercises 36 and 37, Section 3.4.)

19. Adam and three of his friends are playing bridge. (a) If, in a certain hand, Adam announces that he has a king, what is the probability that he has at least one more king? (b) If, in some other hand, Adam announces that he has the king of diamonds, what is the probability that he has at least one more king? Compare parts (a) and (b) and explain why the answers are not the same.

Chapter 4

DISTRIBUTION FUNCTIONS AND DISCRETE RANDOM VARIABLES

4.1   Random Variables

In real-world problems we are often faced with one or more quantities that do not have fixed values. The values of such quantities depend on random actions and they usually change from one experiment to another. For example, the number of babies born in a certain hospital each day is not a fixed quantity. It is a very complicated function of many random phenomena. It varies from one day to another. So are the following quantities: the arrival time of a bus at a station, the sum of the outcomes of two dice when thrown, the amount of rainfall in Seattle during a given year, the number of earthquakes that occur in California per month, the weight of grains of wheat grown on a certain plot of ground (it varies from one grain to another), and the number of misprints per page in a book. In probability, quantities introduced in these diverse examples and similar ones are called *random variables*. The numerical values of random variables are unknown. They depend on random elements developed at the time of the experiment and we have no control over them. For example, if in the experiment of rolling two fair dice, X is the sum, then X can only assume the values 2, 3, 4, ..., 12 with the following probabilities:

$$P(X = 2) = P\{(1, 1)\} = \frac{1}{36},$$

$$P(X = 3) = P\{(1, 2), (2, 1)\} = \frac{2}{36},$$

$$P(X = 4) = P\{(1, 3), (2, 2), (3, 1)\} = \frac{3}{36},$$

and similarly:

Sum, i	5	6	7	8	9	10	11	12
$P(X = i)$	4/36	5/36	6/36	5/36	4/36	3/36	2/36	1/36

Clearly, $\{2, 3, 4, \ldots, 12\}$ is the set of possible values of X. Since $X \in \{2, 3, 4, \ldots, 12\}$, we should have $\sum_{i=2}^{12} P(X = i) = 1$, which is easily verified. The numerical value of a random variable depends on the outcome of the experiment. In this example, for instance, if the outcome is $(2, 3)$, then X is 5, and if it is $(5, 6)$, then X is 11. X is not defined for points that do not belong to S, the sample space of the experiment. Thus X is a real-valued function on S. However, not all real-valued functions on S are considered to be random variables. For theoretical reasons, it is necessary that the inverse image of an interval in **R** be an event of S. This motivates the following definition.

DEFINITION   *Let S be the sample space of an experiment. A real-valued function $X : S \to$ **R** is called a random variable of the experiment if for each interval $I \subseteq$ **R**, $\{s : X(s) \in I\}$ is an event.*

In probability, the set $\{s : X(s) \in I\}$ is often abbreviated as $\{X \in I\}$.

Example 4.1   Suppose that three cards are drawn from an ordinary deck of 52 cards, one by one, at random, and with replacement. Let X be the number of spades drawn; then X is a random variable. If an outcome of spades is denoted by s and other outcomes are represented by t, then X is a real-valued function defined on the sample space

$$S = \{(s, s, s), (t, s, s), (s, t, s), (s, s, t),$$
$$(s, t, t), (t, s, t), (t, t, s), (t, t, t)\},$$

by $X(s, s, s) = 3$, $X(s, t, s) = 2$, $X(s, s, t) = 2$, $X(s, t, t) = 1$, and so on. Now we must determine the values that X assumes and the probabilities that are associated with them. Clearly, X can take the values 0, 1, 2, and 3. The probabilities associated with these values are calculated as follows:

$$P(X = 0) = P\{(t, t, t)\} = \frac{3}{4} \times \frac{3}{4} \times \frac{3}{4} = \frac{27}{64},$$

$$P(X = 1) = P\{(s, t, t), (t, s, t), (t, t, s)\}$$

$$= \left(\frac{1}{4} \times \frac{3}{4} \times \frac{3}{4}\right) + \left(\frac{3}{4} \times \frac{1}{4} \times \frac{3}{4}\right) + \left(\frac{3}{4} \times \frac{3}{4} \times \frac{1}{4}\right) = \frac{27}{64},$$

$$P(X = 2) = P\{(s, s, t), (s, t, s), (t, s, s)\}$$

$$= \left(\frac{1}{4} \times \frac{1}{4} \times \frac{3}{4}\right) + \left(\frac{1}{4} \times \frac{3}{4} \times \frac{1}{4}\right) + \left(\frac{3}{4} \times \frac{1}{4} \times \frac{1}{4}\right) = \frac{9}{64},$$

$$P(X = 3) = P\{(s, s, s)\} = \frac{1}{64}.$$

If the cards are drawn without replacement, the probabilities associated with the values 0, 1, 2, and 3 are $P(X = 0) = \binom{39}{3} \Big/ \binom{52}{3}$, $P(X = 1) = \left[\binom{13}{1}\binom{39}{2}\right] \Big/ \binom{52}{3}$, $P(X = 2) = \left[\binom{13}{2}\binom{39}{1}\right] \Big/ \binom{52}{3}$, and $P(X = 3) = \binom{13}{3} \Big/ \binom{52}{3}$. Therefore,

$$P(X = i) = \frac{\binom{13}{i}\binom{39}{3-i}}{\binom{52}{3}}, \qquad i = 0, 1, 2, 3. \quad \blacklozenge$$

Example 4.2 A bus stops at a station every day at some random time between 11:00 A.M. and 11:30 A.M. If X is the actual arrival time of the bus, X is a random variable. It is a function defined on the sample space $S = \{t : 11 < t < 11\frac{1}{2}\}$ by $X(t) = t$. As we know from Section 1.7, $P\{X = t\} = 0$ for any $t \in S$ and $P\{X \in (\alpha, \beta)\} = \dfrac{\beta - \alpha}{11\frac{1}{2} - 11} = 2(\beta - \alpha)$ for any subinterval (α, β) of $(11, 11\frac{1}{2})$. $\blacklozenge$

Example 4.3 In the United States, the number of twin births is approximately 1 in 90. Let X be the number of births in a certain hospital until the first twins are born. X is a random variable. Denote twin births by T and single births by N. Then X is a real-valued function defined on the sample space $S = \{T, NT, NNT, NNNT, \ldots\}$ by $X(\underbrace{NNN \cdots N}_{i-1} T) = i$. The set of all possible values of X is $\{1, 2, 3, \ldots\}$ and

$$P(X = i) = P(\underbrace{NNN \cdots N}_{i-1} T) = \left(\frac{89}{90}\right)^{i-1} \left(\frac{1}{90}\right). \quad \blacklozenge$$

Example 4.4   In a certain country, the draft-status priorities of eligible men are determined according to their birthdays. Suppose that numbers 1 to 31 are assigned to men with birthdays on January 1 to January 31, numbers 32 to 60 to men with birthdays on February 1 to February 29, numbers 61 to 91 to men with birthdays on March 1 to March 31, ..., and finally, numbers 336 to 366 to those with birthdays on December 1 to December 31. Then numbers are selected at random, one by one, and without replacement from 1 to 366 until all of them are chosen. Those with birthdays corresponding to the first number drawn would have the highest draft priority, those with birthdays corresponding to the second number drawn would have the second-highest priority, and so on. Let X be the largest of the first 10 numbers selected. Then X is a random variable that assumes the values 10, 11, 12, ..., 366. The event $X = i$ occurs if the largest number among the first 10 is i, that is, if one of the first 10 numbers is i and the other 9 are from 1 through $i - 1$. Thus

$$P(X = i) = \frac{\binom{i-1}{9}}{\binom{366}{10}}, \qquad i = 10, 11, 12, \ldots, 366.$$

As an application, let us calculate $P\{X \geq 336\}$, the probability that among persons having one of the 10 highest draft priorities, there are people with birthdays in December. We have that

$$P(X \geq 336) = \sum_{i=336}^{366} \frac{\binom{i-1}{9}}{\binom{366}{10}} \approx 0.592. \quad \blacklozenge$$

REMARK: On December 1, 1969 the Selective Service headquarters in Washington, D.C. determined the draft priorities of 19-year-old males using a method very similar to that of Example 4.4. $\blacklozenge$

We will now explain how, from given random variables, new ones can be formed. Let X and Y be two random variables over the same sample space S; then $X: S \to \mathbf{R}$ and $Y: S \to \mathbf{R}$ are real-valued functions having the same domain. Therefore, we can form the functions $X + Y$; $X - Y$; $aX + bY$, where a and b are constants; XY; and X/Y, where $Y \neq 0$. Since the domain of these real-valued functions is S, they are also random variables defined on S. Hence the sum, the difference, linear combinations, the product, and the quotients (if they exist) of random variables are themselves random variables. Similarly, if $f: \mathbf{R} \to \mathbf{R}$ is an ordinary real-valued function, the composition of f and X, $f \circ X: S \to \mathbf{R}$, is also a random variable. For example, let $f: \mathbf{R} \to \mathbf{R}$ be defined by $f(x) = x^2$; then $f \circ X: S \to \mathbf{R}$ is X^2. Hence X^2 is a random variable as well. Similarly, functions such as $\sin X$, $\cos X^2$, e^X, and $X^3 - 2X$ are random variables. So are the following functions: $X^2 + Y^2$; $\sin X + \cos Y$; $(X^2 + Y^2)/(2Y + 1)$, $Y \neq -1/2$; $\sqrt{X^2 + Y^2}$; and so on. Functions of random variables appear naturally in probability problems. As an example, suppose that we choose a point at random from the unit disk in the plane. If X and Y are the x and the y coordinates of the point, X and Y are random variables. The distance of (X, Y) from the origin is $\sqrt{X^2 + Y^2}$, a random variable that is a function of both X and Y.

Example 4.5   The diameter of a flat metal disk manufactured by a factory is a random number between 4 and 4.5. What is the probability that the area of such a flat disk chosen at random is at least 4.41π?

SOLUTION: Let D be the diameter of the metal disk selected at random. D is a random variable, and the area of the metal disk, $\pi(D/2)^2$, which is a function of D, is also a random variable. We are interested in the probability of the event $\pi D^2/4 > 4.41\pi$, which is

$$P\left(\frac{\pi D^2}{4} > 4.41\pi\right) = P(D^2 > 17.64) = P(D > 4.2).$$

Now to calculate $P(D > 4.2)$, note that since the length of D is a random number in the interval $(4, 4.5)$, the probability that it falls into the subinterval $(4.2, 4.5)$ is $(4.5 - 4.2)/(4.5 - 4) = 3/5$. Hence $P(\pi D^2/4 > 4.41\pi) = 3/5$. ◆

Example 4.6   A random number is selected from the interval $(0, \pi/2)$. What is the probability that its sine is greater than its cosine?

SOLUTION: Let the number selected be X; then X is a random variable and therefore $\sin X$ and $\cos X$, which are functions of X, are also random variables. We are interested in the probability of the event $\sin X > \cos X$:

$$P(\sin X > \cos X) = P(\tan X > 1) = P\left(X > \frac{\pi}{4}\right) = \frac{\frac{\pi}{2} - \frac{\pi}{4}}{\frac{\pi}{2} - 0} = \frac{1}{2},$$

where the first equality holds since in the interval $(0, \pi/2)$, $\cos X > 0$, and the second equality holds since in this interval $\tan X$ is strictly increasing. ◆

4.2   Distribution Functions

Random variables are often introduced in connection with the calculation of the probabilities of events. For example, in the experiment of throwing two dice, if we are interested in a sum of at least 8, we define X to be the sum and calculate $P(X > 8)$. Other examples are the following:

1. If a bus arrives at a random time between 10:00 A.M. and 10:30 A.M. at a station and X is the arrival time, then $X < 10\frac{1}{6}$ is the event that the bus arrives before 10:10 A.M.
2. If X is the price of gold per troy ounce on a random day, then $X \leq 400$ is the event that the price of gold remains at or below \$400 per troy ounce.
3. If X is the number of votes that the next Democratic presidential candidate will get, then $X \geq 5 \times 10^7$ is the event that he or she will get at least 50 million votes.
4. If X is the number of heads in 100 tosses of a coin, then $40 < X \leq 60$ is the event that the number of heads is at least 41 and at most 60.

Usually when dealing with a random variable X, for constants a and b ($b < a$), computation of one or several of the probabilities $P(X = a)$, $P(X < a)$, $P(X \leq a)$, $P(X > b)$, $P(X \geq b)$, $P(b \leq X \leq a)$, $P(b < X \leq a)$, $P(b \leq X < a)$, and $P(b < X < a)$ is our ultimate goal. For this reason we calculate $P(X \leq t)$ for all $t \in (-\infty, +\infty)$. As we will show shortly, if $P(X \leq t)$ is known for all $t \in \mathbf{R}$, then for any a and b, all of the probabilities that are mentioned above can be calculated easily. In fact, since the real-valued function $P(X \leq t)$ characterizes X, it tells us almost everything about X. This function is called the distribution function of X.

DEFINITION   *If X is a random variable, then the function F defined on $(-\infty, +\infty)$ by $F(t) = P(X \leq t)$ is called the distribution function of X.*

Since F "accumulates" all of the probabilities of the values of X up to and including t, sometimes it is called the *cumulative distribution function of X*. The most important properties of the distribution functions are:

1. *F is nondecreasing; that is, if $t < u$, then $F(t) \leq F(u)$.* To see this, note that the occurrence of the event $\{X \leq t\}$ implies the occurrence of the event $\{X \leq u\}$. Thus $\{X \leq t\} \subseteq \{X \leq u\}$ and hence $P\{X \leq t\} \leq P\{X \leq u\}$. That is, $F(t) \leq F(u)$.

2. $\lim_{t\to\infty} F(t) = 1$. To prove this, it suffices to show that for any increasing sequence $\{t_n\}$ of real numbers that converges to ∞, $\lim_{n\to\infty} F(t_n) = 1$. This follows from the continuity property of the probability function (see Theorem 1.7). The events $\{X \le t_n\}$ form an increasing sequence that converges to the event $\bigcup_{n=1}^{\infty}\{X \le t_n\} = \{X < \infty\}$; that is, $\lim_{n\to\infty}\{X \le t_n\} = \{X < \infty\}$. Hence

$$\lim_{n\to\infty} P\{X \le t_n\} = P\{X < \infty\} = 1,$$

which means that

$$\lim_{n\to\infty} F(t) = 1.$$

3. $\lim_{t\to-\infty} F(t) = 0$. The proof of this is similar to the proof that $\lim_{t\to\infty} F(t) = 1$.

4. *F is right continuous. That is, for every $t \in \mathbf{R}$, $F(t+) = F(t)$. This means that if t_n is a decreasing sequence of real numbers converging to t, then*

$$\lim_{n\to\infty} F(t_n) = F(t).$$

To prove this, note that since t_n decreases to t, the events $\{X \le t_n\}$ form a decreasing sequence that converges to the event $\bigcap_{n=1}^{\infty}\{X \le t_n\} = \{X \le t\}$. Thus by the continuity property of the probability function,

$$\lim_{n\to\infty} P\{X \le t_n\} = P\{X \le t\},$$

which means that

$$\lim_{n\to\infty} F(t_n) = F(t).$$

As mentioned above, by means of F, the distribution function of a random variable X, a wide range of probabilistic questions concerning X can be answered. Here are some examples.

1. To calculate $P(X > a)$, note that $P(X > a) = 1 - P(X \le a)$, thus

$$P(X > a) = 1 - F(a).$$

2. To calculate $P(a < X \le b)$, $b > a$, note that $\{a < X \le b\} = \{X \le b\} - \{X \le a\}$ and $\{X \le a\} \subseteq \{X \le b\}$. Hence, by Theorem 1.5,

$$P(a < X \le b) = P(X \le b) - P(X \le a) = F(b) - F(a).$$

3. To calculate $P(X < a)$, note that the sequence of the events $\{X \le a - 1/n\}$ is an increasing sequence that converges to $\bigcup_{n=1}^{\infty}\{X \le a - 1/n\} = \{X < a\}$. Therefore, by the continuity property of the probability function,

$$\lim_{n \to \infty} P\left\{X \le a - \frac{1}{n}\right\} = P\{X < a\},$$

which means that

$$P(X < a) = \lim_{n \to \infty} F\left(a - \frac{1}{n}\right).$$

Hence $P(X < a)$ is the left-hand limit of the function F as $x \to a$, that is,

$$P(X < a) = F(a-).$$

4. To calculate $P(X \ge a)$, note that $P(X \ge a) = 1 - P(X < a)$. Thus

$$P(X \ge a) = 1 - F(a-).$$

5. Since $\{X = a\} = \{X \le a\} - \{X < a\}$ and $\{X < a\} \subseteq \{X \le a\}$, we can write

$$P\{X = a\} = P\{X \le a\} - P\{X < a\} = F(a) - F(a-).$$

Note that since F is right continuous, $F(a)$ is the right-hand limit of F. Thus $P\{X = a\}$ is the difference between the right- and left-hand limits of F at a. If the function F is continuous at a, these limits are the same and equal to $F(a)$. Hence $P\{X = a\} = 0$. Otherwise, F has a *jump* at a, and the magnitude of the jump, $F(a) - F(a-)$, is the probability that $X = a$.

As in cases 1 to 5, we can easily establish similar cases to obtain the following table.

Event concerning X	Probability of the event in terms of F	Event concerning X	Probability of the event in terms of F
$X \le a$	$F(a)$	$a < X \le b$	$F(b) - F(a)$
$X > a$	$1 - F(a)$	$a < X < b$	$F(b-) - F(a)$
$X < a$	$F(a-)$	$a \le X \le b$	$F(b) - F(a-)$
$X \ge a$	$1 - F(a-)$	$a \le X < b$	$F(b-) - F(a-)$
$X = a$	$F(a) - F(a-)$		

Example 4.7 The distribution function of a random variable X is given by

$$F(x) = \begin{cases} 0 & x < 0 \\ \dfrac{x}{4} & 0 \le x < 1 \\ \dfrac{1}{2} & 1 \le x < 2 \\ \dfrac{1}{12}x + \dfrac{1}{2} & 2 \le x < 3 \\ 1 & x \ge 3, \end{cases}$$

where the graph of F is presented in Figure 4.1. Compute the following quantities: (a) $P(X < 2)$; (b) $P(X = 2)$; (c) $P(1 \le X < 3)$; (d) $P(X > 3/2)$; (e) $P(X = 5/2)$; (f) $P(2 < X \le 7)$.

SOLUTION:

(a) $P(X < 2) = F(2-) = 1/2$.

(b) $P(X = 2) = F(2) - F(2-) = (2/12 + 1/2) - 1/2 = 1/6$.

(c) $P(1 \le X < 3) = P(X < 3) - P(X < 1) = F(3-) - F(1-) = (3/12 + 1/2) - 1/4 = 1/2$.

(d) $P(X > 3/2) = 1 - F(3/2) = 1 - 1/2 = 1/2$.

(e) $P(X = 5/2) = 0$ since F is continuous at $5/2$ and has no jumps.

(f) $P(2 < X \le 7) = F(7) - F(2) = 1 - (2/12 + 1/2) = 1/3$. ◆

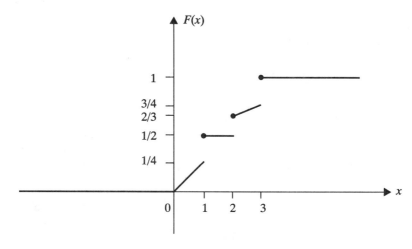

Figure 4.1 Distribution function of Example 4.7.

Example 4.8   For the experiment of flipping a fair coin twice, let X be the number of tails and calculate $F(t)$, the distribution function of X, and then sketch its graph.

SOLUTION: Since X assumes only the values 0, 1, and 2, we have $F(t) = P(X \leq t) = 0$, if $t < 0$. If $0 \leq t < 1$, then

$$F(t) = P(X \leq t) = P(X = 0) = P(\text{HH}) = \frac{1}{4}.$$

If $1 \leq t < 2$, then

$$F(t) = P(X \leq t) = P(X = 0 \text{ or } X = 1) = P\{\text{HH, HT, TH}\} = \frac{3}{4},$$

and if $t \geq 2$, then $P(X \leq t) = 1$. Hence

$$F(t) = \begin{cases} 0 & t < 0 \\ \dfrac{1}{4} & 0 \leq t < 1 \\ \dfrac{3}{4} & 1 \leq t < 2 \\ 1 & t \geq 2. \end{cases}$$

Figure 4.2 presents the graph of F.   ◆

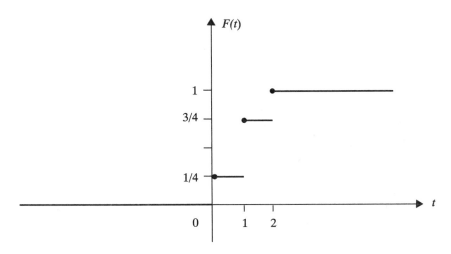

Figure 4.2   Distribution function of Example 4.8.

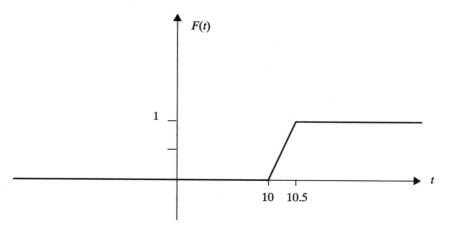

Figure 4.3   Distribution function of Example 4.9.

Example 4.9   Suppose that a bus arrives at a station every day sometime between 10:00 A.M. and 10:30 A.M., at random. Let X be the arrival time; find the distribution function of X and sketch its graph.

SOLUTION: The bus arrives at the station at random between 10 and $10\frac{1}{2}$, so if $t \leq 10$, $F(t) = P(X \leq t) = 0$.   Now if $t \in (10, 10\frac{1}{2})$, then $F(t) = P(X \leq t) = (t - 10)/(10\frac{1}{2} - 10) = 2(t - 10)$ and if $t \geq 10\frac{1}{2}$, then $F(t) = P(X \leq t) = 1$. Thus

$$F(t) = \begin{cases} 0 & t < 10 \\ 2(t - 10) & 10 \leq t < 10\frac{1}{2} \\ 1 & t \geq 10\frac{1}{2}. \end{cases}$$

The graph of F is shown in Figure 4.3.   ◆

Example 4.10   The sales of a convenience store on a randomly selected day are X thousand dollars, where X is a random variable with a distribution function of the following form:

$$F(t) = \begin{cases} 0 & t < 0 \\ \dfrac{1}{2}t^2 & 0 \leq t < 1 \\ k(4t - t^2) & 1 \leq t < 2 \\ 1 & t \geq 2. \end{cases}$$

Suppose that this convenience store's total sales on any given day are less than $2000.

(a) Find the value of k.

(b) Let A and B be the events that tomorrow the store's total sales are between 500 and 1500 dollars, and over 1000 dollars, respectively. Find $P(A)$ and $P(B)$.

(c) Are A and B independent events?

SOLUTION: **(a)** Since $X < 2$, we have that $P(X < 2) = 1$, so $F(2-) = 1$. This gives $k(8 - 4) = 1$, so $k = 1/4$.

(b)
$$P(A) = P\left(\frac{1}{2} \le X \le \frac{3}{2}\right) = F\left(\frac{3}{2}\right) - F\left(\frac{1}{2}-\right)$$

$$= F\left(\frac{3}{2}\right) - F\left(\frac{1}{2}\right) = \frac{15}{16} - \frac{1}{8} = \frac{13}{16},$$

$$P(B) = P(X > 1) = 1 - F(1) = 1 - \frac{3}{4} = \frac{1}{4}.$$

(c)
$$P(AB) = P\left(1 < X \le \frac{3}{2}\right) = F\left(\frac{3}{2}\right) - F(1) = \frac{15}{16} - \frac{3}{4} = \frac{3}{16}.$$

Since $P(AB) \neq P(A)P(B)$, A and B are not independent. ◆

REMARK: Suppose that F is a right-continuous, nondecreasing function on $(-\infty, \infty)$ that satisfies $\lim_{t \to \infty} F(t) = 1$ and $\lim_{t \to -\infty} F(t) = 0$. It can be shown that there exists a sample space S with a probability function and a random variable X over S such that the distribution function of X is F. Therefore, a function is a distribution function if it satisfies the conditions specified in this remark. ◆

EXERCISES

A

1. Two fair dice are rolled and the absolute value of the difference of the outcomes is denoted by X. What are the possible values of X and the probabilities associated with them?

2. From an urn that contains five red, five white, and five blue chips, we draw two chips at random. For each blue chip we win $1, for each white

chip we win $2, but for each red chip we lose $3. If X represents the amount that either we win or we lose, what are the possible values of X and probabilities associated with them?

3. In a society of population N, the probability is p that a person has a certain rare disease. Let X be the number of people who should be tested until a person with the disease is found, $X = 0$ if no one with the disease is found. What are the possible values of X? Determine the probabilities associated with these values.

4. The length of the side of a plastic die manufactured by a factory is a random number between 1 and $1\frac{1}{4}$ centimeters. What is the probability that the volume of a randomly selected die manufactured by this company is greater than 1.424? Assume that the die will always be a cube.

5. F, the distribution function of a random variable X, is given by

$$
F(t) = \begin{cases}
0 & t < -1 \\
\dfrac{1}{4}t + \dfrac{1}{4} & -1 \le t < 0 \\
\dfrac{1}{2} & 0 \le t < 1 \\
\dfrac{1}{12}t + \dfrac{7}{12} & 1 \le t < 2 \\
1 & t \ge 2.
\end{cases}
$$

(a) Sketch the graph of F.
(b) Calculate the following quantities: $P(X < 1)$, $P(X = 1)$, $P(1 \le X < 2)$, $P(X > 1/2)$, $P(X = 3/2)$, and $P(1 < X \le 6)$.

6. From families with three children a *family* is chosen at random. Let X be the number of girls in the family. Calculate and sketch the distribution function of X. Assume that in a three-child family all gender distributions are equally probable.

7. A grocery store sells X hundred kilograms of rice every day, where the distribution of the random variable X is of the following form:

$$
F(x) = \begin{cases}
0 & x < 0 \\
kx^2 & 0 \le x < 3 \\
k(-x^2 + 12x - 3) & 3 \le x < 6 \\
1 & x \ge 6.
\end{cases}
$$

Suppose that this grocery store's total sales of rice do not reach 600 kilograms on any given day.
(a) Find the value of k.
(b) What is the probability that the store sells between 200 and 400 kilograms of rice next Thursday?
(c) What is the probability that the store sells over 300 kilograms of rice next Thursday?
(d) We are given that the store sold at least 300 kilograms of rice last Friday. What is the probability that it did not sell more than 400 kilograms on that day?

8. Let X be a random variable with distribution function F.   For p $(0 < p < 1)$, Q_p is said to be a *quantile of order p* if

$$F(Q_p-) \le p \le F(Q_p).$$

In a certain country, the rate at which the price of oil per gallon changes from one year to another has the following distribution function:

$$F(x) = \frac{1}{1+e^{-x}}, \qquad -\infty < x < \infty.$$

Find $Q_{0.50}$, called *the median* of F; $Q_{0.25}$, called the *first quartile* of F; and $Q_{0.75}$, called the *third quartile* of F. Interpret these quantities.

9. A random variable X is called *symmetric* if for all $x \in \mathbf{R}$,

$$P(X \ge x) = P(X \le -x).$$

Prove that if X is symmetric, then for all $t > 0$, its distribution function F satisfies the following relations:
(a) $P(|X| \le t) = 2F(t) - 1$.
(b) $P(|X| > t) = 2[1 - F(t)]$.
(c) $P(X = t) = F(t) + F(-t) - 1$.

10. Determine if the following is a distribution function.

$$F(t) = \begin{cases} 1 - \dfrac{1}{\pi}e^{-t} & \text{if } t \ge 0 \\ 0 & \text{if } t < 0. \end{cases}$$

11. Determine if the following is a distribution function.

$$F(t) = \begin{cases} \dfrac{t}{1+t} & \text{if } t \geq 0 \\ 0 & \text{if } t < 0. \end{cases}$$

12. Airline A has commuter flights every 45 minutes from San Francisco airport to Fresno. A passenger who wants to take one of these flights arrives at the airport at a random time. Suppose that X is the waiting period for this passenger; find the distribution function of X. Assume that for these flights seats are always available.

B

13. In a small town there are 40 taxis numbered 1 to 40. Three taxis arrive at random at a station to pick up passengers. What is the probability that the number of at least one of them is less than 5?

14. Let X be a randomly selected point from the interval $(0, 3)$. What is the probability that $X^2 - 5X + 6 > 0$?

15. Let X be a random point selected from the interval $(0, 1)$. Calculate F, the distribution function of $Y = X/(1 + X)$, and sketch its graph.

16. In the United States, the number of twin births is approximately 1 in 90. Let X be the number of births in a certain hospital until the first twins are born. Find the first quartile, the median, and the third quartile of X. See Exercise 8 for the definitions of these quantities.

17. Let the time until a new car breaks down be denoted by X, and let

$$Y = \begin{cases} X & \text{if } X \leq 5 \\ 5 & \text{if } X > 5. \end{cases}$$

Then Y is the life of the car, if it lasts less than five years, and is 5 if it lasts longer than five years. Calculate the distribution function of Y in terms of F, the distribution function of X.

4.3   Discrete Random Variables

In Section 4.1 we observed that the set of possible values of a random variable might be finite, infinite but countable, or uncountable. For example, let X, Y, and Z be three random variables representing the respective number of tails in flipping a coin twice, the number of flips of a coin until the first heads, and the amount of next year's rainfall. Then the sets of possible values for X, Y, and Z are the finite set $\{0, 1, 2\}$, the countable set $\{1, 2, 3, 4, \ldots\}$, and the uncountable set $\{x : x \geq 0\}$, respectively. Whenever the set of possible values that a random variable X can assume is at most countable, X is called *discrete*. Therefore, X is discrete if either the set of its possible values is finite or it is countably infinite. To each discrete random variable, a real-valued function $p: \mathbf{R} \to \mathbf{R}$, defined by $p(x) = P\{X = x\}$, is assigned and is called the *probability function* of X. (It is also called the *discrete probability function of X* or the *probability mass function of X*.) Since the set of values of X is countable, $p(x)$ is positive at most for a countable set. It is zero elsewhere, that is, if possible values of X are $x_1, x_2, x_3, \ldots$, then $p(x_i) \geq 0$ $(i = 1, 2, 3, \ldots)$ and $p(x) = 0$ if $x \notin \{x_1, x_2, x_3, \ldots\}$. Now, clearly, the occurrence of one of the events $X = x_1$, $X = x_2$, $X = x_3$, $\ldots$, is certain. Therefore, $\sum_{i=1}^{\infty} P\{X = x_i\} = 1$ or, equivalently, $\sum_{i=1}^{\infty} p(x_i) = 1$.

DEFINITION *The probability function p of a random variable X whose set of possible values is $\{x_1, x_2, x_3, \ldots\}$ is a function from $\mathbf{R}$ to $\mathbf{R}$ that satisfies the following properties.*

(a) $p(x) = 0$ *if* $x \notin \{x_1, x_2, x_3, \ldots\}$.
(b) $p(x_i) = P\{X = x_i\}$ *and hence* $p(x_i) \geq 0$ $(i = 1, 2, 3, \ldots)$.
(c) $\sum_{i=1}^{\infty} p(x_i) = 1$.

Because of this definition, if for a set $\{x_1, x_2, x_3, \ldots\}$, there exists a function $p: \mathbf{R} \to \mathbf{R}$ such that $p(x_i) \geq 0$ $(i = 1, 2, 3, \ldots)$, $p(x) = 0$, $x \notin \{x_1, x_2, x_3, \ldots\}$, and $\sum_{i=1}^{\infty} p(x_i) = 1$, then p is called a probability function.

The probability function of a random variable is often demonstrated geometrically by a set of vertical lines connecting the points $(x_i, 0)$ and $(x_i, p(x_i))$. For example, if X is the number of heads in two flips of a fair coin, then $X = 0, 1, 2$ with $p(0) = 1/4$, $p(1) = 1/2$, and $p(2) = 1/4$. Hence the graphical representation of p is as shown in Figure 4.4.

The distribution function F of a discrete random variable X with the set of possible values $\{x_1, x_2, x_3, \ldots\}$ is a step function. Assuming that $x_1 < x_2 < x_3 < \cdots$, we have that if $t < x_1$, then

$$F(t) = 0;$$

if $x_1 \leq t < x_2$, then

$$F(t) = P(X \leq t) = P(X = x_1) = p(x_1);$$

if $x_2 \leq t < x_3$, then

$$F(t) = P(X \leq t) = P(X = x_1 \text{ or } X = x_2) = p(x_1) + p(x_2);$$

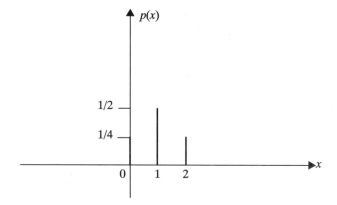

Figure 4.4 Graph of the number of heads in two flips of a fair coin.

and in general, if $x_{n-1} \leq t < x_n$, then

$$F(t) = \sum_{i=1}^{n-1} p(x_i).$$

Thus F is constant in the intervals $[x_{n-1}, x_n)$ with jumps at $x_1, x_2, x_3, \ldots$. The magnitude of the jump at x_i is $p(x_i)$.

Example 4.11 In the experiment of rolling a balanced die twice, let X be the maximum of the two numbers obtained. Determine and sketch the probability function and the distribution function of X.

SOLUTION: The possible values of X are 1, 2, 3, 4, 5, and 6. The sample space of this experiment consists of 36 sample points. Hence the probability of any of them is 1/36. Thus

$$p(1) = P\{X = 1\} = P\{(1, 1)\} = \frac{1}{36},$$

$$p(2) = P\{X = 2\} = P\{(1, 2), (2, 2), (2, 1)\} = \frac{3}{36},$$

$$p(3) = P\{X = 3\} = P\{(1, 3), (2, 3), (3, 3), (3, 2), (3, 1)\} = \frac{5}{36}.$$

Similarly, $p(4) = 7/36$, $p(5) = 9/36$, and $p(6) = 11/36$; $p(x) = 0$ for $x \notin \{1, 2, 3, 4, 5, 6\}$. The graphical representation of p is shown in Figure 4.5. The distribution function of X, F, is as follows (its graph is shown in Figure 4.6):

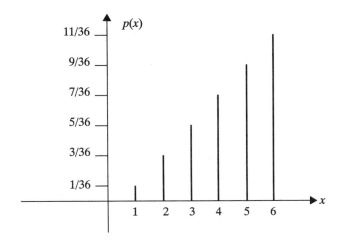

Figure 4.5   Probability function of Example 4.11.

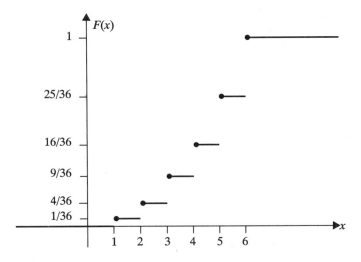

Figure 4.6   Distribution function of Example 4.11.

$$F(x) = \begin{cases} 0 & x < 1 \\[6pt] \dfrac{1}{36} & 1 \le x < 2 \\[10pt] \dfrac{4}{36} & 2 \le x < 3 \\[10pt] \dfrac{9}{36} & 3 \le x < 4 \\[10pt] \dfrac{16}{36} & 4 \le x < 5 \\[10pt] \dfrac{25}{36} & 5 \le x < 6 \\[10pt] 1 & x \ge 6. \quad \blacklozenge \end{cases}$$

Example 4.12   Can a function of the form

$$p(x) = \begin{cases} c\left(\dfrac{2}{3}\right)^x & x = 1, 2, 3, \ldots \\[10pt] 0 & \text{elsewhere} \end{cases}$$

be a probability function?

SOLUTION: A probability function should have three properties: (1) $p(x)$ must be zero at all points except on a finite or a countable set. Clearly, this property is satisfied. (2) $p(x)$ should be nonnegative. This is satisfied if and only if $c \ge 0$. (3) $\sum p(x_i) = 1$. This condition is satisfied if and only if $\sum_{i=1}^{\infty} c(2/3)^i = 1$. This happens precisely when

$$c = \frac{1}{\displaystyle\sum_{i=1}^{\infty}(2/3)^i} = \frac{1}{\dfrac{2/3}{1-2/3}} = \frac{1}{2},$$

where the second equality follows from the geometric series theorem. Thus only for $c = 1/2$, a function of the given form is a probability function. $\blacklozenge$

Example 4.13   Let X be the number of births in a hospital until the first girl is born. Determine the probability function and the distribution function of X. Assume that the probability that a baby born is a girl is $1/2$.

SOLUTION: X is a random variable that can assume any positive integer i. $p(i) = P\{X = i\}$, and $X = i$ occurs if the first $i - 1$ births are all boys and the ith birth is a girl. Thus $p(i) = (1/2)^{i-1}(1/2) = (1/2)^i$ for $i = 1, 2, 3, \ldots$, and $p(x) = 0$ if $x \neq 1, 2, 3, \ldots$. To determine $F(t)$, note that for $t < 1$, $F(t) = 0$; for $1 \leq t < 2$, $F(t) = 1/2$; for $2 \leq t < 3$, $F(t) = 1/2 + 1/4 = 3/4$; for $3 \leq t < 4$, $F(t) = 1/2 + 1/4 + 1/8 = 7/8$; and in general for $n - 1 \leq t < n$,

$$F(t) = \frac{1}{2} + \frac{1}{2^2} + \frac{1}{2^3} + \cdots + \frac{1}{2^{n-1}} = \sum_{i=1}^{n-1} \left(\frac{1}{2}\right)^i$$

$$= \frac{1 - (1/2)^n}{1 - 1/2} - 1 = 1 - \left(\frac{1}{2}\right)^{n-1},$$

by the partial sum formula for geometric series. Thus

$$F(t) = \begin{cases} 0 & t < 1 \\ 1 - \left(\dfrac{1}{2}\right)^{n-1} & n - 1 \leq t < n, \quad n = 2, 3, 4, \ldots. \end{cases} \quad \blacklozenge$$

EXERCISES

A

1. Let $p(x) = x/15$, $x = 1, 2, 3, 4, 5$ be probability function of a random variable X. Determine F, the distribution function of X, and sketch its graph.

2. In the experiment of rolling a balanced die twice, let X be the minimum of the two numbers obtained. Determine the probability function and the distribution function of X and sketch their graphs.

3. The distribution function of a random variable X is given by

$$F(x) = \begin{cases} 0 & \text{if } x < -2 \\ \dfrac{1}{2} & \text{if } -2 \leq x < 2 \\ \dfrac{3}{5} & \text{if } 2 \leq x < 4 \\ \dfrac{8}{9} & \text{if } 4 \leq x < 6 \\ 1 & \text{if } x \geq 6. \end{cases}$$

Determine the probability function of X and sketch its graph.

4. Let X be the number of random numbers selected from $\{0, 1, 2, \ldots, 9\}$ independently until 0 is chosen. Find the probability functions of X and $Y = 2X + 1$.

5. A value i is said to be the *mode* of a discrete random variable X if it maximizes $p(x)$, the probability function of X. Find the modes of random variables X and Y with probability functions

$$p(x) = \left(\frac{1}{2}\right)^x, \qquad x = 1, 2, 3, \ldots,$$

and

$$q(y) = \frac{4!}{y!\,(4-y)!}\left(\frac{1}{4}\right)^y\left(\frac{3}{4}\right)^{4-y}, \qquad y = 0, 1, 2, 3, 4,$$

respectively.

6. For each of the following, determine the value(s) of k for which p is a probability function.
   (a) $p(x) = kx$, $x = 1, 2, 3, 4, 5$.
   (b) $p(x) = k(1 + x)^2$, $x = -2, 0, 1, 2$.
   (c) $p(x) = k(1/9)^x$, $x = 1, 2, 3, \ldots$
   (d) $p(x) = kx$, $x = 1, 2, 3, \ldots, n$
   (e) $p(x) = kx^2$, $x = 1, 2, 3, \ldots, n$,
   where in parts (d) and (e), n is a positive integer.
   HINT: Recall that

$$\sum_{i=1}^{n} i = \frac{n(n+1)}{2}, \qquad \sum_{i=1}^{n} i^2 = \frac{n(n+1)(2n+1)}{6}.$$

7. From 18 potential women jurors and 28 potential men jurors, a jury of 12 is chosen at random. Let X be the number of women selected. Find the probability function of X.

8. Let $p(x) = 3/4(1/4)^x$, $x = 0, 1, 2, 3, \ldots$, be probability function of a random variable X. Find F, the distribution function of X, and sketch its graph.

9. In successive rolls of a fair die, let X be the number of rolls until the first 6. Determine the probability function and the distribution function of X.

4.4   Expectations of Discrete Random Variables

To motivate the concept of expectation, consider a casino game in which the probability of losing $1 per game is 0.6 and the probabilities of winning $1, $2, and $3 per game are 0.3, 0.08, and 0.02, respectively. The gain or loss of a gambler who plays this game only a few times depends on his luck more than anything else. For example, in one play of the game, a lucky gambler might win $3, but he has a 60% chance of losing $1. However, if a gambler decides to play the game a large number of times, his loss or gain depends more on the number of plays than on his luck. A calculating player argues that if he plays the game n times, for a large n, then in approximately $(0.6)n$ games he will lose $1 per game, and in approximately $(0.3)n$, $(0.08)n$, and $(0.02)n$ games he will win $1, $2, and $3, respectively. Therefore, his total gain is

$$(0.6)n \cdot -1 + (0.3)n \cdot 1 + (0.08)n \cdot 2 + (0.02)n \cdot 3 = (-0.08)n. \qquad (4.1)$$

This gives an average of $ -0.08 or about 8 cents of loss per game. The more the gambler plays, the less luck interferes and the closer his loss comes to $0.08 per game. If X is the random variable denoting the gain in one play, then the number -0.08 is called the *expected value* of X. We write $E(X) = -0.08$. $E(X)$ is the average value of X. That is, if we play the game n times and find the average of the values of X, then as $n \to \infty$, $E(X)$ is obtained. Since for this game, $E(X) < 0$, on the average, the more we play, the more we lose. If for some game $E(X) = 0$, then in long run the player neither loses nor wins. Such games are called *fair*. In this example, X is a discrete random variable with the set of possible values $\{-1, 1, 2, 3\}$. The probability function of X, $p(x)$, is given by

i	-1	1	2	3
$p(i) = P(X = i)$	0.6	0.3	0.08	0.02

and $p(x) = 0$ if $x \notin \{-1, 1, 2, 3\}$. Dividing both sides of (4.1) by n, we obtain

$$(0.6)(-1) + (0.3)(1) + (0.08)(2) + (0.02)(3) = -0.08.$$

Hence

$$-1 \cdot p(-1) + 1 \cdot p(1) + 2 \cdot p(2) + 3 \cdot p(3) = -0.08,$$

a relation showing that the expected value of X can be calculated directly by summing up the product of possible values of X by their probabilities. This and similar examples motivate the following general definition, which was first used casually by Pascal but introduced formally by Huygens in the late seventeenth century.

DEFINITION   *The expected value of a discrete random variable X with the set of possible values A and probability function $p(x)$ is defined by*

$$E(X) = \sum_{x \in A} xp(x).$$

We say that $E(X)$ exists if this sum converges absolutely.

The expected value of a random variable X is also called the *mean*, or the *mathematical expectation*, or simply the *expectation* of X. It is also occasionally denoted by $E[X]$, EX, μ_X, or μ.

Note that if each value x of X is weighted by $p(x) = P(X = x)$, then $\sum_{x \in A} xp(x)$ is nothing but the weighted average of X. Similarly, if we think of a unit mass distributed along the real line at the points of A so that the mass at $x \in A$ is $P(X = x)$, then $E(X)$ is the center of gravity.

To illuminate the notion of expected value, a fundamental concept in probability and statistics, we now present some examples.

Example 4.14   We flip a fair coin twice and let X be the number of heads obtained. What is the expected value of X?

SOLUTION: The possible values of X are 0, 1, and 2 and the probability function of X is given by $p(0) = P\{X = 0\} = 1/4$, $p(1) = P\{X = 1\} = 1/2$, $p(2) = P\{X = 2\} = 1/4$, and $p(x) = 0$ if $x \notin \{0, 1, 2\}$. Thus

$$E(X) = 0 \cdot p(0) + 1 \cdot p(1) + 2 \cdot p(2) = 0 \cdot \frac{1}{4} + 1 \cdot \frac{1}{2} + 2 \cdot \frac{1}{4} = 1.$$

Therefore, on the average, we expect one head in every two flips.   ♦

Example 4.15   We write the numbers $a_1, a_2, \ldots, a_n$ on n identical balls and mix them in a box. What is the expected value of a ball selected at random?

SOLUTION: The set of possible values of X, the numbers written on the balls selected, is $\{a_1, a_2, \ldots, a_n\}$. The probability function of X is given by

$$p(a_1) = p(a_2) = \cdots = p(a_n) = \frac{1}{n},$$

and $p(x) = 0$ if $x \notin \{a_1, a_2, \ldots, a_n\}$. Thus

$$E(X) = \sum_{x \in \{a_1, a_2, \ldots, a_n\}} xp(x) = a_1 \frac{1}{n} + a_2 \frac{1}{n} + \cdots + a_n \frac{1}{n}$$

$$= \frac{a_1 + a_2 + \cdots + a_n}{n}.$$

Therefore, as expected, $E(X)$ coincides with the average of the values a_1, $a_2, \ldots, a_n$. That is, if for a large number of times we draw balls at random and with replacement, record their values, and then find their average, the result obtained is the arithmetic mean of the numbers $a_1, a_2, \ldots, a_n$. ◆

Example 4.16   The department of mathematics of a college sends eight to 12 professors to the annual meeting of the American Mathematical Society, which lasts five days. The hotel at which the conference is held offers a bargain rate of a dollars per day per person if reservations are made 45 or more days in advance but charges a cancellation fee of $2a$ dollars per person. The department is not certain how many professors will go. However, from past experience it is known that the probability of the attendance of i professors is $1/5$ for $i = 8, 9, 10, 11$, and 12. If the regular rate of the hotel is $2a$ dollars per day per person, should the department make any reservations? If so, how many?

SOLUTION: For $i = 8, 9, 10, 11$, and 12, let X_i be the total cost in dollars if the department makes reservations for i professors. We should compute $E(X_i)$ for $i = 8, 9, 10, 11$, and 12. If $E(X_j)$ is the smallest of all, the department should make reservations for j professors. To calculate $E(X_8)$, note that X_8 only assumes the values $40a$, $50a$, $60a$, $70a$, and $80a$, which correspond to the cases that 8, 9, 10, 11, and 12 professors attend, respectively, while reservations are made only for eight professors. Since the probability of any of these is $1/5$, we have

$$E(X_8) = (40a)\frac{1}{5} + (50a)\frac{1}{5} + (60a)\frac{1}{5} + (70a)\frac{1}{5} + (80a)\frac{1}{5} = 60a.$$

Similarly,

$$E(X_9) = (42a)\frac{1}{5} + (45a)\frac{1}{5} + (55a)\frac{1}{5} + (65a)\frac{1}{5} + (75a)\frac{1}{5} = 56.4a,$$

$$E(X_{10}) = (44a)\frac{1}{5} + (47a)\frac{1}{5} + (50a)\frac{1}{5} + (60a)\frac{1}{5} + (70a)\frac{1}{5} = 54.2a,$$

$$E(X_{11}) = (46a)\frac{1}{5} + (49a)\frac{1}{5} + (52a)\frac{1}{5} + (55a)\frac{1}{5} + (65a)\frac{1}{5} = 53.4a,$$

$$E(X_{12}) = (48a)\frac{1}{5} + (51a)\frac{1}{5} + (54a)\frac{1}{5} + (57a)\frac{1}{5} + (60a)\frac{1}{5} = 54a.$$

We see that X_{11} has the smallest expected value. Thus making 11 reservations is the most reasonable policy. ♦

Example 4.17   In the lottery of a certain state, players pick six different integers between 1 and 49, the order of selection being irrelevant. The lottery commission then selects six of these numbers at random as the winning numbers. A player wins the grand prize of $1,200,000 if all six numbers that he has selected match the winning numbers. He wins the second and third prizes of $800 and $35, respectively, if exactly five and four of his six selected numbers match the winning numbers. What is the expected value of the amount a player wins in one game?

SOLUTION: Let X be the amount that a player wins in one game. Then the possible values of X are 1,200,000, 800, 35, and 0. The probabilities associated with these values are

$$P(X = 1,200,000) = \frac{1}{\binom{49}{6}} \approx 0.000,000,072,$$

$$P(X = 800) = \frac{\binom{6}{5}\binom{43}{1}}{\binom{49}{6}} \approx 0.000,018,$$

$$P(X = 35) = \frac{\binom{6}{4}\binom{43}{2}}{\binom{49}{6}} \approx 0.000,97,$$

$$P(X = 0) = 1 - 0.000,000,072 - 0.000,018 - 0.000,97$$
$$= 0.999,011,928.$$

Therefore,

$$E(X) \approx 1,200,000(0.000,000,072) + 800(0.000,018) + 35(0.000,97)$$
$$+ 0(0.999,011,928) \approx 0.13.$$

This shows that on the average players will win 13 cents per game. If the cost per game is 50 cents, then, on the average, a player will lose 37 cents per game. Therefore, a player who plays 10,000 games over several years will lose approximately $3700. ◆

Let X be a discrete random variable with a set of possible values A and probability function p. We say that $E(X)$ exists if the sum $\sum_{x \in A} xp(x)$ converges; that is, if $\sum_{x \in A} xp(x) < \infty$. We now present two examples of random variables whose mathematical expectations do not exist.

Example 4.18 (St. Petersburg Paradox) In a game, the player flips a fair coin successively until he gets heads. If this occurs on the kth flip, the player wins 2^k dollars. Therefore, if the outcome of the first flip is heads, the player wins $2. If the outcome of the first flip is tails but that of the second flip is heads, he wins $4. If the outcomes of the first two are tails but the third one heads, he will win $8, and so on. The question is: How much should a person who is willing to play a fair game pay to play this game? To answer this, let X be the amount of money the player wins. Then X is a random variable with the set of possible values $\{2, 4, 8, \ldots, 2^k, \ldots\}$ and

$$P(X = 2^k) = \left(\frac{1}{2}\right)^k, \qquad k = 1, 2, 3, \ldots.$$

Therefore,

$$E(X) = \sum_{k=1}^{\infty} 2^k \left(\frac{1}{2}\right)^k = \sum_{k=1}^{\infty} 1 = 1 + 1 + 1 + \cdots = \infty.$$

This shows that the game remains unfair even if a person pays the largest possible amount to play it. In other words, this is a game in which one always wins no matter how expensive it is to play. To see what the flaw is, note that theoretically this game is not feasible to play because it requires an enormous amount of money. In practice, however, the probability that a gambler wins 2^k for a large k is close to 0. Even for small values of k, winning is highly unlikely. For example, to win $2^{30} = 1,073,741,824$, you should get 29 tails in a row followed by a head. The chance of this happening is 1 in 1,073,741,824, much less than 1 in a billion. ◆

The following example serves to illuminate further the concept of expected value. At the same time, it shows the inadequacy of this concept as a central measure.

Example 4.19 Let X_0 be the amount of rain that will fall in the United States on the following Christmas day. For $n > 0$, let X_n be the amount of rain that will fall in the United States on Christmas n years later. Let N be the smallest number of years that elapse before we get a Christmas rainfall greater than X_0. Suppose that $P(X_i = X_j) = 0$ if $i \neq j$, the events concerning the amount of rain on Christmas days of different years are all independent, and the X_n's are identically distributed. Find the expected value of N.

SOLUTION: Since N is the first value for n for which $X_n > X_0$,

$$P(N > n) = P(X_0 > X_1, X_0 > X_2, \ldots, X_0 > X_n)$$

$$= P\{\max(X_0, X_1, X_2, \ldots, X_n) = X_0\} = \frac{1}{n+1};$$

where the last equality follows from the symmetry, there is no more reason for the maximum to be at X_0 than there is for it to be at X_i, $0 \leq i \leq n$. Therefore,

$$P(N = n) = P(N > n - 1) - P(N > n) = \frac{1}{n} - \frac{1}{n+1} = \frac{1}{n(n+1)}.$$

From this it follows that

$$E(N) = \sum_{n=1}^{\infty} nP(N = n) = \sum_{n=1}^{\infty} \frac{n}{n(n+1)} = \sum_{n=1}^{\infty} \frac{1}{n+1} = \infty.$$

Note that $P(N > n - 1) = 1/n$ gives the probability that in the United States, we will have to wait more than, say, three years for a Christmas rainfall that is greater than X_0 is only 1/4, and the probability that we must wait more than nine years is only 1/10. Even with such low probabilities, on average, it will still take infinitely many years before we will have more rain on a Christmas day than we will have on next Christmas day. ◆

Example 4.20 The tanks of a country's army are numbered 1 to N. In a war this country loses n random tanks to the enemy, who discovers that the captured tanks are numbered. If $X_1, X_2, \ldots, X_n$ are the numbers of the captured tanks, what is $E(\max X_i)$? How can the enemy use $E(\max X_i)$ to find an estimation of N, the total number of this country's tanks?

SOLUTION: Let $Y = \max X_i$; then

$$P(Y = k) = \frac{\binom{k-1}{n-1}}{\binom{N}{n}} \qquad \text{for } k = n, n+1, n+2, \ldots, N,$$

because if the maximum of X_i's is k, the numbers of the remaining $n - 1$ tanks are from 1 to $k - 1$. Now

$$E(Y) = \sum_{k=n}^{N} kP(Y = k) = \sum_{k=n}^{N} \frac{k\binom{k-1}{n-1}}{\binom{N}{n}}$$

$$= \frac{1}{\binom{N}{n}} \sum_{k=n}^{N} \frac{k\,(k-1)!}{(n-1)!\,(k-n)!} = \frac{n}{\binom{N}{n}} \sum_{k=n}^{N} \frac{k!}{n!\,(k-n)!}$$

$$= \frac{n}{\binom{N}{n}} \sum_{k=n}^{N} \binom{k}{n}. \tag{4.2}$$

To calculate $\displaystyle\sum_{k=n}^{N} \binom{k}{n}$, note that $\binom{k}{n}$ is the coefficient of x^n in the polynomial $(1 + x)^k$. Therefore, $\displaystyle\sum_{k=n}^{N} \binom{k}{n}$ is the coefficient of x^n in the polynomial $\displaystyle\sum_{k=n}^{N}(1 + x)^k$. Since

$$\sum_{k=n}^{N}(1 + x)^k = (1 + x)^n \sum_{k=0}^{N-n}(1 + x)^k$$

$$= (1 + x)^n \frac{(1 + x)^{N-n+1} - 1}{(1 + x) - 1} = \frac{1}{x}\left[(1 + x)^{N+1} - (1 + x)^n\right],$$

and the coefficient of x^n in the polynomial $\dfrac{1}{x}[(1 + x)^{N+1} - (1 + x)^n]$ is $\binom{N+1}{n+1}$, we have

$$\sum_{k=n}^{N} \binom{k}{n} = \binom{N+1}{n+1}.$$

Substituting this in (4.2), we obtain

$$E(Y) = \frac{n\binom{N+1}{n+1}}{\binom{N}{n}} = \frac{n\dfrac{(N+1)!}{(n+1)!\,(N-n)!}}{\dfrac{N!}{n!\,(N-n)!}} = \frac{n(N+1)}{n+1}.$$

To estimate N, the total number of this country's tanks, we solve $E(Y) = \frac{n(N+1)}{n+1}$ for N. We obtain $N = \frac{n+1}{n}E(Y) - 1$. Therefore, if, for example, the enemy captures 12 tanks and the maximum of the numbers of the tanks captured is 117, then assuming that $E(Y)$ is approximately equal to the value of Y observed, we get $N \approx (13/12) \times 117 - 1 \approx 126$.   ◆

We now discuss some elementary properties of the expectation of a discrete random variable. Further properties of expectations are discussed in future chapters.

THEOREM 4.1   *If X is a constant random variable, that is, if $P(X = c) = 1$ for a constant c, then $E(X) = c$.*

PROOF: There is only one possible value for X and that is c; hence $E(X) = c \cdot P(X = c) = c \cdot 1 = c$.   ◆

Let $f : \mathbf{R} \to \mathbf{R}$ be a real-valued function and X be a discrete random variable with set of possible values A and probability function $p(x)$. Similar to $E(X) = \sum_{x \in A} xp(x)$, there is the extremely important relation $E[f(X)] = \sum_{x \in A} f(x)p(x)$, known as *the law of the unconscious statistician*, which we now prove. This relation enables us to calculate the expected value of the random variable $f(X)$ without deriving its probability function. It implies that, for example, $E(X^2) = \sum_{x \in A} x^2 p(x)$, $E(X^2 - 2X + 4) = \sum_{x \in A} (x^2 - 2x + 4)p(x)$, $E(X \cos X) = \sum_{x \in A} (x \cos x)p(x)$, and $E(e^X) = \sum_{x \in A} e^x p(x)$.

THEOREM 4.2   *Let X be a discrete random variable with set of possible values A and probability function $p(x)$, and let f be a real-valued function. Then $f(X)$ is a random variable with*

$$E[f(X)] = \sum_{x \in A} f(x)p(x).$$

PROOF: Let S be the sample space. We are given that $f : \mathbf{R} \to \mathbf{R}$ is a real-valued function and $X : S \to A \subseteq \mathbf{R}$ is a random variable with the set of possible values A. As we know, $f(X)$, the composition of f and X, is a function from S to the set $f(A) = \{f(x) : x \in A\}$. Hence $f(X)$ is a random variable with the possible set of values $f(A)$. Now by the definition of expectation,

$$E[f(X)] = \sum_{z \in f(A)} zP\{f(X) = z\}.$$

Let $f^{-1}(\{z\}) = \{x : f(x) = z\}$, and notice that we are not claiming that f has an inverse function. We are simply considering the set $\{x : f(x) = z\}$, which is called *the inverse image* of z and is denoted by $f^{-1}(\{z\})$. Now

$$P\{f(X) = z\} = P\{X \in f^{-1}(\{z\})\} = \sum_{\{x : x \in f^{-1}(\{z\})\}} P(X = x)$$

$$= \sum_{\{x : f(x) = z\}} p(x).$$

Thus

$$E[f(X)] = \sum_{z \in f(A)} zP\{f(X) = z\} = \sum_{z \in f(A)} z \sum_{\{x : f(x) = z\}} p(x)$$

$$= \sum_{z \in f(A)} \sum_{\{x : f(x) = z\}} zp(x) = \sum_{z \in f(A)} \sum_{\{x : f(x) = z\}} f(x)p(x)$$

$$= \sum_{x \in A} f(x)p(x),$$

where the last equality follows from the fact that the sum over A can be performed in two stages: We can first sum over all x with $f(x) = z$ and then over all z. ◆

COROLLARY   *Let X be a discrete random variable; $f_1, f_2, \ldots, f_n$ be real-valued functions, and let $\alpha_1, \alpha_2, \ldots, \alpha_n$ be real numbers. Then*

$$E[\alpha_1 f_1(X) + \alpha_2 f_2(X) + \cdots + \alpha_n f_n(X)]$$
$$= \alpha_1 E[f_1(X)] + \alpha_2 E[f_2(X)] + \cdots + \alpha_n E[f_n(X)].$$

PROOF: Let the set of possible values of X be A and its probability function be $p(x)$. Then, by Theorem 4.2,

$$E[\alpha_1 f_1(X) + \alpha_2 f_2(X) + \cdots + \alpha_n f_n(X)]$$

$$= \sum_{x \in A} [\alpha_1 f_1(x) + \alpha_2 f_2(x) + \cdots + \alpha_n f_n(x)] p(x)$$

$$= \alpha_1 \sum_{x \in A} f_1(x) p(x) + \alpha_2 \sum_{x \in A} f_2(x) p(x) + \cdots + \alpha_n \sum_{x \in A} f_n(x) p(x)$$

$$= \alpha_1 E[f_1(X)] + \alpha_2 E[f_2(X)] + \cdots + \alpha_n E[f_n(X)]. \quad \blacklozenge$$

By this corollary, we have that, for example

$$E(2X^3 + 5X^2 + 7X + 4) = 2E(X^3) + 5E(X^2) + 7E(X) + 4$$

and

$$E(e^X + 2\sin X + \log X) = E(e^X) + 2E(\sin X) + E(\log X).$$

Moreover, this corollary implies that $E(X)$ is linear. That is, if $\alpha, \beta \in \mathbf{R}$, then

$$E(\alpha X + \beta) = \alpha E(X) + \beta.$$

Example 4.21 The probability function of a discrete random variable X is given by

$$p(x) = \begin{cases} \dfrac{x}{15} & x = 1, 2, 3, 4, 5 \\ 0 & \text{otherwise.} \end{cases}$$

What is the expected value of $X(6 - X)$?

SOLUTION: By Theorem 4.2:

$$E[X(6 - X)] = 5 \cdot \frac{1}{15} + 8 \cdot \frac{2}{15} + 9 \cdot \frac{3}{15} + 8 \cdot \frac{4}{15} + 5 \cdot \frac{5}{15} = 7. \quad \blacklozenge$$

Example 4.22 A box contains 10 disks of radii $1, 2, \ldots$, and 10, respectively. What is the expected value of the area of a disk selected at random from this box?

SOLUTION: Let the radius of the disk be R; then R is a random variable with the probability function $p(x) = 1/10$ if $x = 1, 2, \ldots, 10$, and $p(x) = 0$ otherwise. $E(\pi R^2)$, the desired quantity is calculated as follows:

$$E(\pi R^2) = \pi E(R^2) = \pi \left(\sum_{i=1}^{10} i^2 \frac{1}{10} \right) = 38.5\pi. \quad \blacklozenge$$

EXERCISES

A

1. In a certain location of downtown Baltimore it costs $7 per day to park at parking lots. A car that is illegally parked on the street will be fined $25 if caught, and the chance of being caught is 60%. If money is the only concern of a citizen who must park in this location every day, should he park at a lot or park illegally?

2. In a lottery every week 2,000,000 tickets are sold for $1 apiece. If 4000 of these tickets pay off $30 each, 500 pay off $800 each, one ticket pays off $1,200,000, and no ticket pays off more than one prize, what is the expected value of the winning amount for a player with a single ticket?

3. In a lottery a player pays $1 and selects four distinct numbers from 0 to 9. Then from an urn containing 10 identical balls numbered from 0 to 9, four balls are drawn at random and without replacement. If the numbers of three or all four of these balls matches the player's numbers, he wins $5 and $10, respectively. Otherwise, he loses. On the average, how much money does the player *gain* per game? (Gain = win − loss.)

4. An urn contains five balls, two of which are marked $1, two $5, and one $15. A game is played by paying $10 and winning the sum of the amounts marked on two balls selected randomly from the urn. Is this a fair game?

5. A box contains 20 fuses, of which five are defective. What is the expected number of defective items among three fuses selected randomly?

6. The number of requests for a certain weekly magazine from a newsstand is a random variable with probability function $p(i) = (10 - i)/18$, $i = 4, 5, 6, 7$. If the magazine sells for a and costs $2a/3$ to the owner and the unsold magazines cannot be returned, how many magazines should be ordered every week to maximize the profit in the long run?

7. It is well known that $\sum_{x=1}^{\infty} 1/x^2 = \pi^2/6$.
   (a) Show that $p(x) = 6/(\pi x)^2$, $x = 1, 2, 3, \ldots$ is the probability function of a random variable X.
   (b) Prove that $E(X)$ does not exist.

8. (a) Show that $p(x) = (|x| + 1)^2/27$, $x = -2, -1, 0, 1, 2$, is the probability function of a random variable X.

(b) Calculate $E(X)$, $E(|X|)$, and $E(2X^2 - 5X + 7)$.

9. A box contains 10 disks of radii $1, 2, \ldots, 10$, respectively. What is the expected value of the circumference of a disk selected at random from this box?

10. The distribution function of a random variable X is given by

$$F(x) = \begin{cases} 0 & \text{if } x < -3 \\[2mm] \dfrac{3}{8} & \text{if } -3 \le x < 0 \\[2mm] \dfrac{1}{2} & \text{if } 0 \le x < 3 \\[2mm] \dfrac{3}{4} & \text{if } 3 \le x < 4 \\[2mm] 1 & \text{if } x \ge 4. \end{cases}$$

Calculate $E(X)$, $E(X^2 - 2|X|)$, and $E(X|X|)$.

11. If X is a random number selected from the first 10 positive integers, what is $E[X(11 - X)]$?

12. Let X be the number of different birthdays among four persons selected randomly. Find $E(X)$.

B

13. Suppose that there exist N families on the earth and that the maximum number of children a family has is c. For $j = 0, 1, 2, \ldots, c$, let α_j be the fraction of families with j children ($\sum_{j=0}^{c} \alpha_j = 1$). A *child* is selected at random from the set of all children in the world. Let this child be the Kth born of his or her family; then K is a random variable. Find $E(K)$.

14. A newly married couple decides to continue having children until they have one of each sex. If the events of having a boy and a girl are independent and equiprobable, how many children should this couple expect?

15. An ordinary deck of 52 cards is well shuffled and then the cards are turned face up one by one until an ace appears. Find the expected number of cards that are face up.

16. Suppose that n random integers are selected from $\{1, 2, \ldots, N\}$ with replacement. What is the expected value of the largest number selected? Show that for large N the answer is approximately $nN/(n+1)$.

17. (a) Show that

$$p(n) = \frac{1}{n(n+1)}, \qquad n \geq 1,$$

is a probability function.

   (b) Let X be a random variable with probability function p given in part (a); find $E(X)$.

4.5   Variances of Discrete Random Variables

So far, through many examples, we have explained the importance of mathematical expectation in detail. For instance, in Example 4.16, we have shown how expectation is applied in decision making. Also in Example 4.17, concerning lottery, we showed that the expectation of the winning amount per game gives an excellent estimation for the total amount a player will win if he or she plays a large number of times. In these and many other situations, mathematical expectation is the only quantity one needs to calculate. However, very frequently we face situations in which the expectation by itself does not say much. In such cases more information should be extracted from the probability function. As an example, suppose that we are interested in measuring a certain quantity. Let X be the true value [†] of the quantity minus the value obtained by measurement. Then X is the error of measurement. It is a random variable with expected value zero, the reason being that in measuring a quantity a very large number of times, positive and negative errors of the same magnitudes occur with equal probabilities. Now consider an experiment in which a quantity is measured several times and the average of the errors is obtained to be a number close to zero. Can we conclude that the measurements are very close to the true value and thus are accurate? The answer is no because they might differ from the true value by relatively large quantities but be scattered both in positive and negative directions resulting in zero expectation. Thus in this and similar cases, expectation by itself does not give adequate information. Therefore, additional measures for decision making are needed. One such quantity is the *variance* of a random variable. Variance measures the average magnitude of the fluctuations of a random variable from its expectation. This is particularly important because random variables fluctuate about their expectations. To define the variance of a random variable X mathematically, the first temptation is to consider the expectation of the difference of X from

---

[†] True value is a nebulous concept. Here we shall use it to mean the average of a *large* number of measurements.

its expectation, that is, $E[X - E(X)]$. But the difficulty with this quantity is that the positive and negative deviations of X from $E(X)$ cancel each other and we always get 0. This can be seen mathematically from the corollary of Theorem 4.2: Let $E(X) = \mu$; then

$$E[X - E(X)] = E(X - \mu) = E(X) - \mu = E(X) - E(X) = 0.$$

Hence $E[X - E(X)]$ is not an appropriate measure for the variance. However, if we consider $E(|X - E(X)|)$ instead, the problem of negative and positive deviations canceling each other disappears. Since this quantity is the true average magnitude of the fluctuations of X from $E(X)$, it seems that it is the best candidate for an expression for the variance of X. But mathematically, $E(|X - E(X)|)$ is difficult to handle; for this reason the quantity $E[(X - E(X))^2]$, analogous to Euclidean distance in geometry, is used instead and is called the *variance of X*. The square root of $E[(X - E(X))^2]$ is called the *standard deviation* of X.

DEFINITION   *Let X be a discrete random variable with set of possible values A, probability function p(x), and $E(X) = \mu$. Then σ_X and Var(X), called the standard deviation and the variance of X, respectively, are defined by*

$$\sigma_X = \sqrt{E\left[(X - \mu)^2\right]} \quad and \quad Var(X) = E\left[(X - \mu)^2\right].$$

Note that by this definition and Theorem 4.2,

$$\text{Var}(X) = E[(X - \mu)^2] = \sum_{x \in A}(x - \mu)^2 p(x).$$

Let X be a discrete random variable with the set of possible values A and probability function $p(x)$. Suppose that the prediction of the value of X is in order, and if the value t is predicted for X, then based on the error $X - t$, a penalty is charged. To minimize the penalty, it seems reasonable to minimize $E(X - t)^2$. But

$$E(X - t)^2 = \sum_{x \in A}(x - t)^2 p(x).$$

Assuming that this series converges [i. e., $E(X^2) < \infty$], we differentiate it to find the minimum value of $E(X - t)^2$:

$$\frac{d}{dt}E(X - t)^2 = \frac{d}{dt}\sum_{x \in A}(x - t)^2 p(x) = \sum_{x \in A}-2(x - t)p(x) = 0.$$

This gives

$$\sum_{x \in A} xp(x) = t \sum_{x \in A} p(x) = t.$$

Therefore, $E(X - t)^2$ is a minimum for $t = \sum_{x \in A} xp(x) = E(X)$ and the minimum value is $E[X - E(X)]^2 = \text{Var}(X)$. So the smaller that $\text{Var}(X)$ is, the better $E(X)$ predicts X. We have that

$$\text{Var}(X) = \min_t E(X - t)^2.$$

Earlier we mentioned that if we think of a unit mass distributed along the real line at the points of A so that the mass at $x \in A$ is $p(x) = P(X = x)$, then $E(X)$ is the center of gravity. As we know, since center of gravity does not provide any information about how the mass is distributed around this center, the concept of moment of inertia is introduced. Moment of inertia is a measure of dispersion (spread) of the mass distribution about the center of gravity. $E(X)$ is analogous to the center of gravity and it too does not provide any information about the distribution of X about this center of location. However, variance, the analog of the moment of inertia, measures the dispersion or spread of a distribution about its expectation.

RELATED HISTORICAL REMARK: In 1900, the Wright brothers, inventors of the first airplane, were looking for a private location with consistent wind to test their gliders. The data they got from the U.S. Weather Bureau indicated that Kill Devil Hill, near Kitty Hawk, North Carolina, had, on average, suitable winds, so they chose this location for their tests. However, in practice, consistent winds were not the case. There were many calm days and many days with strong winds that were not suitable for their tests. The summary of the data was obtained by averaging undesirable extreme wind conditions. The Weather Bureau statistics were misleading because they failed to utilize the standard deviation. If the Wright brothers had been provided with the standard deviation of the wind speed at Kitty Hawk, they would not have chosen this location for their tests. ♦

Example 4.23   Karen is interested in two games, Keno and Bolita. To play Bolita, she buys a ticket for \$1, draws a ball at random from a box of 100 balls numbered 1 to 100. If the ball drawn matches the number on her ticket, she wins \$75; otherwise, she loses. To play Keno, Karen bets \$1 on a single number that has a 25% chance to win. If she wins, they will give her dollar back plus two dollars more; otherwise, they keep the dollar. Let B and K be the amounts that Karen gains in one play of Bolita and Keno, respectively. Then

$$E(B) = (74)(0.01) + (-1)(0.99) = -0.25,$$
$$E(K) = (2)(0.25) + (-1)(0.75) = -0.25.$$

Therefore, in the long run, it does not matter which of the two games Karen plays. Her gain would be about the same. However, by virtue of

$$\text{Var}(B) = E[(B - \mu)^2] = (74 + 0.25)^2(0.01)$$
$$+ (-1 + 0.25)^2(0.99) = 55.69$$

and

$$\text{Var}(K) = E[(K - \mu)^2] = (2 + 0.25)^2(0.25)$$
$$+ (-1 + 0.25)^2(0.75) = 2.4375,$$

we can say that in Bolita, on average, the deviation of the gain from the expectation is much higher than in Keno. In other words, the risk in Keno is far less than the risk in Bolita. In Bolita the probability of winning is very small, but the amount won is high. In Keno players win more often but in smaller amounts. ◆

Another useful formula for Var(X) is obtained as follows:

$$\text{Var}(X) = E[(X - \mu)^2] = E(X^2 - 2\mu X + \mu^2)$$
$$= E(X^2) - 2\mu E(X) + \mu^2$$
$$= E(X^2) - 2\mu^2 + \mu^2 = E(X^2) - \mu^2$$
$$= E(X^2) - [E(X)]^2.$$

One immediate application of this formula is that since for any discrete random variable X, Var(X) ≥ 0,

$$[E(X)]^2 \leq E(X^2).$$

The formula Var(X) $= E(X^2) - [E(X)]^2$ is usually an easier alternative for computing the variance of X. Here is an example.

Example 4.24   What is the variance of the random variable X, the outcome of rolling a fair die?

SOLUTION: The probability function of X is given by $p(x) = 1/6$; $x = 1, 2, 3, 4, 5, 6$, and $p(x) = 0$, otherwise. Hence

$$E(X) = \sum_{x=1}^{6} xp(x) = \frac{1}{6}\sum_{x=1}^{6} x = \frac{1}{6}(1+2+3+4+5+6) = \frac{7}{2}$$

$$E(X^2) = \sum_{x=1}^{6} x^2 p(x) = \frac{1}{6}\sum_{x=1}^{6} x^2 = \frac{1}{6}(1+4+9+16+25+36) = \frac{91}{6}.$$

Thus

$$\text{Var}(X) = E(X^2) - [E(X)]^2 = \frac{91}{6} - \frac{49}{4} = \frac{35}{12}. \quad \blacklozenge$$

Suppose that a random variable X is constant; then $E(X) = X$ and the deviations of X from $E(X)$ are 0. Therefore, the average deviation of X from $E(X)$ is also 0. We have the following theorem.

THEOREM 4.3   *Let X be a discrete random variable with the set of possible values A. Then $\text{Var}(X) = 0$ if and only if X is a constant.*

PROOF: If $\text{Var}(X) = 0$, then by definition

$$\text{Var}(X) = E[(X-\mu)^2] = \sum_{x \in A}(x-\mu)^2 p(x) = 0.$$

But all the terms of this series are nonnegative. Hence $x - \mu = 0$ for all $x \in A$. This means that A consists of only one number, namely μ. Thus $X = \mu$. Conversely, if X is constant, then $X = E(X) = \mu$. Hence $\text{Var}(X) = E[(X-\mu)^2] = 0$. $\blacklozenge$

For constants a and b, a linear relation similar to $E(aX + b) = aE(X) + b$ does not exist for variance and for standard deviation. However, other important relations exist and are given by the following theorem.

THEOREM 4.4   *Let X be a discrete random variable; then for constants a and b we have that*

$$Var(aX + b) = a^2 Var(X),$$

$$\sigma_{aX+b} = |a|\sigma_X.$$

PROOF: To see this, note that

$$\text{Var}(aX + b) = E[(aX+b) - E(aX+b)]^2$$
$$= E[(aX+b) - (aE(X)+b)]^2$$

$$= E\left[a(X - E(X))\right]^2$$
$$= E\left[a^2(X - E(X))^2\right]$$
$$= a^2 E\left[(X - E(X))^2\right]$$
$$= a^2 \text{Var}(X).$$

Taking the square roots of both sides of this relation, we find that $\sigma_{aX+b} = |a|\sigma_X$. ◆

Example 4.25   Suppose that for a discrete random variable X, $E(X) = 2$ and $E[X(X - 4)] = 5$. Find the variance and the standard deviation of $-4X + 12$.

SOLUTION: By Theorem 4.4:

$$\text{Var}(-4X + 12) = 16\text{Var}(X) = 16E\left[(X - \mu)^2\right] = 16E\left[(X - 2)^2\right]$$
$$= 16E(X^2 - 4X + 4) = 16\left[E(X^2 - 4X) + 4\right]$$
$$= 16E\left[X(X - 4)\right] + 64 = 16 \times 5 + 64 = 144.$$

The standard deviation of $-4X + 12$ is therefore equal to $\sqrt{144} = 12$. ◆

EXERCISES

1. Mr. Jones is about to purchase a business. There are two businesses available. The first has a daily expected profit of \$150 with standard deviation \$30, and the second has a daily expected profit of \$150 with standard deviation \$55. If Mr. Jones is interested in a business with a steady income, which should he choose?

2. The temperature of a material is measured by two devices. Using the first device the expected temperature is t with standard deviation 0.8; using the second device the expected temperature is t with standard deviation 0.3. Which device measures the temperature more accurately?

3. Find the variance of X, the random variable with probability function

$$p(x) = \begin{cases} \dfrac{|x - 3| + 1}{28} & x = -3, -2, -1, 0, 1, 2, 3 \\ 0 & \text{otherwise.} \end{cases}$$

4. Find the variance and the standard deviation of a random variable X with distribution function

$$F(x) = \begin{cases} 0 & x < -3 \\ \dfrac{3}{8} & -3 \le x < 0 \\ \dfrac{3}{4} & 0 \le x < 6 \\ 1 & x \ge 6. \end{cases}$$

5. Let X be a random integer from the set $\{1, 2, 3, \ldots, N\}$. Find $E(X)$, $\mathrm{Var}(X)$, and σ_X.

6. What are the expected number, the variance, and the standard deviation of the number of spades in a poker hand? (A poker hand is a set of five cards that are randomly selected from an ordinary deck of 52 cards.)

7. Suppose that X is a discrete random variable with $E(X) = 1$ and $E[X(X - 2)] = 3$. Find $\mathrm{Var}(-3X + 5)$.

8. A drunk man who has n keys wants to open the door of his office. He tries the keys at random, one by one, and independently. Compute the mean and the variance of the number of trials required to open the door if the unsuccessful keys (a) are not eliminated; (b) are eliminated.

4.6   Standardized Random Variables

Let X be a random variable with mean μ and standard deviation σ. The random variable $X^* = (X - \mu)/\sigma$ is called the *standardized* X. We have that

$$E(X^*) = E\left(\frac{1}{\sigma}X - \frac{\mu}{\sigma}\right) = \frac{1}{\sigma}E(X) - \frac{\mu}{\sigma} = \frac{\mu}{\sigma} - \frac{\mu}{\sigma} = 0,$$

$$\mathrm{Var}(X^*) = \mathrm{Var}\left(\frac{1}{\sigma}X - \frac{\mu}{\sigma}\right) = \frac{1}{\sigma^2}\mathrm{Var}(X) = \frac{\sigma^2}{\sigma^2} = 1.$$

When standardizing a random variable X, we change the origin to μ and the scale to the units of standard deviation. The value that is obtained for X^* is independent of the units in which X is measured. It is the number of standard deviation units by which X differs from $E(X)$. For example, let X be a random variable with mean 10 feet and standard deviation 2 feet. Suppose that in a random observation it is obtained that $X = 16$; then $X^* = (16 - 10)/2 = 3$. This shows that the distance of X from its mean is 3 standard deviation units regardless of the scale of measurement. That is, if the same quantities are measured, say, in inches (12 inches = 1 foot), then we will get the same standardized value:

$$X^* = \frac{16 \times 12 - 10 \times 12}{2 \times 12} = 3.$$

Standardization is particularly useful if two or more random variables with different distributions must be compared. Suppose that, for example, the grade of a student in a probability test is 72 and that her grade in a history test is 85. At first glance these grades suggest that the student is doing much better in the history course than in the probability course. However, this might not be true—the relative grade of the student in probability might be better than that in history. To show this, suppose that the mean and standard deviation of all grades in the history test are 82 and 7, respectively, while these quantities in the probability test are 68 and 4. If we convert the student's grades to their standard deviation units, we find that her standard scores on the probability and history tests are given by $(72 - 68)/4 = 1$ and $(85 - 82)/7 = 0.43$, respectively. These show that the student's grade in probability is 1 and in history is 0.43 standard deviation unit higher than their respective averages. Therefore, the student is doing relatively better in the probability course than in the history course. This comparison is most useful when only the means and standard deviations of the random variables being studied are known. If the distribution functions of these random variables are given, better comparisons might be possible.

We now prove that for a random variable X, the standardized X, denoted by X^*, is independent of the units in which X is measured. To do so, let X_1 be the observed value of X when a different scale of measurement is used. Then for some $\alpha > 0$, we have that $X_1 = \alpha X + \beta$, and

$$X_1^* = \frac{X_1 - E(X_1)}{\sigma_{X_1}} = \frac{(\alpha X + \beta) - [\alpha E(X) + \beta]}{\sigma_{\alpha X + \beta}}$$

$$= \frac{\alpha[X - E(X)]}{\alpha \sigma_X} = \frac{X - E(X)}{\sigma_X} = X^*.$$

EXERCISES

1. Mr. Norton has two appliance stores. In store 1 the number of TV sets sold by a salesperson is, on average, 13 per week with a standard deviation of five. In store 2 the number of TV sets sold by a salesperson is, on average, seven with a standard deviation of four. Mr. Norton has a position open for a person to sell TV sets. There are two applicants.

Mr. Norton asked one of them to work in store 1 and the other in store 2, each for one week. The salesperson in store 1 sold 10 sets, and the salesperson in store 2 sold six sets. Based on this information, which one should Mr. Norton hire?

2. The mean and standard deviation in midterm tests of a probability course are 72 and 12, respectively. These quantities for final tests are 68 and 15. What final grade is comparable to Velma's 82 in the midterm.

Review Problems

1. An urn contains 10 chips numbered from 0 to 9. Two chips are drawn at random and without replacement. What is the probability function of their total?

2. A word is selected at random from the following poem of the great Persian poet and mathematician Omar Khayyām (1048–1131), translated by Edward Fitzgerald (1808–1883), a prominent English poet. Find the expected value of the length of the word.

> And when thyself with shining Foot shall pass
> Among the Guests Star-scatter'd on the Grass,
> And in thy joyous Errand reach the Spot
> Where I made one-turn down and empty Glass!

3. A statistical survey shows that only 2% of secretaries know how to use the highly sophisticated word processor TEX. If a certain mathematics department prefers to hire a secretary who knows TEX, what is the least number of applicants that should be interviewed to have at least a 50% chance to find one such secretary?

4. An electronic system fails if both of its components fail. Let X be the time (in hours) until the system fails. Experience has shown that

$$P(X > t) = \left(1 + \frac{t}{200}\right) e^{-t/200}, \qquad t \geq 0.$$

What is the probability that the system lasts at least 200 but not more than 300 hours?

5. A professor has made 30 exams of which eight are hard, 12 are reasonable, and 10 are easy. The exams are mixed up and the professor selects four of them at random to give to four sections of the course he is teaching. How many sections would be expected to get a hard test?

6. The annual amount of rainfall in a certain area in centimeters is a random variable with distribution function

$$F(x) = \begin{cases} 0 & x < 5 \\ 1 - \dfrac{5}{x^2} & x \geq 5. \end{cases}$$

What is the probability that next year it rains (a) at least 6 centimeters; (b) at most 9 centimeters; (c) at least 2 and at most 7 centimeters?

7. Let X be the amount (in fluid ounces) of soft drink in a randomly chosen bottle from company A, and Y be the amount of soft drink in a randomly chosen bottle from company B. A study has shown that the probability distributions of X and Y are as follows:

x	15.85	15.9	16	16.1	16.2
$P(X = x)$	0.15	0.21	0.35	0.15	0.14
$P(Y = x)$	0.14	0.05	0.64	0.08	0.09

Find $E(X)$, $E(Y)$, $\text{Var}(X)$, and $\text{Var}(Y)$ and interpret them.

8. The fasting blood glucose levels of 30 children are as follows.

```
58  62  80  58  64  76  80  80  80  58
62  64  76  76  58  64  62  80  58  58
80  64  58  62  76  62  64  80  62  76
```

Let X be the fasting blood glucose level of a child chosen randomly from this group. Find the distribution function of X.

9. Experience shows that X, the number of customers entering a post office during any period of length t, is a random variable whose probability function is of the form

$$p(i) = k \frac{(2t)^i}{i!}, \qquad i = 0, 1, 2, \ldots.$$

(a) Determine the value of k, (b) compute $P(X < 4)$ and $P(X > 1)$.

10. From the set of families with three children a *family* is selected at random and the number of its boys is denoted by the random variable X. Find the probability function and the probability distribution functions of X. Assume that in a three-child family all gender distributions are equally probable.

Chapter 5

SPECIAL DISCRETE DISTRIBUTIONS

In this chapter we study a few important discrete random variables. These random variables have enormous applications in many diverse sciences.

5.1 Bernoulli and Binomial Random Variables

Bernoulli trials, named after the Swiss mathematician James Bernoulli, are perhaps the simplest type of random variable. They have only two possible outcomes. One outcome is usually called a *success* and is denoted by s. The other outcome is called *failure* and is denoted by f. The experiment of flipping a coin is a Bernoulli trial. Its only outcomes are "heads" and "tails." If we are interested in heads, we may call it a success; tails is then a failure. The experiment of tossing a die is a Bernoulli trial if, for example, we are interested in knowing whether the outcome is odd or even. An even outcome may be called a success and hence an odd outcome a failure, or vice versa. If a fuse is inspected, either it is "defective" or it is "good." So the experiment of inspecting fuses is a Bernoulli trial. A good fuse may be called a success and a defective fuse a failure.

The sample space of a Bernoulli trial contains two points, s and f. The random variable defined by $X(s) = 1$ and $X(f) = 0$ is called a *Bernoulli random variable*. Therefore, a Bernoulli random variable takes on the value 1 when the outcome of the Bernoulli trial is a success and 0 when it is a failure. If p is the probability of a success, then $1 - p$ (sometimes denoted by q) is the probability of a failure. Hence the probability function of X is given by

$$p(x) = \begin{cases} 1 - p \equiv q & \text{if } x = 0 \\ p & \text{if } x = 1 \\ 0 & \text{otherwise.} \end{cases} \qquad (5.1)$$

Note that the same symbol p is used for the probability function and the Bernoulli parameter. This should not be confusing since the p's used for the probability function often appear in the form $p(x)$.

An accurate mathematical definition for Bernoulli random variables is as follows.

DEFINITION   *A random variable is called Bernoulli with parameter p if its probability function is given by (5.1).*

From (5.1) it follows that the expected value of a Bernoulli random variable X with parameter p is p. This is because

$$E(X) = 0 \cdot P(X = 0) + 1 \cdot P(X = 1) = P(X = 1) = p.$$

Also since

$$E(X^2) = 0 \cdot P(X = 0) + 1 \cdot P(X = 1) = p,$$

we have that

$$\text{Var}(X) = E(X^2) - [E(X)]^2 = p - p^2 = p(1 - p).$$

Example 5.1   If in a throw of a fair die the event of obtaining 4 or 6 is called a success, and the event of obtaining 1, 2, 3, or 5 is called a failure, then

$$X = \begin{cases} 1 & \text{if 4 or 6 is obtained} \\ 0 & \text{otherwise,} \end{cases}$$

is a Bernoulli random variable with the parameter $p = 1/3$. Therefore, its probability function is

$$p(x) = \begin{cases} \dfrac{2}{3} & \text{if } x = 0 \\[2mm] \dfrac{1}{3} & \text{if } x = 1 \\[2mm] 0 & \text{elsewhere.} \end{cases}$$

The expected value of X is given by $E(X) = p = 1/3$ and its variance by $\text{Var}(X) = 1/3(1 - 1/3) = 2/9$. ◆

Let $X_1, X_2, X_3, \ldots$ be a sequence of Bernoulli random variables. If for all $j_i \in \{0, 1\}$, the sequence of events $\{X_1 = j_1\}, \{X_2 = j_2\}, \{X_3 = j_3\}$, $\ldots$ are independent, we say that $\{X_1, X_2, X_3, \ldots\}$ and the corresponding Bernoulli trials are independent.

Although Bernoulli trials are very simple, if they are repeated independently, they may pose interesting and even sometimes complicated questions. If n Bernoulli trials all with probability of success p are performed independently, then X, the number of successes, is one of the most important random variables. It is called *binomial with parameters n and p*. The set of possible values of X is $\{0, 1, 2, \ldots, n\}$ and its probability function is given by the following theorem.

THEOREM 5.1 *Let X be a binomial random variable with parameters n and p. Then p(x), the probability function of X, is given by*

$$p(x) = P(X = x) = \begin{cases} \binom{n}{x} p^x (1 - p)^{n-x} & \text{if } x = 0, 1, 2, \ldots, n \\[2mm] 0 & \text{elsewhere.} \end{cases} \qquad (5.2)$$

PROOF: Observe that the number of ways that in n Bernoulli trials x ($x = 0, 1, 2, \ldots, n$) successes can occur is equal to the number of different sequences of length n with x successes (s's) and ($n - x$) failures (f's.) But the number of such sequences is $\binom{n}{x}$ because the number of distinguishable permutations of n objects of two different types where x are alike and $n - x$ are alike is $\dfrac{n!}{x! \, (n - x)!} = \binom{n}{x}$ (see Theorem 2.4). Since by the independence of the trials the probability of each of these sequences is $p^x (1 - p)^{n-x}$, we have that $P(X = x) = \binom{n}{x} p^x (1 - p)^{n-x}$. Hence (5.2) follows. ◆

DEFINITION   *The function $p(x)$ given by (5.2) is called the binomial probability function with parameters (n, p).*

The reason for this name is that the binomial expansion theorem (Theorem 2.5) guarantees that p is a probability function:

$$\sum_{x=0}^{n} p(x) = \sum_{x=0}^{n} \binom{n}{x} p^x (1 - p)^{n-x} = [p + (1 - p)]^n = 1^n = 1.$$

Example 5.2   A restaurant serves eight entrées of fish, 12 of beef, and 10 of poultry. If customers select from these entrées randomly, what is the probability that two of the next four customers order fish entrées?

SOLUTION: Let X denote the number of fish entrées (successes) ordered by the next four customers. Then X is binomial with the parameters $(4, 8/30 = 4/15)$. Thus

$$P(X = 2) = \binom{4}{2} \left(\frac{4}{15}\right)^2 \left(\frac{11}{15}\right)^2 = 0.23 \quad \blacklozenge$$

Example 5.3   In a county hospital 10 babies, of whom six were boys, were born last Thursday. What is the probability that the first six births were all boys? Assume that the events that a child born is a girl or is a boy are equiprobable.

SOLUTION: Let A be the event that the first six births were all boys and the last four all girls. Let X be the number of boys; then X is binomial with parameters 10 and 1/2. The desired probability is

$$P(A \mid X = 6) = \frac{P(A \text{ and } X = 6)}{P(X = 6)} = \frac{P(A)}{P(X = 6)}$$

$$= \frac{\left(\frac{1}{2}\right)^{10}}{\binom{10}{6} \left(\frac{1}{2}\right)^6 \left(\frac{1}{2}\right)^4} = \frac{1}{\binom{10}{6}} \approx 0.0048. \quad \blacklozenge$$

REMARK: We can also argue as follows: *bbbbbbgggg*, the event that the first six were all boys and the remainder girls, is one sequence out of the $\binom{10}{6}$

possible sequences of the sexes of the births. Since each of these sequences has the same probability of occurrence, the answer is $1/\binom{10}{6}$. ♦

Example 5.4   In a small town, out of 12 accidents that occurred in June 1986, four happened on Friday the 13th. Is this a good reason for a superstitious person to argue that Friday the 13th is inauspicious?

SOLUTION: Suppose the probability that each accident occurs on Friday the 13th is 1/30, just as on any other day. Then the probability of at least 4 accidents on Friday the 13th is

$$1 - \sum_{i=0}^{3} \binom{12}{i} \left(\frac{1}{30}\right)^i \left(\frac{29}{30}\right)^{12-i} \approx 0.000, 493.$$

Since the probability of 4 or more of these accidents occurring on Friday the 13th is very small, this is a good reason for superstitious persons to argue that Friday the 13th is inauspicious. ♦

Example 5.5   A realtor claims that only 30% of the houses in a certain neighborhood are appraised at less than $200,000. A random sample of 20 houses from this neighborhood is selected and appraised. The results in (thousands of dollars) are as follows:

$$
\begin{array}{ccccccc}
285 & 156 & 202 & 306 & 276 & 562 & 415 \\
245 & 185 & 143 & 186 & 377 & 225 & 192 \\
510 & 222 & 264 & 198 & 168 & 363 &
\end{array}
$$

Based on these data, is the realtor's claim acceptable?

SOLUTION: Suppose that the realtor's claim is acceptable and $p = 0.30$ is the probability that a randomly selected house is appraised at less than $200,000. In a randomly selected sample of 20 houses, let X be the number that are less than $200,000. Since in the data given there are seven such houses, we will calculate $P(X \geq 7)$. If this probability is too small, we will reject the claim of the realtor; otherwise, we will accept it. To find $P(X \geq 7)$, note that X is binomial with parameters 20 and 0.30. Therefore,

$$P(X \geq 7) = 1 - P(X \leq 6) = 1 - \sum_{i=0}^{6} \binom{20}{i}(0.30)^i(0.70)^{20-i} \approx 0.392.$$

This shows the event that seven or more houses are appraised at less than \$200,000 is highly probable. So the claim of the realtor should be accepted. ◆

Example 5.6   Suppose that jury members decide independently and that each with probability p $(0 < p < 1)$ makes the correct decision. If the decision of the majority is final, which is preferable: a three-person jury or a single juror?

SOLUTION: Let X denote the number of persons who decide correctly among a three-person jury. Then X is a binomial random variable with parameters $(3, p)$. Hence the probability that a three-person jury decides correctly is

$$P(X \geq 2) = P(X = 2) + P(X = 3) = \binom{3}{2} p^2(1 - p) + \binom{3}{3} p^3(1 - p)^0$$

$$= 3p^2(1 - p) + p^3 = 3p^2 - 2p^3.$$

Since the probability that a single juror decides correctly is p, a three-person jury is preferable to a single juror if and only if

$$3p^2 - 2p^3 > p.$$

This is equivalent to $3p - 2p^2 > 1$, so $-2p^2 + 3p - 1 > 0$. But $-2p^2 + 3p - 1 = 2(1 - p)(p - 1/2)$. Since $1 - p > 0$, $2(1 - p)(p - 1/2) > 0$ if and only if $p > 1/2$. Hence a three-person jury is preferable if $p > 1/2$. If $p < 1/2$, the decision of a single juror is preferable. For $p = 1/2$ it makes no difference. ◆

Example 5.7   Let p be the probability that a randomly chosen American is against abortion, and let X be the number of persons against abortion in a random sample of size n. Suppose that in a particular random sample of n persons, k are against abortion. Show that $P(X = k)$ is maximum for $\hat{p} = k/n$. That is, $\hat{p}$ is the value of p that makes the outcome $X = k$ *most probable.*

SOLUTION: By definition of X:

$$P(X = k) = \binom{n}{k} p^k (1 - p)^{n-k}.$$

This gives

$$\frac{d}{dp}P(X = k) = \binom{n}{k}\left[kp^{k-1}(1 - p)^{n-k} - (n - k)p^k(1 - p)^{n-k-1}\right]$$

$$= \binom{n}{k}p^{k-1}(1 - p)^{n-k-1}\left[k(1 - p) - (n - k)p\right].$$

Letting $\frac{d}{dp}P(X = k) = 0$, we obtain $p = k/n$. Now since $\frac{d^2}{dp^2}P(X = k) < 0$, $\hat{p} = k/n$ is the maximum of $P(X = k)$ and hence it is an estimate of p that makes the outcome $x = k$ most probable.   ◆

Let X be a binomial random variable with parameters (n, p), $0 < p < 1$, and probability function $p(x)$. We will now find the value of X at which $p(x)$ is maximum. For any real number t, let $[t]$ denote the largest integer less than or equal to t. We will prove that $p(x)$ is maximum at $x = [(n + 1)p]$. To do so, note that

$$\frac{p(x)}{p(x - 1)} = \frac{\dfrac{n!}{(n - x)! \, x!} p^x(1 - p)^{n-x}}{\dfrac{n!}{(n - x + 1)! \, (x - 1)!} p^{x-1}(1 - p)^{n-x+1}}$$

$$= \frac{(n - x + 1)p}{x(1 - p)} \tag{5.3}$$

$$= \frac{(n + 1)p - xp + x - x}{x(1 - p)} = \frac{[(n + 1)p - x] + x(1 - p)}{x(1 - p)}$$

$$= \frac{(n + 1)p - x}{x(1 - p)} + 1.$$

This equality shows that $p(x) > p(x - 1)$ if and only if $(n + 1)p - x > 0$, or equivalently if and only if $x < (n + 1)p$. Hence *as x changes from 0 to $[(n+1)p]$, $p(x)$ increases. As x changes from $[(n+1)p]$ to n, $p(x)$ decreases. The maximum value of $p(x)$ [the peak of the graphical representation of $p(x)$] occurs at $[(n + 1)p]$.*

What we have just shown is illustrated by the graphical representations of $p(x)$ in Figure 5.1 for binomial random variables with parameters $(5, 1/2)$, $(10, 1/2)$, and $(20, 1/2)$.

Expectations and Variances of Binomial Random Variables

Let X be a binomial random variable with parameters (n, p). Intuitively, we expect that the expected value of X will be np. For example, if we toss

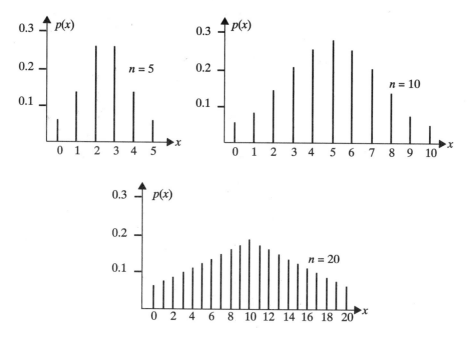

Figure 5.1   Examples of binomial probability functions.

a fair coin 100 times, we expect that the average number of heads will be 50, which is $100 \cdot 1/2 = np$. Also, if we choose 10 fuses from a set with 30% defective fuses, we expect the average number of defective fuses to be $np = 10(0.30) = 3$. The formula $E(X) = np$ can be verified directly from the definition of mathematical expectation as follows:

$$E(X) = \sum_{x=0}^{n} x \binom{n}{x} p^x (1-p)^{n-x} = \sum_{x=1}^{n} x \frac{n!}{x!\,(n-x)!} p^x (1-p)^{n-x}$$

$$= \sum_{x=1}^{n} \frac{n!}{(x-1)!\,(n-x)!} p^x (1-p)^{n-x}$$

$$= np \sum_{x=1}^{n} \frac{(n-1)!}{(x-1)!\,(n-x)!} p^{x-1} (1-p)^{n-x}$$

$$= np \sum_{x=1}^{n} \binom{n-1}{x-1} p^{x-1} (1-p)^{n-x}.$$

Letting $i = x - 1$ (by reindexing this sum), we obtain

$$E(X) = np \sum_{i=0}^{n-1} \binom{n-1}{i} p^i (1-p)^{(n-1)-i}$$

$$= np[p + (1-p)]^{n-1} = np,$$

where the next-to-last equality follows from binomial expansion (Theorem 2.5). To calculate the variance of X, from a procedure similar to the one we used to compute $E(X)$, we obtain (see Exercise 23)

$$E(X^2) = \sum_{x=1}^{n} x^2 \binom{n}{x} p^x (1-p)^{n-x} = n^2 p^2 - np^2 + np.$$

Therefore,

$$\text{Var}(X) = E(X^2) - [E(X)]^2 = -np^2 + np = np(1-p).$$

Example 5.8 A town of 100,000 inhabitants is exposed to a contagious disease. If the probability that a person becomes infected is 0.04, what is the expected number of people who become infected?

SOLUTION: Assuming that people become infected independent of each other, the number of those who will become infected is a binomial random variable with parameters 100,000 and 0.04. Thus the expected number of such people is $100,000 \times 0.04 = 4000$. Note that here "getting infected" is implicitly defined to be a success for mathematical simplicity! Therefore, in the context of Bernoulli trials, what is called a success may be a failure in real life. ♦

Example 5.9 Two proofreaders, Ruby and Myra, read a book independently and found r and m misprints, respectively. Suppose that the probability that a misprint is noticed by Ruby is p and the probability that it is noticed by Myra is q, where these two probabilities are independent. If the number of misprints noticed by *both* Ruby and Myra is b, estimate the number of unnoticed misprints. [This problem was posed and solved by a prominent mathematician of this century, George Pólya (1888–1985), in the January 1976 issue of the *American Mathematical Monthly.*]

SOLUTION: Let M be the total number of misprints in the book. The expected numbers of misprints that may be noticed by Ruby, Myra, and both of them are Mp, Mq, and Mpq, respectively. Assuming that the numbers of misprints found are approximately equal to the expected numbers, we have $Mp \approx r$, $Mq \approx m$, and $Mpq \approx b$. Therefore,

$$M = \frac{(Mp)(Mq)}{Mpq} \approx \frac{rm}{b}.$$

The number of unnoticed misprints is therefore estimated by

$$M - (r + m - b) \approx \frac{rm}{b} - (r + m - b) = \frac{(r - b)(m - b)}{b}. \quad \blacklozenge$$

EXERCISES

A

1. From an ordinary deck of 52 cards, cards are drawn at random and with replacement. What is the probability that of the first eight cards drawn four are spades?

2. The probability that a randomly selected person is female is 1/2. What is the expected number of the girls in the first-grade classes of an elementary school that has 64 first graders? What is the expected number of females in a family with six children?

3. A graduate class consists of six students. What is the probability that exactly three of them are born either in April or in October?

4. In a state where license plates consist of six digits, what is the probability that the license number of a randomly selected car has two 9's? Assume that each digit of the license number is randomly selected from $\{0, 1, \ldots, 9\}$.

5. A box contains 30 balls numbered 1 through 30. Suppose that five balls are drawn at random, one at a time, with replacement. What is the probability that the numbers of two of them are prime?

6. A manufacturer of nails claims that only 3% of its nails are defective. A random sample of 24 nails is selected and it is is found that two of them are defective. Is it fair to reject the manufacturer's claim based on this observation?

7. Only 60% of certain kinds of seeds germinate when planted under normal conditions. Suppose that four such seeds are planted and X denotes the number of those that will germinate. Find the probability functions of X and $Y = 2X + 1$.

8. Suppose that the Internal Revenue Service will audit 20% of income tax returns reporting an annual gross income of over \$80,000. What is the probability that of 15 such returns, at most four will be audited?

9. If two fair dice are rolled 10 times, what is the probability of at least one 6 (on either die) in exactly five of these 10 rolls?

10. From the interval $(0, 1)$, five points are selected at random and independently. What is the probability that (a) at least two of them are less than $1/3$; (b) the first decimal point of exactly two of them is 3?

11. Let X be a binomial random variable with parameters (n, p) and probability function $p(x)$. Prove that if $(n + 1)p$ is an integer, then $p(x)$ is maximum at two different points. Find both of these points.

12. On the average, how many times should Ernie play poker in order to be dealt a straight flush (royal flush included)? (See Exercise 20 of Section 2.4 for definitions of a royal and a straight flush.)

13. Suppose that each day the price of a stock moves up $1/8$ of a point with probability $1/3$ and moves down $1/8$ of a point with probability $2/3$. If the price fluctuations from one day to another are independent, what is the probability that after six days the stock has its original price?

14. From the set $\{x : 0 \le x \le 1\}$, 100 independent numbers are selected at random and rounded to three decimal places. What is the probability that at least one of them is 0.345?

15. A certain basketball player makes a foul shot with probability 0.45. Determine for what value of k the probability of k baskets in 10 shots is maximum, and find this maximum probability.

16. What are the expected value and variance of the number of full houses in n poker hands? (See Exercise 20 of Section 2.4 for a definition of a full house.)

17. What is the probability that at least two of the six members of a family are not born in the fall? Assume that all seasons have the same probability of containing the birthday of a person selected randomly.

18. A certain rare blood type can be found in only 0.05% of people. If the population of a randomly selected group is 3000, what is the probability that at least two people in the group have this rare blood type?

19. Edward's experience shows that 7% of the parcels he mails will not reach their destination. He has bought two books for $20 apiece and wants to mail them to his brother. If he sends them in one parcel, the postage is $5.20, but if he sends them in separate parcels, the postage is $3.30 for each book. To minimize the expected value of his expenses (loss + postage), which way is preferable to send the books, in two separate parcels or in a single parcel?

20. A woman and her husband want to have a 95% chance for at least one boy and at least one girl. What is the minimum number of children that they should plan to have? Assume that the events that a child is a girl and a boy are equiprobable and independent of the gender of other children born in the family.

B

21. In a community, a persons are for abortion, b ($b < a$) are against it, and n ($n > a - b$) are undecided. Suppose that there will be a vote taken to determine the will of the majority with regard to legalizing abortion. If by then all of the undecided persons make up their minds, what is the probability that those against abortion will win? Assume that it is equally likely that an undecided person eventually votes against or for abortion.

22. A game often played in carnivals and gambling houses is called *chuck-a-luck*. In this game, a player bets on any of the numbers 1 through 6. Then three fair dice are tossed. If one, two, or all three land the same as the player's number, then he or she receives one, two, or three times the original stake plus his or her original bet, respectively. Otherwise, the player loses his or her stake. Let X be the net gain of the player per unit of stake. First find the probability function of X; then determine the expected amount that the player will lose per unit of stake.

23. Let X be a binomial random variable with the parameters (n, p). Prove that

$$E(X^2) = \sum_{x=1}^{n} x^2 \binom{n}{x} p^x (1 - p)^{n-x} = n^2 p^2 - np^2 + np.$$

24. Suppose that an engine of an airplane in flight will fail with probability $1 - p$ independent of other engines of the plane. Also suppose that a plane can complete the journey successfully if at least half of its engines remain operating.
   (a) Is it true that a four-engine plane is always preferable to a two-engine plane? Explain.
   (b) Is it true that a five-engine plane is always preferable to a three-engine plane? Explain.

25. How many games of poker are required in order that a preassigned player is dealt at least one straight flush with probability of at least 3/4? (See Exercise 20 of Section 2.4 for a definition of a straight flush.)

26. The post office of a certain small town has only one clerk to wait on customers. The probability that a customer will finish in any given minute is 0.6 regardless of the time that the customer has already taken. The probability of a new customer arriving is 0.45 regardless of the number of customers already in line. The chances of two customers arriving during the same minute are negligible. Similarly, the chances of two customers being served in the same minute are negligible. Suppose that we start with exactly two customers: one at the postal window and one waiting on line. After 4 minutes, what is the probability that there will be exactly four customers: one at the window and three waiting on line?

27. **(a)** What is the probability of an even number of successes in n independent Bernoulli trials?

    HINT: Let r_n be the probability of an even number of successes in n Bernoulli trials. By conditioning on the first trial and using the law of total probability (Theorem 3.3), show that for $n \geq 1$,

    $$r_n = p(1 - r_{n-1}) + (1 - p)r_{n-1}.$$

    Then using this, prove that $r_n = (1/2)[1 + (1 - 2p)^n]$.

    (b) Prove that

    $$\sum_{k=0}^{[n/2]} \binom{n}{2k} p^{2k}(1 - p)^{n-2k} = \frac{1}{2}[1 + (1 - 2p)^n].$$

28. An urn contains n balls whose colors are red or blue with equal probabilities. [For example, the probability that all of the balls are red is $(1/2)^n$.] If in drawing k balls from the urn, successively, with replacement, and randomly, no red balls appear, what is the probability that the urn contains no red balls at all?

    HINT: Use Bayes' theorem.

5.2   Poisson Random Variable

Poisson as an Approximation to Binomial

From Section 5.1, it should have become clear that in many natural phenomena and real-world problems, we are dealing with binomial random variables. Therefore, for possible values of parameters n, x, and p, we need to calculate $p(x) = \binom{n}{x} p^x (1 - p)^{n-x}$. However, in many cases, direct calculation of $p(x)$ from this formula is not possible, because even for moderate values of n, $n!$ exceeds the largest integer that a computer can store. To overcome this difficulty, indirect methods for calculation of $p(x)$ are developed. For example, we may use the recursive formula

$$p(x) = \frac{(n - x + 1)p}{x(1 - p)} p(x - 1),$$

obtained from (5.3), with the initial condition $p(0) = (1 - p)^n$, to evaluate $p(x)$. An important study concerning calculation of $p(x)$ is the one by the prominent French mathematician Simeon-Denis Poisson in 1837 in his book concerning the applications of probability to law. In this book, Poisson introduced the following procedure to obtain the formula that approximates the binomial probability function when the number of trials is large ($n \to \infty$), the probability of success is small ($p \to 0$), and the average number of successes remains a fixed quantity of moderate value ($np = \lambda$ for some constant λ).

Let X be a binomial random variable with parameters (n, p); then

$$
P(X = i) = \binom{n}{i} p^i (1 - p)^{n-i} = \frac{n!}{(n - i)!\, i!} \left(\frac{\lambda}{n}\right)^i \left(1 - \frac{\lambda}{n}\right)^{n-i}
$$

$$
= \frac{n(n - 1)(n - 2) \cdots (n - i + 1)}{n^i} \frac{\lambda^i}{i!} \frac{\left(1 - \dfrac{\lambda}{n}\right)^n}{\left(1 - \dfrac{\lambda}{n}\right)^i}.
$$

Now for large n and appreciable λ, $(1 - \lambda/n)^i$ is approximately 1, $(1 - \lambda/n)^n$ is approximately $e^{-\lambda}$ [from calculus we know that $\lim_{n \to \infty}(1 + x/n)^n = e^x$; thus $\lim_{n \to \infty}(1 - \lambda/n)^n = e^{-\lambda}$], and $[n(n - 1)(n - 2) \cdots (n - i + 1)]/n^i$ is approximately 1 because its numerator and denominator are both polynomials in n of degree i. Thus as $n \to \infty$,

$$
P(X = i) \to \frac{e^{-\lambda} \lambda^i}{i!}.
$$

Three years after his book was published, Poisson died and the importance of this approximation was left unknown until 1889, when the German-Russian mathematician L. V. Bortkiewicz demonstrated how significant it is both in probability theory and its applications. Among other things, Bortkiewicz argued that since

$$
\sum_{i=0}^{\infty} \frac{e^{-\lambda} \lambda^i}{i!} = e^{-\lambda} \sum_{i=0}^{\infty} \frac{\lambda^i}{i!} = e^{-\lambda} e^{\lambda} = 1,
$$

Poisson's approximation by itself is a probability function. This and the introduction of the Poisson processes in the twentieth century made the Poisson probability function one of the three most important probability functions, the other two being the binomial and normal probability functions.

DEFINITION   *A discrete random variable X with possible values 0, 1, 2, 3, ... is called Poisson with parameter* λ, $\lambda > 0$, *if*

$$P(X = i) = \frac{e^{-\lambda}\lambda^i}{i!}, \qquad i = 0, 1, 2, 3, \ldots.$$

Under the conditions specified in our discussion, binomial probabilities can be approximated by Poisson probabilities. Such approximations are generally good if $p < 0.1$ and $np \leq 10$. If $np > 10$, it would be more appropriate to use normal approximation, discussed in Section 7.2.

Since a Poisson probability function is the limit of a binomial probability function, the expected value of a binomial random variable with parameters (n, p) is np, and $np = \lambda$, it is reasonable to expect that the mean of a Poisson random variable with parameter λ is λ. To prove that this is the case, observe that

$$E(X) = \sum_{i=0}^{\infty} i P(X = i) = \sum_{i=1}^{\infty} i \frac{e^{-\lambda}\lambda^i}{i!}$$

$$= \lambda e^{-\lambda} \sum_{i=1}^{\infty} \frac{\lambda^{i-1}}{(i-1)!} = \lambda e^{-\lambda} \sum_{i=0}^{\infty} \frac{\lambda^i}{i!} = \lambda e^{-\lambda} e^{\lambda} = \lambda.$$

The variance of a Poisson random variable X with parameter λ is also λ. To see this, note that

$$E(X^2) = \sum_{i=0}^{\infty} i^2 P(X = i) = \sum_{i=1}^{\infty} i^2 \frac{e^{-\lambda}\lambda^i}{i!}$$

$$= \lambda e^{-\lambda} \sum_{i=1}^{\infty} \frac{i\lambda^{i-1}}{(i-1)!} = \lambda e^{-\lambda} \sum_{i=1}^{\infty} \frac{1}{(i-1)!} \frac{d}{d\lambda}(\lambda^i)$$

$$= \lambda e^{-\lambda} \frac{d}{d\lambda} \left[\sum_{i=1}^{\infty} \frac{\lambda^i}{(i-1)!} \right] = \lambda e^{-\lambda} \frac{d}{d\lambda} \left[\lambda \sum_{i=1}^{\infty} \frac{\lambda^{i-1}}{(i-1)!} \right]$$

$$= \lambda e^{-\lambda} \frac{d}{d\lambda}(\lambda e^{\lambda}) = \lambda e^{-\lambda}(e^{\lambda} + \lambda e^{\lambda}) = \lambda + \lambda^2,$$

and hence

$$\text{Var}(X) = (\lambda + \lambda^2) - \lambda^2 = \lambda.$$

Some examples of binomial random variables that obey Poisson's approximation are as follows:

1. Let X be the number of babies from a community who grow up to at least 190 centimeters. If a baby is called a success provided that he or she grows up to the height of 190 or more centimeters, then X is a binomial random variable. Since n, the total number of babies, is large, p, the probability that a baby grows to the height of 190 centimeters or more, is small, and np, the average number of such babies, is appreciable, X is approximately a Poisson random variable.

2. Let X be the number of winning tickets among the Maryland lottery tickets sold in Baltimore during one week. Then calling winning tickets successes, we have that X is a binomial random variable. Since n, the total number of tickets sold in Baltimore, is large, p, the probability that a ticket wins, is small, and the average number of winning tickets is appreciable, X is approximately a Poisson random variable.

3. Let X be the number of misprints on a page of a document typed by a secretary. Then X is a binomial random variable if a word is called a success provided that it is misprinted! Since misprints are rare events, the number of words is large, and np, the average number of misprints, is of moderate value, X is approximately a Poisson random variable.

In the same way we can argue that random variables such as "the number of defective fuses produced by a factory in a given year," "the number of inhabitants of a town who live at least 85 years," "the number of machine failures per day in a plant," and "the number of drivers arriving at a gas station per hour" are approximately Poisson random variables.

We will now give examples of problems that can be solved by Poisson's approximations. In the course of these examples, new Poisson random variables are also introduced.

Example 5.10　Every week the average number of wrong numbers received by a certain mail-order house is seven. What is the probability that this house receives (a) two wrong calls tomorrow; (b) at least one wrong call tomorrow?

Solution: Assuming that the house receives a lot of calls, X, the number of wrong numbers received tomorrow, is approximately a Poisson random variable with $\lambda = E(X) = 1$. Thus

$$P(X = n) = \frac{e^{-1} \cdot (1)^n}{n!} = \frac{1}{e \cdot n!},$$

and hence the answers to (a) and (b) are, respectively,

$$P(X = 2) = \frac{1}{2e} \approx 0.18$$

and

$$P(X \geq 1) = 1 - P(X = 0) = 1 - \frac{1}{e} \approx 0.63. \quad \blacklozenge$$

Example 5.11   Suppose that, on average, in every three pages of a book there is one typographical error. If the number of typographical errors on a single page of the book is a Poisson random variable, what is the probability of at least one error on a certain page of the book?

SOLUTION: Let X be the number of errors on the page we are interested in. Then X is a Poisson random variable with $E(X) = 1/3$. Hence $\lambda = E(X) = 1/3$, and thus

$$P(X = n) = \frac{(1/3)^n e^{-1/3}}{n!}.$$

Therefore,

$$P(X \geq 1) = 1 - P(X = 0) = 1 - e^{-1/3} \approx 0.28. \quad \blacklozenge$$

Example 5.12   The atoms of a radioactive element are randomly disintegrating. If every gram of this element, on average, emits 3.9 alpha particles per second, what is the probability that during the next second the number of alpha particles emitted from 1 gram is (a) at most six; (b) at least two; (c) at least three and at most six?

SOLUTION: Every gram of radioactive material consists of a large number n of atoms. If for these atoms the event of disintegrating and emitting an alpha particle during the next second is called a success, then X, the number of alpha particles emitted during the next second, is a binomial random variable. Now $E(X) = 3.9$, so $np = 3.9$ and $p = 3.9/n$. Since n is very large, p is very small, and hence, to a very close approximation, X has a Poisson distribution with the appreciable parameter $\lambda = 3.9$. Therefore,

$$P(X = n) = \frac{(3.9)^n e^{-3.9}}{n!},$$

and hence (a), (b), and (c) are calculated as follows.

(a) $P(X \leq 6) = \displaystyle\sum_{n=0}^{6} \frac{(3.9)^n e^{-3.9}}{n!} \approx 0.899.$

(b) $P(X \geq 2) = 1 - P(X = 0) - P(X = 1) = 0.901.$

(c) $P(3 \leq X \leq 6) = \sum_{n=3}^{6} \dfrac{(3.9)^n e^{-3.9}}{n!} \approx 0.646.$ ◆

Example 5.13   Suppose that n raisins are thoroughly mixed in dough. If we bake k raisin cookies of equal sizes from this mixture, what is the probability that a given cookie contains at least one raisin?

SOLUTION: Since the raisins are thoroughly mixed in the dough, the probability that a given cookie contains any particular raisin is $p = 1/k$. If for the raisins the event of ending up in the given cookie is called a success, then X, the number of raisins in the given cookie, is a binomial random variable. For large values of k, $p = 1/k$ is small. If n is also large but n/k has a moderate value, it is reasonable to assume that X is approximately Poisson. Hence

$$P(X = i) = \frac{\lambda^i e^{-\lambda}}{i!},$$

where $\lambda = np = n/k$. Therefore, the probability that a given cookie contains at least one raisin is

$$P(X \neq 0) = 1 - P(X = 0) = 1 - e^{-n/k}. \quad ◆$$

For real-world problems, it is important to know that in most cases numerical examples show that even for small values of n the agreement between the binomial and the Poisson probabilities is surprisingly good (see Exercise 10).

EXERCISES

A

1. Jim buys 60 lottery tickets every week. If only 5% of the lottery tickets win, what is the probability that he wins next week?

2. Suppose that 3% of the families in a large city have an annual income of over $60,000. What is the probability that of 60 random families, at most 3 have an annual income of over $60,000?

3. Suppose that 2.5% of the population of a border town are illegal immigrants. Find the probability that in a theater of this town with 80 random viewers, there are at least two illegal immigrants.

4. By Example 2.20, the probability that a poker hand is a full house is 0.0014. What is the probability that in 500 random poker hands there are at least 2 full houses?

5. On a random day, the number of vacant rooms of a big hotel in New York City is 35, on average. What is the probability that next Saturday this hotel has at least 30 vacant rooms?

6. On average, there are three misprints in every 10 pages of a book. If every chapter of the book contains 35 pages, what is the probability that Chapters 1 and 5 have 10 misprints each?

7. Suppose that X is a Poisson random variable with $P(X = 1) = P(X = 3)$. Find $P(X = 5)$.

8. Suppose that n raisins have been carefully mixed with a batch of dough. If we bake k ($k > 4$) raisin buns of equal size from this mixture, what is the probability that two out of four randomly selected buns contain no raisins?
   HINT: Note that by Example 5.13, the number of raisins in a given bun is approximately Poisson with parameter n/k.

9. The children in a small town all own slingshots. In a recent contest, 4% of them were such poor shots that they did not hit the target even once in 100 shots. If the number of times a randomly selected child has hit the target is approximately a Poisson random variable, determine the percentage of children who have hit the target at least twice.

10. The department of mathematics of a state university has 26 faculty members. For $i = 0, 1, 2, 3$, find p_i, the probability that i of them were born on Independence Day (a) using the binomial distribution; (b) using the Poisson distribution. Assume that the birth rates are constant throughout the year and that each year has 365 days.

B

11. Suppose that in Maryland, on a certain day, N lottery tickets are sold and M win. To have a probability of at least α of winning on that day, approximately how many tickets should be purchased?

12. In a forest, the number of trees that grow in a region of area R has a Poisson distribution with mean λR, where λ is a given positive number.
   (a) Find the probability that the distance from a certain tree to the nearest tree is more than d.
   (b) Find the probability that the distance from a certain tree to the nth nearest tree is more than d.

13. Let X be a Poisson random variable with parameter λ. Show that the maximum of $P(X = i)$ occurs at $[\lambda]$, where $[\lambda]$ is the greatest integer less than or equal to λ.

HINT: Let p be the probability function of X. Prove that $p(i) = (\lambda/i)p(i - 1)$. Use this to find the values of i at which p is increasing and the values of i at which it is decreasing.

Poisson Process

Earlier, the Poisson distribution was introduced to approximate binomial distribution for large n, small p, and appreciable np. On its own merit, Poisson distribution appears in connection with the study of sequences of random events that occur in time. Suppose that certain types of events occur at random times. Let $N(t)$ be the number of events that have occurred in the time interval $[0, t]$. Then the set of random variables $\{N(t), t > 0\}$ is called a *counting process*. This is because $N(t)$ counts the number of events that have occurred at or up to t. Some examples of counting processes follow.

Example 5.14   Let $N(t)$ be the number of accidents at an intersection by time t. $\{N(t), t \geq 0\}$ is a counting process in which the events are accidents that occur at the intersection.   ♦

Example 5.15   Let $N(t)$ be the number of β-particles emitted from a radioactive substance in $[0, t]$. Then $\{N(t), t \geq 0\}$ is a counting process. Every time that a β-particle is emitted an event has occurred.   ♦

Example 5.16   Let $N(t)$ be the number of customers who have entered a post office by time t, and $M(t)$ be the number of customers present in the post office at time t. Then $\{N(t), t \geq 0\}$ is a counting process, whereas $\{M(t), t \geq 0\}$ is not. The reason for $\{M(t), t \geq 0\}$ not being a counting process is that some customers might have entered and then left the post office before t.   ♦

Example 5.17   Suppose that events are earthquakes that occur in California. If $N(t)$ is the total number of earthquakes that have occurred by time t, then $\{N(t), t \geq 0\}$ is a counting process.   ♦

If $\{N(t), t \geq 0\}$ is a counting process, then for all $t \geq 0$, the random variable $N(t)$ is nonnegative and integer valued. Also, it is increasing in t; that is, if $s < t$, then $N(s) < N(t)$. In fact, $N(t) - N(s)$ is the random variable representing the number of events that have occurred in the time interval $(s, t]$.

From these examples it might have become clear that finding the probability distributions of the random variables $N(t)$, $t \geq 0$, is not an easy task. Indeed, for many years this has been, and for many more will be, a major field of research. But perhaps in this area the most striking result so far is that under a few simple physical conditions (to be discussed later), for all t, $N(t)$ is a Poisson random variable. It is because of this phenomenon that the Poisson distribution has become one of the most important and applicable distributions.

For a deeper understanding of the Poisson distribution, we will now describe the model known as the Poisson process. To explain this model, we need to make some definitions.

DEFINITION   *A counting process* $\{N(t) : t \geq 0\}$ *is said to be stationary if for all $n \geq 0$ and for any two time intervals Δ_1 and Δ_2 with the same length, the probability of n events in Δ_1 is equal to the probability of n events in Δ_2.*

Thus for a stationary counting process $\{N(t), t \geq 0\}$, the number of events in the interval $(t_1, t_2]$ has the same distribution as the number of events in $(t_1 + s, t_2 + s]$, $s \geq 0$. This means that the random variables $N(t_2) - N(t_1)$ and $N(t_2 + s) - N(t_1 + s)$ have the same probability function. In other words, the probability of occurrence of n events during the interval of time from t_1 to t_2 is a function of n and $t_2 - t_1$ and not of t_1 and t_2 independently.

If $N(t)$ represents the number of customers who have entered a post office up to time t, then $N(t)$ is stationary if at no time of the day it is more likely that more customers enter the post office. Otherwise, it will not be stationary. The total number of earthquakes that occur in California up to t, the number of γ-particles that are emitted from a radioactive substance, and the number of children born at or prior to t are all stationary counting processes. The number of accidents at an intersection by time t is stationary if at no time of the day is it more likely that more accidents occur. For example, if, during rush hours or late in the evenings, more accidents occur, the process is not stationary.

DEFINITION   *Let $\{N(t), t \geq 0\}$ be a counting process. Let $\{\Delta_1, \Delta_2, \ldots, \Delta_n\}$ be an arbitrary set of n nonoverlapping time intervals and $\{i_1, i_2, \ldots, i_n\}$ be an arbitrary set of nonnegative integers. For $1 \leq j \leq n$, let A_j be the event that i_j events of the process occur in Δ_j. If $A_1, A_2, \ldots,$ and A_n are n independent events, then $\{N(t), t \geq 0\}$ is said to be a process with independent increments.*

Thus for a counting process with independent increments, for $s > 0$, the occurrence of n events during $(t, t + s]$ is independent of the number of events that have occurred in $[0, t]$. In other words, for any $n \geq 0$, given $N(t) = n$ does not change the probability function of $N(t + s) - N(t)$. That is,

$$P\{N(t + s) - N(t) = i \mid N(t) = n\} = P\{N(t + s) - N(t) = i\}$$

for all $i \geq 0$ and $n \geq 0$.

The number of customers who have entered a post office up to t, the number of earthquakes that have occurred in California at or prior to t, the number of α-particles that are emitted from a radioactive substance in $[0, t]$, and the number of accidents in an intersection at or prior to t are examples of counting processes with independent increments. If $N(t)$ denotes the total number of fish caught by time t from a small lake with a small number of fish, $\{N(t), t \geq 0\}$ is a counting process without independent increments because the number of fish caught in $(t, t + s)$ is not independent of the number of fish caught in $[0, t]$. For example, if $N(t)$ is very large, $N(t + s) - N(t)$ will be small. This process will possess independent increments if the number of fish in the lake is very large (theoretically, infinite). Also, if $N(t)$ is the number of fish caught by time t from an ocean, it is reasonable to assume that $\{N(t), t \geq 0\}$ possesses independent increments.

DEFINITION   *A counting process $\{N(t), t \geq 0\}$ is said to be a Poisson process if $N(0) = 0$, it is stationary, possesses independent increments, and satisfies $\lim_{h \to 0} P\{N(h) > 1\}/h = 0$.*

Under the condition $\lim_{h \to 0} P\{N(h) > 1\}/h = 0$, as $h \to 0$, we find $P\{N(h) > 1\}$ approaches 0 faster than h does. Hence for small values of h, $P\{N(h) > 1\}$ is very small. Therefore $\lim_{h \to 0} P\{N(h) > 1\}/h = 0$ is a mathematical expression for the statement that *the probability of occurrence of two or more events in a very small time interval is negligible*. It can be shown that for stationary counting processes, this relation means that *the simultaneous occurrence of two or more events is impossible*. Since for Poisson processes stationarity is always assumed, we can interpret $\lim_{h \to 0} P\{N(h) > 1\}/h = 0$ as the condition under which simultaneous occurrence of two or more events is impossible.

The reason for calling a counting process under study Poisson is the following celebrated theorem, for whose validity we present a probabilistic argument.

THEOREM 5.2   *If $\{N(t), t \geq 0\}$ is a Poisson process, then there exists a positive number λ such that*

$$P\{N(t) = n\} = \frac{(\lambda t)^n e^{-\lambda t}}{n!}.$$

That is, for all $t > 0$, $N(t)$ is a Poisson random variable with parameter λt. Hence $E[N(t)] = \lambda t$ and therefore $\lambda = E[N(1)]$. Sometimes λ is called the rate of the Poisson process $\{N(t), t \geq 0\}$.

JUSTIFICATION: The property that for a Poisson process $\{N(t), t \geq 0\}$, $N(t)$ is a Poisson random variable is not accidental. It is related to the fact that a Poisson random variable is approximately binomial for large n, small p, and moderate np. To see this, divide the interval $[0, t]$ into n subintervals of equal length. Then, as $n \to \infty$, the probability of two or more events in any of these subintervals is 0. Therefore, $N(t)$, the number of events in $[0, t]$, is equal to the number of subintervals in which an event occurs. If a subinterval in which an event occurs is called a success, then $N(t)$ is the number of successes. Moreover, the stationarity and independent increments properties imply that $N(t)$ is a binomial random variable with parameters (n, p), where p is the probability of success (i.e., the probability that an event occurs in a subinterval). Now let λ be the expected number of events in an interval of unit length. Because of stationarity, events occur at a uniform rate over the entire time period. Therefore, the expected number of events in any period of length t is λt. Hence, in particular, the expected number of the events in $[0, t]$ is λt. But by the formula for the expectation of binomial random variables, the expected number of the events in $[0, t]$ is np. Thus $np = \lambda t$ or $p = (\lambda t)/n$. Since n is extremely large ($n \to \infty$), we have that p is extremely small while $np = \lambda t$ is appreciable. Therefore, $N(t)$ is a Poisson random variable with parameter λt. ◆

This theorem is astonishing because it shows how three simple and natural physical conditions on $N(t)$ characterize the probability functions of random variables $N(t)$, $t > 0$. Moreover, λ, the only unknown parameter of the probability functions of $N(t)$'s, equals $E[N(1)]$. That is, it is the average number of events that occur in one unit of time. Hence in practice it can be measured very easily. Historically, Theorem 5.2 is the most elegant and evolved form of several previous theorems. It was discovered in 1955 by the Russian mathematician Alexander Khinchin (1894–1959). The first major work in this direction was done by Thornton Fry (1892–1992?) in 1929. But before Fry, Albert Einstein (1879–1955) and Roman Smoluchowski (1910–) had also discovered important results in connection with their work on the theory of Brownian motions.

Example 5.18   Suppose that children are born at a Poisson rate of five per day in a certain hospital. What is the probability that (a) at least two babies are born during the next six hours; (b) no babies are born during the next two days?

SOLUTION: Let $N(t)$ denote the number of babies born at or prior to t. The assumption that $\{N(t) : t \geq 0\}$ is a Poisson process is reasonable because it is stationary, it has independent increments, $N(0) = 0$, and simultaneous births are impossible. Thus $\{N(t) : t \geq 0\}$ is a Poisson process. If we choose one day as time unit, then $\lambda = E[N(1)] = 5$. Therefore,

$$P\{N(t) = n\} = \frac{(5t)^n e^{-5t}}{n!}.$$

Hence the probability that at least two babies are born during the next six hours is

$$P\left\{N\left(\frac{1}{4}\right) \geq 2\right\} = 1 - P\left\{N\left(\frac{1}{4}\right) = 0\right\} - P\left\{N\left(\frac{1}{4}\right) = 1\right\}$$

$$= 1 - \frac{(5/4)^0 e^{-5/4}}{0!} - \frac{(5/4)^1 e^{-5/4}}{1!} \approx 0.36,$$

where 1/4 is used since six hours is 1/4 of a day. The probability that no babies are born during the next two days is

$$P\{N(2) = 0\} = \frac{(10)^0 e^{-10}}{0!} \approx 4.54 \times 10^{-5}. \quad \blacklozenge$$

Example 5.19 Suppose that in a certain region of California, earthquakes occur in accordance with a Poisson process at a rate of seven per year.

(a) What is the probability of no earthquakes in one year?

(b) What is the probability that in exactly three of the next eight years no earthquakes will occur?

SOLUTION: (a) Let $N(t)$ be the number of earthquakes in this region at or prior to t. We are given that $\{N(t) : t \geq 0\}$ is a Poisson process. If we choose one year as the unit of time, then $\lambda = E[N(1)] = 7$. Thus

$$P\{N(t) = n\} = \frac{(7t)^n e^{-7t}}{n!}, \qquad n = 0, 1, 2, 3, \ldots.$$

Let p be the probability of no earthquakes in one year; then

$$p = P\{N(1) = 0\} = e^{-7} \approx 0.00091.$$

(b) Suppose that a year is called a success if during its course no earthquakes occur. Of the next eight years, let X be the number of years in which no earthquakes will occur. Then X is a binomial random variable with parameters $(8, p)$. Thus

$$P(X = 3) \approx \binom{8}{3}(0.00091)^3(1 - 0.00091)^5 \approx 4.2 \times 10^{-8}. \quad \blacklozenge$$

Example 5.20 A fisherman catches fish at a Poisson rate of two per hour from a large lake with lots of fish. If he starts fishing at 10:00 A.M., what is the probability that he catches one fish by 10:30 and three fish by noon?

SOLUTION: Let $N(t)$ denote the total number of fish caught at or prior to t. $N(0) = 0$ because at 10:00 A.M. no fish is caught. It is reasonable to assume that catching two or more fish simultaneously is impossible. It is also reasonable to assume that $\{N(t) : t \geq 0\}$ is stationary and has independent increments. Thus the assumption that $\{N(t) : t \geq 0\}$ is a Poisson process is well grounded. Choosing 1 hour as the unit of time, we have $\lambda = E[N(1)] = 2$. Thus

$$P\{N(t) = n\} = \frac{(2t)^n e^{-2t}}{n!}, \qquad n = 0, 1, 2, \ldots.$$

We want to calculate the probability of the event $\{N(1/2) = 1 \text{ and } N(2) = 3\}$. But this event is the same as $\{N(1/2) = 1 \text{ and } N(2) - N(1/2) = 2\}$. Thus by the independent increment property,

$$P\left\{N\left(\frac{1}{2}\right) = 1 \text{ and } N(2) - N\left(\frac{1}{2}\right) = 2\right\}$$
$$= P\left\{N\left(\frac{1}{2}\right) = 1\right\} \cdot P\left\{N(2) - N\left(\frac{1}{2}\right) = 2\right\}.$$

Since stationarity implies that

$$P\left\{N(2) - N\left(\frac{1}{2}\right) = 2\right\} = P\left\{N\left(\frac{3}{2}\right) = 2\right\},$$

the desired probability equals

$$P\left\{N\left(\frac{1}{2}\right) = 1\right\} \cdot P\left\{N\left(\frac{3}{2}\right) = 2\right\} = \frac{1^1 e^{-1}}{1!} \cdot \frac{3^2 e^{-3}}{2!} \approx 0.082. \quad \blacklozenge$$

EXERCISES

A

1. Suppose that in a summer evening, shooting stars are observed at a Poisson rate of one every 12 minutes. What is the probability that three shooting stars are observed in 30 minutes?

2. Suppose that in Japan earthquakes occur at a Poisson rate of three per week. What is the probability that the next earthquake occurs after two weeks?

3. Suppose that for a telephone subscriber, the number of wrong numbers is Poisson, having rate $\lambda = 1$ per week. A certain subscriber has not received any wrong numbers from Sunday through Friday. What is the probability that he receives no wrong numbers on Saturday either?

4. In a certain town, crimes occur at a Poisson rate of five per month. What is the probability of having exactly two months (not necessarily consecutive) with no crimes during the next year?

5. Accidents occur at an intersection at a Poisson rate of three per day. What is the probability that during January there are exactly three days (not necessarily consecutive) without any accidents?

6. Customers arrive at a bookstore at a Poisson rate of six per hour. Given that the store opens at 9:30 A.M., what is the probability that exactly one customer arrives by 10:00 A.M. and 10 customers by noon?

B

7. Let $\{N(t) : t \geq 0\}$ be a Poisson process. What is the probability of (a) an even number of events in $(t, t + \alpha)$; (b) an odd number of events in $(t, t + \alpha)$?

8. Let $\{N(t) : t \geq 0\}$ be a Poisson process with rate λ. Suppose that $N(t)$ is the total number of two types of events that have occurred in $[0, t]$. Let $N_1(t)$ and $N_2(t)$ be the total number of events of type 1 and events of type 2 that have occurred in $[0, t]$, respectively. If events of type 1 and type 2 occur independently with probabilities p and $1 - p$, respectively, prove that $\{N_1(t) : t \geq 0\}$ and $\{N_2(t) : t \geq 0\}$ are Poisson processes with respective rates λp and $\lambda(1 - p)$.

   HINT: First calculate $P\{N_1(t) = n \text{ and } N_2(t) = m\}$ using the relation

$$P\{N_1(t) = n \text{ and } N_2(t) = m\}$$

$$= \sum_{i=0}^{\infty} P\{N_1(t) = n \text{ and } N_2(t) = m \mid N(t) = i\} P\{N(t) = i\}.$$

(This is true because of Theorem 3.4.) Then use the relation

$$P\{N_1(t) = n\} = \sum_{m=0}^{\infty} P\{N_1(t) = n \text{ and } N_2(t) = m\}.$$

9. Customers arrive at a grocery store at a Poisson rate of one per minute. If 2/3 of the customers are female and 1/3 are male, what is the probability that 15 females enter the store between 10:30 and 10:45? HINT: Use the result of Exercise 8.

5.3   Other Discrete Random Variables

Geometric Random Variable

Suppose that a sequence of independent Bernoulli trials, each with probability of success p, $0 < p < 1$, are performed. Let X be the number of experiments until the first success occurs. Then X is a discrete random variable called *geometric*. Its set of possible values is $\{1, 2, 3, 4, \ldots\}$ and

$$P(X = n) = (1 - p)^{n-1}p, \qquad n = 1, 2, 3, \ldots.$$

This equation follows since (a) the first $(n - 1)$ trials are all failures, (b) the nth trial is a success, and (c) the successive Bernoulli trials are all independent.

Let $p(x) = (1 - p)^{x-1}p$ for $x = 1, 2, 3, \ldots$, and 0 elsewhere. Then for all values of x in $\mathbf{R}$, $p(x) \geq 0$ and

$$\sum_{x=1}^{\infty} p(x) = \sum_{x=1}^{\infty} (1 - p)^{x-1}p = \frac{p}{1 - (1 - p)} = 1,$$

by the geometric series theorem. Hence $p(x)$ is a probability function.

DEFINITION   *The probability function* $p(x) = (1 - p)^{x-1}p$, $0 < p < 1$, *for* $x = 1, 2, 3, \ldots$, *and* 0 *elsewhere, is called geometric.*

Let X be a geometric random variable with parameter p; then

$$E(X) = \sum_{x=1}^{\infty} xp(1 - p)^{x-1} = \frac{p}{1 - p} \sum_{x=1}^{\infty} x(1 - p)^x$$

$$= \frac{p}{1 - p} \frac{1 - p}{[1 - (1 - p)]^2} = \frac{1}{p},$$

where the third equality follows from the relation $\sum_{x=1}^{\infty} xr^x = r/(1-r)^2$, $|r| < 1$. $E(X) = 1/p$ indicates that to get the first success, on average, $1/p$ independent Bernoulli trials are needed. The relation $\sum_{x=1}^{\infty} x^2 r^x = [r(r+1)]/(1-r)^3$, $|r| < 1$, implies that

$$E(X^2) = \sum_{x=1}^{\infty} x^2 p(1-p)^{x-1} = \frac{p}{1-p} \sum_{x=1}^{\infty} x^2 (1-p)^x = \frac{2-p}{p^2}.$$

Hence

$$\text{Var}(X) = E(X^2) - [E(X)]^2 = \frac{2-p}{p^2} - \left(\frac{1}{p}\right)^2 = \frac{1-p}{p^2}.$$

Example 5.21   From an ordinary deck of 52 cards we draw cards at random, *with replacement*, and successively until an ace is drawn. What is the probability that at least 10 draws are needed?

SOLUTION: Let X be the number of draws until the first ace. The random variable X is geometric with the parameter $p = 1/13$. Thus

$$P(X = n) = \left(\frac{12}{13}\right)^{n-1}\left(\frac{1}{13}\right), \qquad n = 1, 2, 3, \ldots,$$

and so the probability that at least 10 draws are needed is

$$P(X \geq 10) = \sum_{n=10}^{\infty} \left(\frac{12}{13}\right)^{n-1}\left(\frac{1}{13}\right) = \frac{1}{13}\sum_{n=10}^{\infty}\left(\frac{12}{13}\right)^{n-1}$$

$$= \frac{1}{13} \cdot \frac{(12/13)^9}{1 - 12/13} = \left(\frac{12}{13}\right)^9 \approx 0.49. \quad \blacklozenge$$

REMARK: There is a shortcut to the solution of this problem: The probability that at least 10 draws are needed to get an ace is the same as the probability that in the first nine draws there are no aces. This is equal to $(12/13)^9 \approx 0.49$. $\blacklozenge$

Example 5.22   A father asks his sons to cut their backyard lawn. Since he does not specify which of the three sons is to do the job, each boy tosses a coin to determine the odd person, who must then cut the lawn. In the case that all three get heads or tails, they continue tossing until they reach a decision. Let p be the probability of heads and $q = 1 - p$, the probability of tails.

(a) Find the probability that they reach a decision in less than n tosses.

(b) If $p = 1/2$, what is the minimum number of tosses required to reach a decision with probability 0.95?

SOLUTION: (a) The probability that they reach a decision on a certain round of coin tossing is

$$\binom{3}{2}p^2q + \binom{3}{2}q^2p = 3pq(p+q) = 3pq.$$

The probability that they do not reach a decision on a certain round is $1 - 3pq$. Let X be the number of tosses until they reach a decision; then X is a geometric random variable with parameter $3pq$. Therefore

$$P(X < n) = 1 - P(X \geq n) = 1 - (1 - 3pq)^{n-1},$$

where the second equality follows since $X \geq n$ if and only if none of the first $n - 1$ tosses results in a success.

   (b) We want to find the minimum n so that $P(X \leq n) \geq 0.95$. This gives $1 - P(X > n) \geq 0.95$ or $P(X > n) \leq 0.05$. But $P(X > n) = (1 - 3pq)^n = (1 - 3/4)^n = (1/4)^n$. Therefore, we must have $(1/4)^n \leq 0.05$, or $n \ln 1/4 \leq \ln 0.05$. This gives $n \geq 2.16$; hence the smallest n is 3.   ♦

Negative Binomial Random Variable

Negative binomial random variables are generalizations of geometric random variables. Suppose that a sequence of independent Bernoulli trials each with probability of success p, $0 < p < 1$, is performed. Let X be the number of experiments until the rth success occurs. Then X is a discrete random variable called *negative binomial*. Its set of possible values is $\{r, r+1, r+2, r+3, \ldots\}$ and

$$P(X = n) = \binom{n-1}{r-1}p^r(1-p)^{n-r}, \qquad n = r, r+1, \ldots. \qquad (5.4)$$

This equation follows since if the outcome of the nth trial is the rth success, then in the first $(n-1)$ trials exactly $(r-1)$ successes have occurred and the nth trial is a success. The probability of the former event is

$$\binom{n-1}{r-1}p^{r-1}(1-p)^{(n-1)-(r-1)} = \binom{n-1}{r-1}p^{r-1}(1-p)^{n-r},$$

and the probability of the latter is p. Therefore by the independence of the trials, (5.4) follows.

DEFINITION   *The probability function*

$$p(x) = \binom{x-1}{r-1} p^r (1-p)^{x-r},$$

$$0 < p < 1, \quad x = r, r+1, r+2, r+3, \ldots,$$

is called negative binomial with parameters (r, p).

Note that a negative binomial probability function with parameters $(1, p)$ is geometric. In Section 9.3 we show that the expected value of a negative binomial random variable with parameters (r, p) is r/p.

Example 5.23   Sharon and Ann play a series of backgammon games until one of them wins five games. Suppose that the games are independent and the probability that Sharon wins a game is 0.58.

(a) Find the probability that the series ends in seven games.

(b) If the series ends in seven games, what is the probability that Sharon wins?

SOLUTION: (a) Let X be the number of games until Sharon wins five games. Let Y be the number of games until Ann wins five games. The random variables X and Y are negative binomial with parameters $(5, 0.58)$ and $(5, 0.42)$, respectively. The probability that the series ends in seven games is

$$P(X = 7) + P(Y = 7) = \binom{6}{4}(0.58)^5(0.42)^2 + \binom{6}{4}(0.42)^5(0.58)^2$$

$$\approx 0.17 + 0.066 \approx 0.24.$$

(b) Let A be the event that Sharon wins and B be the event that the series ends in seven games. Then the desired probability is

$$P(A \mid B) = \frac{P(AB)}{P(B)} = \frac{P(X = 7)}{P(X = 7) + P(Y = 7)} \approx \frac{0.17}{0.24} \approx 0.71. \quad \blacklozenge$$

The following example, given by Kaigh in January 1979 issue of *Mathematics Magazine*, is a modification of the gambler's ruin problem, Example 3.12.

Example 5.24   (Attrition Ruin Problem)   Two gamblers play a game in which in each play gambler A beats B with probability p, $0 < p < 1$, and loses to B with probability $q = 1 - p$. Suppose that each play results in a forfeiture of \$1 for the loser and in no change for the winner. If player A initially has a dollars and player B has b dollars, what is the probability that B will be ruined?

SOLUTION: Let E_i be the event that in the first $b + i$ plays B loses b times. Let A^* be the event that A wins. Then

$$P(A^*) = \sum_{i=0}^{a-1} P(E_i).$$

If every time that A wins is called a success, E_i is the event that the bth success occurs on the $(b+i)$th play. Using the negative binomial distribution, we have

$$P(E_i) = \binom{i+b-1}{b-1} p^b q^i.$$

Therefore,

$$P(A^*) = \sum_{i=0}^{a-1} \binom{i+b-1}{b-1} p^b q^i.$$

As a numerical illustration, let $a = b = 4$, $p = 0.6$ and $q = 0.4$. We get $P(A^*) = 0.710208$ and $P(B^*) = 0.289792$ exactly.   ♦

The following example is due to Hugo Steinhaus (1887–1972), who brought it up in a conference honoring Stefan Banach (1892–1945), a smoker and one of the greatest mathematicians of this century.

Example 5.25   (Banach Matchbox Problem)   A smoking mathematician carries two matchboxes, one in his right pocket and one in his left pocket. Whenever he wants to smoke, he selects a pocket at random and takes a match from the box in that pocket. If each matchbox initially contains N matches, what is the probability that when the mathematician for the first time *discovers* that one box is empty, there are exactly m matches in the other box, $m = 0, 1, 2, \ldots, N$?

SOLUTION: Every time that the left pocket is selected we say that a success has occurred. When the mathematician discovers that the left box is empty, the right one contains m matches if and only if the $(N + 1)$st success occurs on the $(N - m) + (N + 1) = (2N - m + 1)$st trial. The probability of this event is

$$\binom{(2N - m + 1) - 1}{(N + 1) - 1} \left(\frac{1}{2}\right)^{N+1} \left(\frac{1}{2}\right)^{(2N-m+1)-(N+1)}$$

$$= \binom{2N - m}{N} \left(\frac{1}{2}\right)^{2N-m+1}.$$

By symmetry, when the mathematician discovers that the right box is empty, with probability $\binom{2N - m}{N} \left(\frac{1}{2}\right)^{2N-m+1}$, the left box contains m matches. Therefore, the desired probability is

$$2\binom{2N - m}{N} \left(\frac{1}{2}\right)^{2N-m+1} = \binom{2N - m}{N} \left(\frac{1}{2}\right)^{2N-m}. \quad \blacklozenge$$

Hypergeometric Random Variable

Suppose that from a box containing D defective and $N - D$ nondefective items, n are drawn at random and *without replacement*. Let X be the number of defective items drawn. Then X is a discrete random variable with the set of possible values

$$A = \{x \in \mathbf{N} : \max(0, n + D - N) \leq x \leq \min(D, n)\},$$

and probability function

$$p(x) = P(X = x) = \frac{\binom{D}{x}\binom{N - D}{n - x}}{\binom{N}{n}}, \qquad x \in A.$$

Any random variable X with such a probability function is called a *hypergeometric random variable*. The fact that $p(x)$ is a probability function is easily verified. Clearly, $p(x) \geq 0, \forall x$. To prove that $\sum_{x \in A} p(x) = 1$, assume that $\binom{m}{x} = 0$ for $m < x$. This is then equivalent to

$$\sum_{x=0}^{n} \binom{D}{x}\binom{N - D}{n - x} = \binom{N}{n},$$

which can be shown by a simple combinatorial argument. (See Exercise 37 of Section 2.4; in that exercise let $m = D$, $n = N - D$, and $r = n$.) We will prove in Section 9.3 that $E(X) = nD/N$.

DEFINITION   *Let N, D, and n be positive integers and*

$$p(x) = P(X = x) = \frac{\dbinom{D}{x}\dbinom{N-D}{n-x}}{\dbinom{N}{n}}$$

for $x \in A = \{x \in \mathbf{N} : \max(0, n + D - N) \le x \le \min(D, n)\}$, and 0 elsewhere. Then $p(x)$ is said to be a hypergeometric probability function.

Example 5.26   In 500 independent calculations a scientist has made 25 errors. If a second scientist checks seven of these calculations randomly, what is the probability that he detects two errors? Assume that the second scientist will definitely find the error of a false calculation.

SOLUTION: Let X be the number of errors found by the second scientist. Then X is hypergeometric with $N = 500$, $D = 25$, and $n = 7$. We are interested in $P(X = 2)$, which is given by

$$p(2) = \frac{\dbinom{25}{2}\dbinom{500-25}{7-2}}{\dbinom{500}{7}} \approx 0.04. \quad \blacklozenge$$

Example 5.27   In a community of $a + b$ potential voters, a are for abortion and b ($b < a$) are against it. Suppose that a vote will be taken to determine the will of the majority with regard to legalizing abortion. If n ($n < b$) random persons of these $a + b$ potential voters do not vote, what is the probability that those against abortion will win?

SOLUTION: Let X be the number of those who do not vote and are for abortion. The persons against abortion will win if and only if

$$a - X < b - (n - X).$$

But this is true if and only if $X > (a - b + n)/2$. Since X is a hypergeometric random variable, we have that

$$P\left(X > \frac{a-b+n}{2}\right) = \sum_{i=[\frac{a-b+n}{2}]+1}^{n} P(X=i) = \sum_{i=[\frac{a-b+n}{2}]+1}^{n} \frac{\binom{a}{i}\binom{b}{n-i}}{\binom{a+b}{n}},$$

where $\left[\dfrac{a-b+n}{2}\right]$ is the greatest integer less than or equal to $\dfrac{a-b+n}{2}$. ◆

Example 5.28 Professors Davidson and Johnson from the University of Victoria in Canada gave the following problem to their students in a finite math course: "An urn contains N balls of which B are black and $N - B$ are white, n balls are chosen at random and without replacement from the urn. If X is the number of black balls chosen, find $P(X = i)$." Using the hypergeometric formula, the solution to this problem is

$$P(X = i) = \frac{\binom{B}{i}\binom{N-B}{n-i}}{\binom{N}{n}}, \qquad i \le \min(n, B).$$

However, by mistake, some students interchanged B with n and came up with the following answer

$$P(X = i) = \frac{\binom{n}{i}\binom{N-n}{B-i}}{\binom{N}{B}},$$

which can be verified immediately to be the correct answer for all values of i. Explain if this is accidental or there is a probabilistic reason for its validity.

SOLUTION: In their paper "Interchanging Parameters of Hypergeometric Distributions", *Mathematics Magazine*, December 1993, Volume 66, Davidson and Johnson have shown that what happened is not accidental and that there is a probabilistic reason for it. Their justification is elegant and is as follows:

> We begin with an urn that contains N white balls. When Ms. Painter arrives, she picks B balls at random without replacement from the urn, then paints each of the drawn balls black with instant dry paint, and finally returns these B balls to the urn. When Mr. Carver arrives, he chooses n balls at

random without replacement from the urn, then engraves each of the chosen balls with the letter C, and finally, returns these n balls to the urn. Let the random variable X denote the number of black balls chosen (i.e., painted and engraved) when both Painter and Carver have finished their jobs. Since the tasks of Painter and Carver do not depend on which task is done first, the probability distribution of X is the same whether Painter does her job before or after Carver does his. If Painter goes first, Carver chooses n balls from an urn that contains B black balls and $N - B$ white balls, so

$$P(X = i) = \frac{\binom{B}{i}\binom{N-B}{n-i}}{\binom{N}{n}} \qquad \text{for } x = 0, 1, 2, \ldots, \min(n, B).$$

On the other hand, if Carver goes first, Painter draws B balls from an urn that contains n balls engraved with C and $N - n$ balls not engraved, so

$$P(X = i) = \frac{\binom{n}{i}\binom{N-n}{B-i}}{\binom{N}{B}} \qquad \text{for } x = 0, 1, 2, \ldots, \min(n, B).$$

Thus, changing the order of who goes first provides an urn-drawing explanation of why the probability distribution of X remains the same when B is interchanged with n.   ◆

Let X be a hypergeometric random variable with parameters n, D, and N. Calculation of $E(X)$ and $\text{Var}(X)$ directly from $p(x)$, the probability function of X, is tedious. By the methods developed in Sections 9.3 and 11.1, calculation of these quantities is considerably easier. In Section 9.3 we will prove that $E(X) = nD/N$; it can be shown that

$$\text{Var}(X) = \frac{nD(N - D)}{N^2}\left(1 - \frac{n - 1}{N - 1}\right).$$

(See Exercise 19, Section 11.1.) Using these formulas, for example, we have that the average number of face cards in a bridge hand is $(13 \times 12)/52 = 3$ with variance

$$\frac{13 \times 12(52 - 12)}{52^2}\left(1 - \frac{13 - 1}{52 - 1}\right) = 1.765.$$

REMARK: If the n items that are selected at random from the D defective and $N - D$ nondefective items are chosen *with replacement* rather than *without replacement*, then X, the number of defective items, is a binomial random variable with parameters n and D/N. Thus

$$P(X = x) = \binom{n}{x} \left(\frac{D}{N}\right)^x \left(1 - \frac{D}{N}\right)^{n-x}, \qquad x = 0, 1, 2, \ldots, n.$$

Now if N is very large, it is not that important whether a sample is taken with or without replacement. For example, if we have to choose 1000 persons from a population of 200,000,000, replacing persons that are selected back to the population will change the chance of other persons to be selected only very insignificantly. Therefore, for large N, the binomial probability function is an excellent approximation for the hypergeometric probability function. This can be stated mathematically as follows:

$$\lim_{\substack{N \to \infty \\ D \to \infty \\ D/N \to p}} \frac{\binom{D}{x}\binom{N-D}{n-x}}{\binom{N}{n}} = \binom{n}{x} p^x (1-p)^{n-x}.$$

We leave its proof as an exercise. ◆

EXERCISES

A

1. An absentminded professor does not remember which of his 12 keys will open his office door. If he tries them at random and with replacement:
   (a) On average, how many keys should he try before unlocking his office?
   (b) What is the probability that he opens his office door after only three tries?

2. The probability that Marty hits target M when he fires at it is p. The probability that Alvie hits target A when he fires at it is q. Marty and Alvie fire one shot each at their targets. If both of them hit their targets, they stop; otherwise, they will continue.
   (a) What is the probability that they stop after each has fired r times?
   (b) What is the expected value of the number of the times that each of them has fired before stopping?

3. Suppose that 20% of people have hazel eyes. What is the probability that the eighth passenger boarding a plane is the third one having hazel

eyes? Assume that passengers boarding the plane form a randomly chosen group.

4. A certain basketball player makes a foul shot with probability 0.45. What is the probability that (a) his first basket occurs on the sixth shot; (b) his first and second baskets occur on his fourth and eighth shots, respectively?

5. A store has 50 light bulbs available for sale. Of these, five are defective. A customer buys eight light bulbs randomly from this store. What is the probability that he finds exactly one defective light bulb among them?

6. The probability that a randomly chosen light bulb is defective is p. We screw a bulb into a lamp and switch on the current. If the bulb works, we stop; otherwise, we try another and continue until a good bulb is found. What is the probability that at least n bulbs are required?

7. Suppose that independent Bernoulli trials with parameter p are performed successively. Let N be the number of trials needed to get x successes, and X be the number of successes in the first n trials. Show that

$$P(N = n) = \frac{x}{n} P(X = x).$$

NOTE: By this relation, in coin tossing, for example, we can state that the probability of the fifth head on the seventh toss is 5/7 of the probability of five heads in seven tosses.

8. The digits after the decimal point of a random number between 0 and 1 are numbers selected at random, with replacement, independently, and successively from the set $\{0, 1, \ldots, 9\}$. In a random number from $(0, 1)$, on the average, how many digits are there before the fifth 3?

9. On average, how many games of bridge are necessary before a player is dealt three aces? A bridge hand is 13 randomly selected cards from an ordinary deck of 52 cards.

10. Solve the Banach matchbox problem (Example 5.25) for the case where the matchbox in the right pocket contains M matches, the matchbox in his left pocket contains N matches, and $m \leq \min(M, N)$.

11. Suppose that 15% of the population of a town are senior citizens. Let X be the number of nonsenior citizens who enter a mall before the tenth senior citizen arrives. Find the probability function of X. Assume that each customer who enters the mall is a random person from the entire population.

12. Florence is moving and wishes to sell her package of 100 computer diskettes. Unknown to her, 10 of these diskettes are defective. Sharon

will purchase them if a random sample of 10 contains no more than one defective disk. What is the probability that she buys them?

13. In an annual charity drive, 35% of a population of size 560 make contributions. If, in a statistical survey, 15 people are selected at random and without replacement, what is the probability that at least two persons have contributed?

14. The probability that a message sent over a communication channel is garbled is p. If the message is sent repeatedly until it is transmitted reliably, and if each time it takes 2 minutes to process it, what is the probability that the transmission of a message takes more than t minutes?

15. A vending machine that sells cans of grapefruit juice for 75 cents each is not working properly. The probability that it accepts a coin is 10%. Angela has a quarter and five dimes. Determine the probability that she should try the coins at least 50 times before she gets a can of grapefruit juice.

B

16. A fair coin is flipped repeatedly. What is the probability that the fifth tail occurs before the tenth head?

17. On average, how many independent games of poker are required in order that a preassigned player is dealt a straight? (See Exercise 20 of Section 2.4 for a definition of a straight. The cards have distinct consecutive values that are not of the same suit: for example, 3 of hearts, 4 of hearts, 5 of spades, 6 of hearts, and 7 of hearts.)

18. Let X be a geometric random variable with parameter p, and n and m be nonnegative integers.
  (a) Prove that $P(X > n + m \mid X > m) = P(X > n)$. What is the meaning of this property in terms of Bernoulli trials? This is called the *memoryless property* of geometric random variables.
  (b) For what values of n, is $P(X = n)$ maximum?
  (c) What is the probability that X is even?
  (d) Show that the geometric is the only distribution on the positive integers with the memoryless property.

19. Twelve hundred eggs, of which 200 are rotten, are distributed randomly in 100 cartons, each containing a dozen eggs. These cartons are then sold to a restaurant. How many cartons should we expect the chef of the restaurant to open before finding one without rotten eggs?

20. In the Banach matchbox problem, Example 5.25, find the probability that when the first matchbox is emptied (not found empty) there are exactly *m* matches in the other box.

21. In the Banach matchbox problem, Example 5.25, find the probability that the box which is emptied first is not the one that is first found empty.

22. To estimate the number of trout in a lake, we caught 50 trout, tagged them, and returned them. Later we caught 50 trout and found that four of them were tagged. From this experiment estimate *n*, the total number of trout in the lake.

HINT: Let p_n be the probability of four tagged trout among the 50 trout caught. Find the value of *n* that maximizes p_n.

23. Suppose that from a box containing *D* defective and $N - D$ nondefective items, *n* $(\leq D)$ are drawn one by one, at random, and *without replacement*.

   (a) Find the probability that the *k*th item drawn is defective.

   (b) If the $(k - 1)$st item drawn is defective, what is the probability that the *k*th item is also defective?

Review Problems

---

1. Of the applicants for the police academy, only 25% will pass all the examinations. Suppose that 12 successful candidates are needed. What is the probability that by examining 20 candidates, the academy finds all of the 12 persons needed?

2. The time between the arrival of two consecutive customers at a post office is 3 minutes on average. Assuming that customers arrive in accordance with a Poisson process, find the probability that tomorrow during the lunch hour (between noon and 12:30 P.M.) fewer than seven customers arrive.

3. A restaurant serves eight fish entrées, 12 beef, and 10 poultry. If customers select from these entrées randomly, what is the expected number of fish entrées ordered by the next four customers?

4. A university has *n* students, 70% of whom will finish an aptitude test in less than 25 minutes. If 12 students are selected at random, what is the probability that at most two of them will not finish in 25 minutes?

5. A doctor has five patients with migraine headaches. He prescribes for all five a drug that relieves the headaches of 82% of such patients. What is the probability that the medicine will not relieve the headaches of two of these patients?

6. In a community, the chance of a set of triplets is 1 in 1000 births. Determine the probability that the second set of triplets in this community is before the 2000th birth.

7. From a panel of prospective jurors, 12 are selected at random. If there are 200 men and 160 women in the panel, what is the probability that more than half of the jury selected are women?

8. Suppose that 10 trains arrive independently at a station every day each at a random time between 10:00 A.M. and 11:00 A.M.. What is the expected number and the variance of those that arrive between 10:15 A.M. and 10:28 A.M.?

9. Suppose that a certain bank returns bad checks at a Poisson rate of three per day. What is the probability that this bank returns at most four bad checks during the next two days?

10. The policy of the quality control division of a certain corporation is to reject a shipment if more than 5% of its items are defective. A shipment of 500 items is received, 30 of them are randomly tested, and two have been found defective. Should this shipment be rejected?

11. A fair coin is tossed successively until a head occurs. If N is the number of tosses required, what are the expected value and the variance of N?

12. What is the probability that the sixth toss of a die is the first 6?

13. From the 28 professors of a certain department, 18 drive foreign and 10 drive domestic cars. If five of these professors are selected at random, what is the probability that at least three of them drive foreign cars?

14. A bowl contains 10 red and six blue chips. What is the expected number of blue chips among five randomly selected chips?

15. A certain type of seed when planted fails to germinate with probability 0.06. If 40 of such seeds are planted, what is the probability that at least 36 of them germinate?

16. Suppose that 6% of the claims received by an insurance company are for damages from vandalism. What is the probability that at least three of the 20 claims received on a certain day are for vandalism damages?

17. An insurance company claims that only 70% of the drivers regularly use seat belts. In a statistical survey, it was found that out of 20 randomly selected drivers, 12 regularly used seat belts. Is this sufficient evidence to conclude that the insurance company's claim is false?

18. From the set $\{x : 0 \le x \le 1\}$ numbers are selected at random and independently and rounded to three decimal places.   What is the

probability that 0.345 is obtained (a) for the first time after 1000 selections; (b) for the third time after 3000 selections?

19. The probability that a child of a certain family inherits a certain disease is 0.23 independent of other children inheriting the disease. If the family has five children and the disease is detected in one child, what is the probability that exactly two more children have the disease as well?

20. A bowl contains w white and b blue chips. Chips are drawn at random and with replacement until a blue chip is drawn. What is the probability that (a) exactly n draws are required; (b) at least n draws are required?

21. Experience shows that 75% of certain kinds of seeds germinate when planted under normal conditions. Determine the minimum number of seeds to be planted so that the chance of at least five of them germinating is more than 90%.

22. Suppose that n babies were born at a county hospital last week. Also suppose that the probability of a baby having blonde hair is p. If k of these n babies are blondes, what is the probability that the ith baby born is blonde?

23. A farmer, plagued by insects, seeks to attract birds to his property by distributing seeds over a wide area. Let λ be the average number of seeds per unit area and suppose that the seeds are distributed in a way that the probability of having any number of seeds in a given area depends only on the size of the area and not on its location. Argue that X, the number of seeds that fall on the area A, is approximately a Poisson random variable with parameter λA.

24. Show that if all three of n, N, and $D \to \infty$ so that $n/N \to 0$, D/N converges to a small number, and $nD/N \to \lambda$, then for all x,

$$\frac{\binom{D}{x}\binom{N-D}{n-x}}{\binom{N}{n}} \to \frac{e^{-\lambda}\lambda^x}{x!}.$$

This shows that the Poisson distribution is the limit of the hypergeometric distribution.

Chapter 6

CONTINUOUS RANDOM VARIABLES

6.1 Probability Density Functions

As discussed in Section 4.2, the distribution function of a random variable X is a function F from $(-\infty, +\infty)$ to $\mathbf{R}$ defined by $F(t) = P(X \le t)$. From the definition of F we deduced that it is nondecreasing, right continuous, and satisfies $\lim_{t \to \infty} F(t) = 1$, and $\lim_{t \to -\infty} F(t) = 0$. Furthermore, we showed that for discrete random variables, distributions are step functions. We also proved that if X is a discrete random variable with set of possible values $\{x_1, x_2, \ldots\}$, probability function p, and distribution function F, then F has jump discontinuities at $x_1, x_2, \ldots$, where the magnitude of the jump at x_i is $p(x_i)$ and for $x_{n-1} \le t < x_n$,

$$F(t) = P(X \le t) = \sum_{i=1}^{n-1} p(x_i). \qquad (6.1)$$

In the case of discrete random variables a very small change in t may cause relatively large changes in the values of F. For example, if t changes from $x_n - \varepsilon$ to x_n, $\varepsilon > 0$ being arbitrarily small, then $F(x)$ changes from $\sum_{i=1}^{n-1} p(x_i)$ to $\sum_{i=1}^{n} p(x_i)$, a change of magnitude $p(x_n)$, which might be large. In cases such as the lifetime of a random light bulb, the arrival time of a train at a station, and the weight of a random watermelon grown in a certain field, where the set of possible values of X is uncountable, small

204

changes in x produce correspondingly small changes in the distribution of X. In such cases we expect that F, the distribution function of X, will be a continuous function. Random variables that have continuous distributions can be studied under general conditions. However, for practical reasons and mathematical simplicity, we restrict ourselves to a class of random variables that are called *absolutely continuous* and are defined as follows:

DEFINITION   *Let X be a random variable. Suppose that there exists a nonnegative real-valued function* $f : \mathbf{R} \to [0, \infty)$ *such that for any subset of real numbers A which can be constructed from intervals by a countable number of set operations,*

$$P(X \in A) = \int_A f(x)\, dx. \tag{6.2}$$

Then X is called absolutely continuous or in this book, for simplicity, continuous. Therefore, whenever we say that X is continuous, we mean that it is absolutely continuous and hence satisfies (6.2). The function f is called the probability density function, or simply, the density function of X.

Let f be the density function of a random variable X with distribution function F. Some immediate properties of f are as follows:

(a) $F(t) = \int_{-\infty}^{t} f(x)\, dx$.

Let $A = (-\infty, t]$, then using (6.2), we can write

$$F(t) = P(X \le t) = P(X \in A) = \int_A f(x)\, dx = \int_{-\infty}^{t} f(x)\, dx.$$

Comparing (a) with (6.1), we see that f, the probability density function of a continuous random variable X, can be regarded as an extension of the idea of p, the probability function of a discrete random variable. As we study f further, the resemblance between f and p becomes clearer. For example, recall that if X is a discrete random variable with the set of possible values A and probability function p, then $\sum_{x \in A} p(x) = 1$. In the continuous case, the relation analogous to this is the following:

(b) $\int_{-\infty}^{\infty} f(x)\, dx = 1$.

This is true by (a) and the fact that $F(\infty) = 1$ [i.e., $\lim_{t \to \infty} F(t) = 1$]. It also follows from (6.2) with $A = \mathbf{R}$. Because of this property, in general, if a function $g : \mathbf{R} \to [0, \infty)$ satisfies $\int_{-\infty}^{\infty} g(x) \, dx = 1$, we say that g is a probability density function, or simply a density function.

(c) *If f is continuous, then* $F'(x) = f(x)$.

This follows from (a) and the fundamental theorem of calculus: $\dfrac{d}{dt} \displaystyle\int_{a}^{t} f(x) \, dx = f(t)$. Note that even if f is not continuous, still $F'(x) = f(x)$ for every x at which f is continuous. *Therefore, the distribution functions of continuous random variables are integrals of their derivatives.*

(d) *For real numbers* $a \leq b$, $P(a \leq X \leq b) = \int_{a}^{b} f(t) \, dt$.

Let $A = [a, b]$ in (6.2); then

$$P(a \leq X \leq b) = P(X \in A) = \int_{A} f(x) \, dx = \int_{a}^{b} f(x) \, dx.$$

Property (d) states that the probability of X being between a and b is equal to the area under the graph of f from a to b. Letting $a = b$ in (d), we obtain

$$P(X = a) = \int_{a}^{a} f(x) \, dx = 0.$$

This means that for any real number a, $P(X = a) = 0$. That is, the probability that a continuous random variable assumes a certain value is 0. From $P(X = a) = 0$, $\forall a \in \mathbf{R}$, we have that the probability of X lying in an interval does not depend on the endpoints of the interval. Therefore,

(e) $P(a < X < b) = P(a \leq X < b) = P(a < X \leq b)$

$$= P(a \leq X \leq b) = \int_{a}^{b} f(t) \, dt.$$

Another implication of $P(X = a) = 0$, $\forall a \in \mathbf{R}$, is that *the value of the density function f at no point represents a probability.* The probabilistic significance of f is that its integral over any subset of real numbers B gives the probability that X lies in B. Intuitively, $f(a)$ is a measure that determines how likely it is for X to be close to a. To see this, note that if $\varepsilon > 0$ is very small, then $P(a - \varepsilon < X < a + \varepsilon)$ is the probability that X is close to a. Now

$$P(a - \varepsilon < X < a + \varepsilon) = \int_{a-\varepsilon}^{a+\varepsilon} f(t) \, dt$$

is the area under $f(t)$ from $a - \varepsilon$ to $a + \varepsilon$. This area is almost equal to the area of a rectangle with sides of lengths $(a + \varepsilon) - (a - \varepsilon) = 2\varepsilon$ and $f(a)$. Thus

$$P(a - \varepsilon < X < a + \varepsilon) = \int_{a-\varepsilon}^{a+\varepsilon} f(t)\, dt \approx 2\varepsilon f(a). \qquad (6.3)$$

Similarly, for any other real number b (in the domain of f),

$$P(b - \varepsilon < X < b + \varepsilon) \approx 2\varepsilon f(b). \qquad (6.4)$$

Relations (6.3) and (6.4) show that for small fixed $\varepsilon > 0$, if $f(a) < f(b)$, then

$$P(a - \varepsilon < X < a + \varepsilon) < P(b - \varepsilon < X < b + \varepsilon).$$

That is, if $f(a) < f(b)$, the probability that X is close to b is higher than the probability that X is close to a. Thus the larger $f(x)$, the higher the probability will be that X is close to x.

Example 6.1   Experience has shown that while walking in a certain park, the time X, in minutes, between seeing two people smoking has a density function of the form

$$f(x) = \begin{cases} \lambda x e^{-x} & x > 0 \\ 0 & \text{otherwise.} \end{cases}$$

(a) Calculate the value of λ.
(b) Find the probability distribution function of X.
(c) What is the probability that Jeff, who has just seen a person smoking, will see another person smoking in 2 to 5 minutes? In at least 7 minutes?

SOLUTION: (a) To determine λ, we use the property $\int_{-\infty}^{\infty} f(x)\, dx = 1$:

$$\int_{-\infty}^{\infty} f(x)\, dx = \int_{0}^{\infty} \lambda x e^{-x}\, dx = \lambda \int_{0}^{\infty} x e^{-x}\, dx.$$

Now by integration by parts,

$$\int x e^{-x}\, dx = -(x + 1)e^{-x}.$$

Thus

$$\lambda \left[-(x+1)e^{-x}\right]_0^\infty = 1. \tag{6.5}$$

But as $x \to \infty$, using L'Hôpital's rule we get

$$\lim_{x\to\infty} (x+1)e^{-x} = \lim_{x\to\infty} \frac{x+1}{e^x} = \lim_{x\to\infty} \frac{1}{e^x} = 0.$$

Therefore, (6.5) implies that $\lambda[0 - (-1)] = 1$ or $\lambda = 1$.

(b) To find F, the distribution function of X, note that $F(t) = 0$ if $t < 0$. For $t \geq 0$,

$$F(t) = \int_{-\infty}^t f(x)\, dx = \left[-(x+1)e^{-x}\right]_0^t = -(t+1)e^{-t} + 1.$$

Thus

$$F(t) = \begin{cases} 0 & \text{if } t < 0 \\ -(t+1)e^{-t} + 1 & \text{if } t \geq 0. \end{cases}$$

(c) The desired probabilities $P(2 < X < 5)$ and $P(X \geq 7)$ are calculated as follows:

$$P(2 < X < 5) = P(2 < X \leq 5) = P(X \leq 5) - P(X \leq 2) = F(5) - F(2)$$
$$= (1 - 6e^{-5}) - (1 - 3e^{-2}) = 3e^{-2} - 6e^{-5} \approx 0.37.$$
$$P(X \geq 7) = 1 - P(X < 7) = 1 - P(X \leq 7) = 1 - F(7)$$
$$= 8e^{-7} \approx 0.007.$$

NOTE: To calculate these quantities, we can use $P(2 < X < 5) = \int_2^5 f(t)\, dt$ and $P(X \geq 7) = \int_7^\infty f(t)\, dt$ as well.　◆

Example 6.2

(a) Sketch the graph of the function

$$f(x) = \begin{cases} \dfrac{1}{2} - \dfrac{1}{4}|x - 3| & 1 \leq x \leq 5 \\ 0 & \text{otherwise,} \end{cases}$$

and show that it is the probability density function of a random variable X.

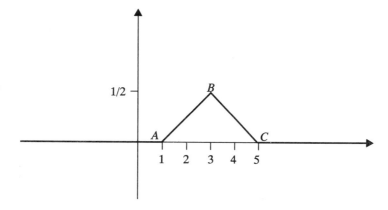

Figure 6.1   Density function of Example 6.2.

(b) Find F, the distribution function of X, and show that it is continuous.
(c) Sketch the graph of F.

SOLUTION: (a) Note that

$$
f(x) = \begin{cases}
0 & x < 1 \\
\dfrac{1}{2} + \dfrac{1}{4}(x - 3) & 1 \le x < 3 \\
\dfrac{1}{2} - \dfrac{1}{4}(x - 3) & 3 \le x < 5 \\
0 & x \ge 5.
\end{cases}
$$

Therefore, the graph of f is as shown in Figure 6.1. Now since $f(x) \ge 0$ and the area under f from 1 to 5, being the area of the triangle ABC (as seen from Figure 6.1) is $1/2(4 \times 1/2) = 1$, f is a density function of some random variable X.

(b) To calculate the distribution function of X, we use the formula $F(t) = \int_{-\infty}^{t} f(x)\,dx$. For $t < 1$,

$$
F(t) = \int_{-\infty}^{t} 0\,dx = 0.
$$

for $1 \le t < 3$,

$$
F(t) = \int_{-\infty}^{t} f(x)\,dx = \int_{1}^{t} \left[\frac{1}{2} + \frac{1}{4}(x - 3)\right] dx = \frac{1}{8}t^2 - \frac{1}{4}t + \frac{1}{8};
$$

for $3 \le t < 5$,

$$F(t) = \int_{-\infty}^{t} f(x)dx = \int_{1}^{3} \left[\frac{1}{2} + \frac{1}{4}(x-3) \right] dx + \int_{3}^{t} \left[\frac{1}{2} - \frac{1}{4}(x-3) \right] dx$$

$$= \frac{1}{2} + \left(-\frac{1}{8}t^2 + \frac{5}{4}t - \frac{21}{8} \right) = -\frac{1}{8}t^2 + \frac{5}{4}t - \frac{17}{8};$$

and for $t \ge 5$,

$$F(t) = \int_{-\infty}^{t} f(x)\, dx = \int_{1}^{3} \left[\frac{1}{2} + \frac{1}{4}(x-3) \right] dx + \int_{3}^{5} \left[\frac{1}{2} - \frac{1}{4}(x-3) \right] dx$$

$$= \frac{1}{2} + \frac{1}{2} = 1.$$

Therefore,

$$F(t) = \begin{cases} 0 & t < 1 \\ \dfrac{1}{8}t^2 - \dfrac{1}{4}t + \dfrac{1}{8} & 1 \le t < 3 \\ -\dfrac{1}{8}t^2 + \dfrac{5}{4}t - \dfrac{17}{8} & 3 \le t < 5 \\ 1 & t \ge 5. \end{cases}$$

F is continuous because

$$\lim_{t \to 1-} F(t) = 0 = F(1),$$

$$\lim_{t \to 3-} F(t) = \frac{1}{8}(3)^2 - \frac{1}{4}(3) + \frac{1}{8} = \frac{1}{2} = -\frac{1}{8}(3)^2 + \frac{5}{4}(3) - \frac{17}{8} = F(3),$$

$$\lim_{t \to 5-} F(t) = -\frac{1}{8}(5)^2 + \frac{5}{4}(5) - \frac{17}{8} = 1 = F(5).$$

(c) The graph of F is shown in Figure 6.2. ◆

REMARK: It is important to realize that there are random variables that are neither continuous nor discrete. Examples 4.7 and 4.10 give two such random variables. ◆

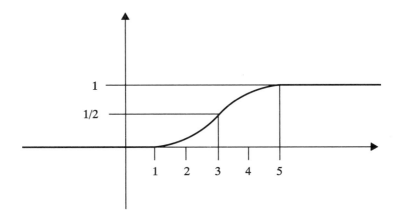

Figure 6.2 Distribution function of Example 6.2.

EXERCISES

A

1. When a certain car breaks down, the time that it takes to fix it (in hours) is a random variable with the density function

$$f(x) = \begin{cases} ce^{-3x} & \text{if } 0 \le x < \infty \\ 0 & \text{otherwise.} \end{cases}$$

(a) Calculate the value of c.
(b) Find the probability that when this car breaks down, it takes at most 30 minutes to fix it.

2. The distribution function for the duration of a certain soap opera (in tens of hours) is

$$F(x) = \begin{cases} 1 - \dfrac{16}{x^2} & x \ge 4 \\ 0 & x < 4. \end{cases}$$

(a) Calculate f, the probability density function of the soap opera.
(b) Sketch the graphs of F and f.
(c) What is the probability that the soap opera takes at most 50 hours? At least 60 hours? Between 50 and 70 hours? Between 10 and 35 hours?

3. The time it takes for a student to finish an aptitude test (in hours) has a density function of the form

$$f(x) = \begin{cases} c(x-1)(2-x) & \text{if } 1 < x < 2 \\ 0 & \text{elsewhere.} \end{cases}$$

(a) Determine the constant c.
(b) Calculate the distribution function of the time it takes for a randomly selected student to finish the aptitude test.
(c) What is the probability that a student will finish the aptitude test in less than 75 minutes? Between $1\frac{1}{2}$ and 2 hours?

4. The lifetime of a tire selected randomly from a used tire shop is $10,000X$ miles, where X is a random variable with the density function

$$f(x) = \begin{cases} \dfrac{2}{x^2} & \text{if } 1 < x < 2 \\ 0 & \text{elsewhere.} \end{cases}$$

(a) What percentage of the tires of this shop last fewer than 15,000 miles?
(b) What percentage of those having lifetimes fewer than 15,000 miles last between 10,000 and 12,500 miles?

5. The probability density function of a random variable X is given by

$$f(x) = \begin{cases} \dfrac{c}{\sqrt{1-x^2}} & \text{if } -1 < x < 1 \\ 0 & \text{elsewhere.} \end{cases}$$

(a) Calculate the value of c.
(b) Find the probability distribution function of X.

6. Let X be a continuous random variable with density and distribution functions f and F, respectively. Assuming that $\alpha \in \mathbf{R}$ is a point at which $P(X \le \alpha) < 1$, prove that

$$h(x) = \begin{cases} \dfrac{f(x)}{1 - F(\alpha)} & \text{if } x \ge \alpha \\ 0 & \text{if } x < \alpha \end{cases}$$

is also a probability density function.

7. Let X be a continuous random variable with density function f. We say that X is symmetric about α if for all x

$$P(X \geq \alpha + x) = P(X \leq \alpha - x).$$

(a) Prove that X is symmetric about α if and only if for all x, we have $f(\alpha - x) = f(\alpha + x)$.

(b) Let X be a continuous random variable with probability density function

$$f(x) = \frac{1}{\sqrt{2\pi}} e^{-(x-3)^2/2}, \qquad x \in \mathbf{R},$$

and Y be a continuous random variable with probability density function

$$g(x) = \frac{1}{\pi[1 + (x-1)^2]}, \qquad x \in \mathbf{R}.$$

Find the points about which X and Y are symmetric.

8. Suppose that the loss in a certain investment, in thousands of dollars, is a continuous random variable X that has a density function of the form

$$f(x) = \begin{cases} k(2x - 3x^2) & -1 < x < 0 \\ 0 & \text{elsewhere.} \end{cases}$$

(a) Calculate the value of k.

(b) Find the probability that the loss is at most \$500.

9. Let X denote the lifetime of a radio, in years, manufactured by a certain company. The density function of X is given by

$$f(x) = \begin{cases} \dfrac{1}{15} e^{-x/15} & \text{if } 0 \leq x < \infty \\ 0 & \text{elsewhere.} \end{cases}$$

What is the probability that, of eight such radios, at least four last more than 15 years?

10. Prove that if f and g are two probability density functions, then for $\alpha \geq 0$, $\beta \geq 0$, and $\alpha + \beta = 1$, $\alpha f + \beta g$ is also a probability density function.

11. The distribution function of a random variable X is given by

$$F(x) = \alpha + \beta \arctan \frac{x}{2}, \qquad -\infty < x < \infty.$$

Determine the constants α and β and the density function of X.

6.2 Density Function of a Function of a Random Variable

In the theory and applications of probability, we are often faced with situations in which f, the density function of a random variable X, is known but the density function for some function of X, $h(X)$, is needed. In such cases there are two common methods for the evaluation of the density function of $h(X)$ that we will explain in this section. One method is to find the density of $h(X)$ by calculating its distribution function. Another method is to find it directly from the density function of X. We now explain the first method, called *the method of distribution functions*. To calculate the probability density function of $h(X)$, we calculate the distribution function of $h(X)$, $P\{h(X) \leq t\}$, by finding a set A for which $h(X) \leq t$ if and only if $X \in A$ and then computing $P(X \in A)$ from the distribution function of X. When $P\{h(X) \leq t\}$ is found, the density function of $h(X)$, if it exists, is obtained by differentiation. Some examples follow.

Example 6.3 Let X be a continuous random variable with the probability density function

$$f(x) = \begin{cases} \dfrac{2}{x^2} & \text{if } 1 < x < 2 \\ 0 & \text{elsewhere.} \end{cases}$$

Find the distribution and the density functions of $Y = X^2$.

SOLUTION: Let G and g be the distribution function and the density function of Y, respectively. By definition,

$$G(t) = P(Y \leq t) = P(X^2 \leq t) = P(-\sqrt{t} \leq X \leq \sqrt{t})$$

$$= \begin{cases} 0 & t < 1 \\ P(1 \leq X \leq \sqrt{t}) & 1 \leq t < 4 \\ 1 & t \geq 4. \end{cases}$$

Now

$$P(1 \leq X \leq \sqrt{t}) = \int_1^{\sqrt{t}} \frac{2}{x^2}\, dx = \left[-\frac{2}{x}\right]_1^{\sqrt{t}} = 2 - \frac{2}{\sqrt{t}}.$$

Therefore,

$$G(t) = \begin{cases} 0 & t < 1 \\ 2 - \dfrac{2}{\sqrt{t}} & 1 \leq t < 4 \\ 1 & t \geq 4. \end{cases}$$

g, the density function of Y, is found by differentiation of G:

$$g(t) = G'(t) = \begin{cases} \dfrac{1}{t\sqrt{t}} & \text{if } 1 \leq t \leq 4 \\ 0 & \text{elsewhere.} \end{cases} \quad \blacklozenge$$

Example 6.4   Let X be a continuous random variable with distribution function F and probability density function f. Find the distribution and the density functions of $Y = X^3$.

SOLUTION: Let G and g be the distribution function and the density function of Y. Then

$$G(t) = P(Y \leq t) = P(X^3 \leq t) = P(X \leq \sqrt[3]{t}) = F(\sqrt[3]{t}).$$

Hence

$$g(t) = G'(t) = \frac{1}{3\sqrt[3]{t^2}} F'(\sqrt[3]{t}) = \frac{1}{3\sqrt[3]{t^2}} f(\sqrt[3]{t}),$$

where g is defined at all points $t \neq 0$ at which f is defined. $\blacklozenge$

Example 6.5   The error of a measurement has the density function

$$f(x) = \begin{cases} \dfrac{1}{2} & \text{if } -1 < x < 1 \\ 0 & \text{otherwise.} \end{cases}$$

Find the distribution and the density functions of the magnitude of the error.

SOLUTION: Let X be the error of the measurement. We want to find G, the distribution, and g, the density functions of $|X|$, the magnitude of the error. By definition,

$$G(t) = P(|X| \le t) = P(-t \le X \le t)$$

$$= \begin{cases} 0 & t < 0 \\ \int_{-t}^{t} \frac{1}{2} \, dx = t & 0 \le t < 1 \\ 1 & t \ge 1. \end{cases}$$

The density of Y, $g(t)$, is obtained by differentiation of G:

$$g(t) = G'(t) = \begin{cases} 1 & \text{if } 0 \le t \le 1 \\ 0 & \text{elsewhere.} \end{cases} \quad \blacklozenge$$

In Examples 6.3, 6.4, and 6.5, the method used to find the density function of $h(X)$, a function of the random variable X, is to determine the distribution of $h(X)$ and then differentiate it. It is also possible to find the density function of $h(X)$ without obtaining its distribution function. We now explain this method, called *the method of transformations*.

Recall that a real-valued function $h : \mathbf{R} \to \mathbf{R}$ defined by $y = h(x)$ is *invertible* if and only if when solving $y = h(x)$ for x, we find that $x = g(y)$ is itself a function. That is, $y = h(x)$ if and only if $x = g(y)$, where $x = g(y)$ is a function. For example, the function $y = h(x) = x^3$ is invertible because if we solve it for x, we obtain $x = g(y) = \sqrt[3]{y}$, which is itself a function. The function $y = h(x) = x^2$ is not invertible if its domain is $(-\infty, \infty)$. When solving for x we obtain $x = \pm\sqrt{y}$, which is not a function. The function $y = x^2$ will be invertible if we restrict its domain to $[0, \infty)$, the set of nonnegative numbers. This follows since solving $y = x^2$ for x gives $x = \sqrt{y}$, which is a function. Other examples of invertible functions are $h(x) = e^x$, $x \in \mathbf{R}$; $h(x) = \ln x$, $x \in (0, \infty)$; and $h(x) = \sin x$, $x \in (0, \pi/2)$. Examples of functions that are not invertible are $h(x) = \sin x$, $x \in \mathbf{R}$; $h(x) = x^2 + 5$, $x \in \mathbf{R}$; and $h(x) = \exp(x^2)$, $x \in \mathbf{R}$.

In calculus, we have seen that a continuous function is invertible if and only if it is either strictly increasing or strictly decreasing. If it is strictly increasing, its inverse is also strictly increasing. If it is strictly decreasing, its inverse is strictly decreasing as well. We use these to prove the next theorem, which enables us to find the density function of $h(X)$ mainly by finding the inverse of h. For this application, all we need to know is how to find the inverse of a function, if it exists. Studying different criteria for invertibility does not serve our purpose.

THEOREM 6.1   (Method of Transformations)   *Let X be a continuous random variable with density function f_X and set of possible values A. For the invertible function $h : A \to \mathbf{R}$, let $Y = h(X)$ be a random variable with the set of possible values $B = h(A) = \{h(a) : a \in A\}$.*

Suppose that the inverse of $y = h(x)$ is the function $x = g(y)$, which is differentiable for all values of $y \in B$. Then f_Y, the density function of Y, is given by

$$f_Y(y) = f_X(g(y))|g'(y)|, \qquad y \in B.$$

PROOF: Let F_X and F_Y be distribution functions of X and $Y = h(X)$, respectively. Differentiability of g implies that it is continuous. Since a continuous invertible function is strictly monotone, g is either strictly increasing or strictly decreasing. If it is strictly increasing, $g'(y) > 0$, and hence $g'(y) = |g'(y)|$. Moreover, in this case h is also strictly increasing, so

$$F_Y(y) = P\{h(X) \le y\} = P\{X \le g(y)\} = F_X(g(y)).$$

Differentiating this by chain rule, we obtain

$$F_Y'(y) = g'(y)F_X'(g(y)) = |g'(y)|f_X(g(y)),$$

which gives the theorem. If g is strictly decreasing, $g'(y) < 0$ and hence $|g'(y)| = -g'(y)$. In this case h is also strictly decreasing and we get

$$F_Y(y) = P\{h(X) \le y\} = P\{X \ge g(y)\} = 1 - F_X(g(y)).$$

Differentiating this by chain rule, we find that

$$F_Y'(y) = -g'(y)F_X'(g(y)) = |g'(y)|f_X(g(y)),$$

showing that the theorem is valid in this case as well. ♦

Example 6.6   Let X be a random variable with the density function

$$f_X(x) = \begin{cases} 2e^{-2x} & \text{if } x > 0 \\ 0 & \text{otherwise.} \end{cases}$$

Using the method of transformations, find the probability density function of $Y = \sqrt{X}$.

SOLUTION: The set of possible values of X is $A = (0, \infty)$. Let $h : (0, \infty) \to \mathbf{R}$ be defined by $h(x) = \sqrt{x}$. We want to find the density function of $Y = h(X) = \sqrt{X}$. The set of possible values of $h(X)$ is $B = \{h(a) : a \in A\} = (0, \infty)$. The function h is invertible with the inverse $x = g(y) = y^2$, which is differentiable and its derivative is $g'(y) = 2y$. Therefore, by Theorem 6.1,

$$f_Y(y) = f_X(g(y))|g'(y)| = 2e^{-2y^2}|2y|, \qquad y \in B = (0, \infty).$$

Since $y \in (0, \infty)$, $|2y| = 2y$ and we find that

$$f_Y(y) = \begin{cases} 4ye^{-2y^2} & \text{if } y > 0 \\ 0 & \text{otherwise,} \end{cases}$$

is the density function of $Y = \sqrt{X}$. ◆

Example 6.7   Let X be a continuous random variable with the probability density function

$$f_X(x) = \begin{cases} 4x^3 & \text{if } 0 < x < 1 \\ 0 & \text{otherwise.} \end{cases}$$

Using the method of transformations, find the probability density function of $Y = 1 - 3X^2$.

SOLUTION: The set of possible values of X is $A = (0, 1)$. Let $h: (0, 1) \to \mathbf{R}$ be defined by $h(x) = 1 - 3x^2$. We want to find the density function of $Y = h(X) = 1 - 3X^2$. The set of possible values of $h(X)$ is $B = \{h(a) : a \in A\} = (-2, 1)$. Since the domain of the function h is $(0, 1)$, h is invertible and its inverse is found by solving $1 - 3x^2 = y$ for x, which gives $x = \sqrt{(1 - y)/3}$. Therefore, the inverse of h is $x = g(y) = \sqrt{(1 - y)/3}$, which is differentiable and its derivative is $g'(y) = -1/(2\sqrt{3(1 - y)})$. Using Theorem 6.1, we find that

$$f_Y(y) = f_X(g(y))|g'(y)| = 4\left(\sqrt{\frac{1-y}{3}}\right)^3 \left| -\frac{1}{2\sqrt{3(1-y)}} \right| = \frac{2}{9}(1 - y)$$

is the density function of Y when $y \in (-2, 1)$. ◆

EXERCISES

A

1. Let X be a continuous random variable with the density function

$$f(x) = \begin{cases} \dfrac{1}{4} & \text{if } x \in (-2, 2) \\ 0 & \text{otherwise.} \end{cases}$$

Using the method of distribution functions, find the probability density functions of $Y = X^3$ and $Z = X^4$.

2. Let X be a continuous random variable with distribution function F and density function f. Calculate the density function of the random variable $Y = e^X$.

3. Let the density function of X be

$$f(x) = \begin{cases} e^{-x} & \text{if } x > 0 \\ 0 & \text{elsewhere.} \end{cases}$$

Using the method of transformations, find the density functions of $Y = X\sqrt{X}$ and $Z = e^{-X}$.

4. Let X be a continuous random variable with the density function

$$f(x) = \begin{cases} 3e^{-3x} & \text{if } x > 0 \\ 0 & \text{otherwise.} \end{cases}$$

Using the method of transformations, find the probability density function of $Y = \log_2 X$.

5. Let the probability density function of X be

$$f(x) = \begin{cases} \lambda e^{-\lambda x} & \text{if } x \geq 0 \\ 0 & \text{otherwise,} \end{cases}$$

for some $\lambda > 0$. Using the method of distribution functions, calculate the probability density function of $Y = \sqrt[3]{X^2}$.

6. Let f be the probability density function of a random variable X. In terms of f, calculate the probability density function of X^2.

7. Let X be a random variable with the density function

$$f(x) = \frac{1}{\pi(1 + x^2)}, \qquad -\infty < x < \infty.$$

(X is called a Cauchy random variable.) Find the density function of $Z = \arctan X$.

B

8. Let X be a random variable with the probability density function given by

$$f(x) = \begin{cases} e^{-x} & \text{if } x \geq 0 \\ 0 & \text{elsewhere.} \end{cases}$$

Let

$$Y = \begin{cases} X & \text{if } X \leq 1 \\ \dfrac{1}{X} & \text{if } X > 1. \end{cases}$$

Find the probability density function of Y.

6.3   Expectations and Variances

Expectations of Continuous Random Variables

Let X be a continuous random variable with probability density and distribution functions f and F, respectively. To define $E(X)$, the average or expected value of X, first suppose that X only takes values from the interval $[a, b]$ and divide $[a, b]$ into n subintervals of equal lengths. Let $h = (b - a)/n$, $x_0 = a$, $x_1 = a + h$, $x_2 = a + 2h$, ..., $x_n = a + nh = b$. Then $a = x_0 < x_1 < x_2 < \cdots < x_n = b$ is a partition of $[a, b]$. Since F is continuous, let us assume that it is differentiable on (a, b). By the mean value theorem (of calculus), there exists $t_i \in (x_{i-1}, x_i)$ such that

$$F(x_i) - F(x_{i-1}) = F'(t_i)(x_i - x_{i-1}), \qquad 1 \leq i \leq n,$$

or equivalently,

$$P(x_{i-1} < X \leq x_i) = f(t_i)h, \qquad 1 \leq i \leq n. \tag{6.6}$$

If n is sufficiently large ($n \to \infty$), then h, the width of the intervals, is sufficiently small ($h \to 0$) and $f(x)$ does not vary appreciably over any subinterval of $(x_{i-1}, x_i]$, $1 \leq i \leq n$. Thus t_i is an approximate value of X in the interval $(x_{i-1}, x_i]$. Now $\sum_{i=1}^{n} t_i P(x_{i-1} < X \leq x_i)$ finds the product of an approximate value of X when it is in $(x_{i-1}, x_i]$ and the probability that it is in $(x_{i-1}, x_i]$, and then sums over all these intervals. From the concept of expectation of a discrete random variable, it is clear that as the lengths of these intervals get smaller and smaller, $\sum_{i=1}^{n} t_i P(x_{i-1} < X \leq x_i)$ gets closer and closer to the "average" value of X. So it is desirable to define $E(X)$ as

$$\lim_{n \to \infty} \sum_{i=1}^{n} t_i P(x_{i-1} < X \leq x_i).$$

But by (6.6) this is the same as $\lim_{n\to\infty} \sum_{i=1}^{n} t_i f(t_i)h$, where this limit is equal to $\int_a^b x f(x)\,dx$ as known from calculus. If X is not restricted to an interval $[a, b]$, a definition for $E(X)$ is motivated in the same way, but at the end the limit is taken as $a \to \infty$ and $b \to \infty$.

DEFINITION *If X is a continuous random variable with probability density function f, the expected value of X is defined by*

$$E(X) = \int_{-\infty}^{\infty} x f(x)\,dx.$$

The expected value of X is also called the mean, or mathematical expectation, or simply the expectation of X, and as in the discrete case, sometimes it is denoted by EX, $E[X]$, μ, or μ_X.

To get a geometric feeling for mathematical expectation, consider a piece of cardboard of uniform density on which the graph of the density function f of a random variable X is drawn. Suppose that the cardboard is cut along the graph of f and we are asked to balance it on a given edge perpendicular to the x-axis. Then to have it in equilibrium, we must balance the cardboard on the given edge at the point $x = E(X)$ on the x-axis.

Example 6.8 In a group of adult males, the difference between the uric acid value and 6, the standard value, is a random variable X with the following probability density function:

$$f(x) = \begin{cases} -\dfrac{1}{2}(2x - 3x^2) & \text{if } -1 < x < 1 \\ 0 & \text{elsewhere.} \end{cases}$$

Calculate the mean of these differences for the group.

SOLUTION: By definition,

$$E(X) = \int_{-\infty}^{\infty} x f(x)\,dx = \int_{-1}^{1} -\frac{1}{2}(2x^2 - 3x^3)\,dx$$

$$= -\frac{1}{2}\left[\frac{2}{3}x^3 - \frac{3}{4}x^4\right]_{-1}^{1} = -\frac{2}{3}. \quad \blacklozenge$$

REMARK 1: If X is a continuous random variable with density function f, X is said to have a *finite expected value* if

$$\int_{-\infty}^{\infty} |x| f(x)\, dx < \infty;$$

that is, X has a finite expected value if the integral of $xf(x)$ converges absolutely. Otherwise, we say that the expected value of X is not finite. We now justify why the absolute convergence of the integral is required. Note that

$$E(X) = \int_{-\infty}^{\infty} xf(x)\, dx = \int_{-\infty}^{0} xf(x)\, dx + \int_{0}^{\infty} xf(x)\, dx$$

$$= -\int_{-\infty}^{0} (-x)f(x)\, dx + \int_{0}^{\infty} xf(x)\, dx,$$

where $\int_{-\infty}^{0} (-x)f(x)\, dx \geq 0$ and $\int_{0}^{\infty} xf(x)\, dx \geq 0$. Thus $E(X)$ is well defined if $\int_{-\infty}^{0} (-x)f(x)\, dx$ and $\int_{0}^{\infty} xf(x)\, dx$ are not both ∞. Moreover, $E(X) < \infty$ if neither of these two integrals is $+\infty$. Hence a necessary and sufficient condition for $E(X)$ to exist and to be finite is that $\int_{-\infty}^{0} (-x)f(x)\, dx < \infty$ and $\int_{0}^{\infty} xf(x)\, dx < \infty$. Since both of these integrals are finite if and only if

$$\int_{-\infty}^{\infty} |x| f(x)\, dx = \int_{-\infty}^{0} (-x)f(x)\, dx + \int_{0}^{\infty} xf(x)\, dx$$

is finite, we have that $E(X)$ is well defined and finite if and only if the integral $\int_{-\infty}^{\infty} xf(x)\, dx$ is absolutely convergent. ◆

Example 6.9   A random variable X with density function

$$f(x) = \frac{c}{1 + x^2}, \qquad -\infty < x < \infty,$$

is called a Cauchy random variable.

(a) Find c.
(b) Show that $E(X)$ does not exist.

SOLUTION: (a) Since f is a density function $\int_{-\infty}^{\infty} f(x)\, dx = 1$. Thus

$$\int_{-\infty}^{\infty} \frac{c}{1 + x^2}\, dx = c \int_{-\infty}^{\infty} \frac{dx}{1 + x^2} = 1.$$

Now

$$\int \frac{dx}{1+x^2} = \arctan x.$$

Since the range of $\arctan x$ is $(-\pi/2, +\pi/2)$, we get

$$1 = c \int_{-\infty}^{\infty} \frac{dx}{1+x^2} = c \left[\arctan x\right]_{-\infty}^{\infty} = c \left[\frac{\pi}{2} - \left(-\frac{\pi}{2}\right)\right] = c\pi.$$

Thus $c = 1/\pi$.

(b) To show that $E(X)$ does not exist, note that

$$\int_{-\infty}^{\infty} |x| f(x) \, dx = \int_{-\infty}^{\infty} \frac{|x| \, dx}{\pi(1+x^2)} = 2 \int_{0}^{\infty} \frac{x \, dx}{\pi(1+x^2)}$$

$$= \frac{1}{\pi} \left[\ln(1+x^2)\right]_{0}^{\infty} = \infty. \quad \blacklozenge$$

REMARK 2: In this book, unless otherwise specified, it is implicitly assumed that the expectation of a random variable is finite. $\blacklozenge$

The following theorem directly relates the distribution function of a random variable to its expectation. It enables us to find the expected value of a continuous random variable without calculating its probability density function. It also has important theoretical applications.

THEOREM 6.2   *For any continuous random variable X with probability distribution function F and density function f,*

$$E(X) = \int_{0}^{\infty} [1 - F(t)] \, dt - \int_{0}^{\infty} F(-t) \, dt.$$

PROOF: Note that

$$E(X) = \int_{-\infty}^{\infty} xf(x) \, dx = \int_{-\infty}^{0} xf(x) \, dx + \int_{0}^{\infty} xf(x) \, dx$$

$$= -\int_{-\infty}^{0} \left(\int_{0}^{-x} dt\right) f(x) \, dx + \int_{0}^{\infty} \left(\int_{0}^{x} dt\right) f(x) \, dx$$

$$= -\int_{0}^{\infty} \left(\int_{-\infty}^{-t} f(x) \, dx\right) dt + \int_{0}^{\infty} \left(\int_{t}^{\infty} f(x) \, dx\right) dt,$$

where the last equality is obtained by changing the order of integration. The theorem follows since $\int_{-\infty}^{-t} f(x) \, dx = F(-t)$ and $\int_{t}^{\infty} f(x) \, dx = P(X > t) = 1 - F(t)$. $\blacklozenge$

REMARK 3: In the proof of this theorem we assumed that the random variable X is continuous. Even without this condition the theorem is still valid. Also note that since $1 - F(t) = P(X > t)$, this theorem may be stated as follows.

For any random variable X,

$$E(X) = \int_0^\infty P(X > t)\, dt - \int_0^\infty P(X \le -t)\, dt.$$

In particular, if X is nonnegative, that is, $P(X < 0) = 0$, this theorem states that

$$E(X) = \int_0^\infty P(X > t)\, dt. \quad \blacklozenge$$

As an important application of Theorem 6.2, we now prove the law of unconscious statistician, Theorem 4.2, for continuous random variables.

THEOREM 6.3   *Let X be a continuous random variable with probability density function $f(x)$; then for any function $h: \mathbf{R} \to \mathbf{R}$,*

$$E[h(X)] = \int_{-\infty}^\infty h(x) f(x)\, dx.$$

PROOF: Let

$$h^{-1}(t, \infty) = \{x : h(x) \in (t, \infty)\} = \{x : h(x) > t\}$$

with similar representation for $h^{-1}(-\infty, -t)$. Notice that we are not claiming that h has an inverse function. We are simply considering the set $\{x : h(x) \in (t, \infty)\}$, which is called the *inverse image* of (t, ∞) and is denoted by $h^{-1}(t, \infty)$. By Theorem 6.2,

$$E[h(X)] = \int_0^\infty P\{h(X) > t\}\, dt - \int_0^\infty P\{h(X) \le -t\}\, dt$$

$$= \int_0^\infty P\{X \in h^{-1}(t, \infty)\}\, dt - \int_0^\infty P\{X \in h^{-1}(-\infty, -t]\}\, dt$$

$$= \int_0^\infty \left(\int_{\{x : x \in h^{-1}(t, \infty)\}} f(x)\, dx \right) dt$$

$$\quad - \int_0^\infty \left(\int_{\{x : x \in h^{-1}(-\infty, -t]\}} f(x)\, dx \right) dt$$

$$= \int_0^\infty \left(\int_{\{x : h(x) > t\}} f(x)\, dx \right) dt - \int_0^\infty \left(\int_{\{x : h(x) \le -t\}} f(x)\, dx \right) dt.$$

Now we change the order of integration for both of these double integrals. Since

$$\{(t, x) : 0 < t < \infty, \ h(x) > t\} = \{(t, x) : h(x) > 0, \ 0 < t < h(x)\},$$

and

$$\{(t, x) : 0 < t < \infty, \ h(x) \leq -t\} = \{(t, x) : h(x) < 0, \ 0 < t \leq -h(x)\},$$

we get

$$
\begin{aligned}
E[h(X)] &= \int_{\{x:h(x)>0\}} \left(\int_0^{h(x)} dt \right) f(x)\, dx \\
&\quad - \int_{\{x:h(x)<0\}} \left(\int_0^{-h(x)} dt \right) f(x)\, dx \\
&= \int_{\{x:h(x)>0\}} h(x) f(x)\, dx + \int_{\{x:h(x)<0\}} h(x) f(x)\, dx \\
&= \int_{-\infty}^{\infty} h(x) f(x)\, dx.
\end{aligned}
$$

Note that the last equality follows because $\int_{\{x:h(x)=0\}} h(x) f(x)\, dx = 0.$ ◆

COROLLARY   *Let X be a continuous random variable with probability density function $f(x)$. Let $h_1, h_2, \ldots, h_n$ be real-valued functions, and $\alpha_1, \alpha_2, \ldots, \alpha_n$ be real numbers. Then*

$$
\begin{aligned}
E[\alpha_1 h_1(X) + \alpha_2 h_2(X) + \cdots + \alpha_n h_n(X)] \\
= \alpha_1 E[h_1(X)] + \alpha_2 E[h_2(X)] + \cdots + \alpha_n E[h_n(X)].
\end{aligned}
$$

PROOF: In the discrete case, in the proof of the corollary of Theorem 4.2, replace $\sum$ by $\int$, and $p(x)$ by $f(x)\, dx$. ◆

Just as in the discrete case, by this corollary we can write that, for example,

$$E(3X^4 + \cos X + 3e^X + 7) = 3E(X^4) + E(\cos X) + 3E(e^X) + 7.$$

Moreover, this corollary implies that if α and β are constants, then

$$E(\alpha X + \beta) = \alpha E(X) + \beta.$$

Example 6.10   A point X is selected from the interval $(0, \pi/4)$ randomly. Calculate $E(\cos 2X)$ and $E(\cos^2 X)$.

SOLUTION: First we calculate the distribution function of X. Clearly,

$$F(t) = P(X \leq t) = \begin{cases} 0 & t < 0 \\[2mm] \dfrac{t-0}{\dfrac{\pi}{4} - 0} = \dfrac{4t}{\pi} & 0 \leq t < \dfrac{\pi}{4} \\[4mm] 1 & t \geq \dfrac{\pi}{4}. \end{cases}$$

Thus f, the probability density function of X, is

$$f(t) = \begin{cases} \dfrac{4}{\pi} & 0 < t < \dfrac{\pi}{4} \\[3mm] 0 & \text{otherwise.} \end{cases}$$

Now, by Theorem 6.3,

$$E(\cos 2X) = \int_0^{\pi/4} \frac{4}{\pi} \cos 2x \, dx = \left[\frac{2}{\pi} \sin 2x \right]_0^{\pi/4} = \frac{2}{\pi}.$$

To calculate $E(\cos^2 X)$, note that $\cos^2 X = (1 + \cos 2X)/2$. So by the corollary of Theorem 6.3,

$$E(\cos^2 X) = E\left(\frac{1}{2} + \frac{1}{2} \cos 2X \right) = \frac{1}{2} + \frac{1}{2} E(\cos 2X)$$

$$= \frac{1}{2} + \frac{1}{2} \cdot \frac{2}{\pi} = \frac{1}{2} + \frac{1}{\pi}. \quad \blacklozenge$$

Variances of Continuous Random Variables

DEFINITION   *If X is a continuous random variable with $E(X) = \mu$, then σ_X and $Var(X)$, called the standard deviation and the variance of X, respectively, are defined by*

$$\sigma_X = \sqrt{E\left[(X - \mu)^2 \right]}$$

$$Var(X) = E\left[(X - \mu)^2 \right].$$

Therefore, if f is the density function of X, then by Theorem 6.3,

$$\text{Var(X)} = E\left[(X - \mu)^2\right] = \int_{-\infty}^{\infty} (x - \mu)^2 f(x)\, dx.$$

Also, as before, we have the following important relations whose proofs are analogous to those in the discrete case.

$$\text{Var}(X) = E(X^2) - [E(X)]^2,$$

$$\text{Var}(aX + b) = a^2\text{Var}(X), \quad \sigma_{aX+b} = |a|\sigma_X, \qquad a \text{ and } b \text{ being constants.}$$

Example 6.11 The time elapsed, in minutes, between the placement of an order of pizza and its delivery is random with the density function

$$f(x) = \begin{cases} \dfrac{1}{15} & \text{if } 25 < x < 40 \\ 0 & \text{otherwise.} \end{cases}$$

(a) Determine the mean and standard deviation of the time it takes for the pizza shop to deliver pizza.

(b) Suppose that it takes 12 minutes for the pizza shop to bake pizza. Determine the mean and standard deviation of the time it takes for the delivery person to deliver pizza.

SOLUTION: **(a)** Let the time elapsed between the placement of an order and its delivery be X minutes. Then

$$E(X) = \int_{25}^{40} x\frac{1}{15}dx = 32.5,$$

$$E(X^2) = \int_{25}^{40} x^2\frac{1}{15}dx = 1075.$$

Therefore, $\text{Var}(X) = 1075 - (32.5)^2 = 18.75$ and hence $\sigma_X = \sqrt{\text{Var}(X)} = 4.33$.

(b) The time it takes for the delivery person to deliver pizza is $X - 12$. Therefore, the desired quantities are

$$E(X - 12) = E(X) - 12 = 32.5 - 12 = 20.5$$

$$\sigma_{X-12} = |1|\sigma_X = \sigma_X = 4.33. \quad \blacklozenge$$

EXERCISES

A

1. The distribution function for the duration of a certain soap opera (in tens of hours) is

$$F(x) = \begin{cases} 1 - \dfrac{16}{x^2} & \text{if } x \geq 4 \\ 0 & \text{if } x < 4. \end{cases}$$

   (a) Find $E(X)$.
   (b) Show that $\text{Var}(X)$ does not exist.

2. The time it takes for a student to finish an aptitude test (in hours) has the density function

$$f(x) = \begin{cases} 6(x - 1)(2 - x) & \text{if } 1 < x < 2 \\ 0 & \text{otherwise.} \end{cases}$$

   Determine the mean and standard deviation of the time it takes for a randomly selected student to finish the aptitude test.

3. The mean and standard deviation of the lifetime of a car muffler manufactured by company A are five and two years, respectively. These quantities for car mufflers manufactured by company B are, respectively, four years and 18 months. Brian buys one muffler from company A and one from company B. That of company A lasts four years and three months, and that of company B lasts three years and nine months. Determine which of these mufflers has performed relatively better?
   HINT: Find the standardized lifetimes of the mufflers and compare (see Section 4.6).

4. A random variable X has the density function

$$f(x) = \begin{cases} 3e^{-3x} & \text{if } 0 \leq x < \infty \\ 0 & \text{otherwise.} \end{cases}$$

   Calculate $E(e^X)$.

5. Find the expected value of a random variable X with the density function

$$f(x) = \begin{cases} \dfrac{1}{\pi\sqrt{1 - x^2}} & \text{if } -1 < x < 1 \\ 0 & \text{otherwise.} \end{cases}$$

6. Let the probability density function of tomorrow's Celsius temperature be h. In terms of h, calculate the corresponding probability density function and its expectation for Fahrenheit temperature.

   HINT: Let C and F be tomorrow's temperature in Celsius and Fahrenheit, respectively. Then $F = 1.8C + 32$.

7. Let X be a continuous random variable with probability density function

$$f(x) = \begin{cases} \dfrac{2}{x^2} & \text{if } 1 < x < 2 \\ 0 & \text{elsewhere.} \end{cases}$$

   Find $E(\ln X)$.

8. A right triangle has a hypotenuse of length 9. If the probability density function of one side's length is given by

$$f(x) = \begin{cases} \dfrac{x}{6} & \text{if } 2 < x < 4 \\ 0 & \text{otherwise,} \end{cases}$$

   what is the expected value of the length of the other side?

9. Let X be a random variable with probability density function

$$f(x) = \frac{1}{2}e^{-|x|}, \qquad -\infty < x < \infty.$$

   Calculate $\text{Var}(X)$.

B

10. Let X be a random variable with the probability density function

$$f(x) = \frac{1}{\pi(1 + x^2)}, \qquad -\infty < x < \infty.$$

    Prove that $E(|X|^\alpha)$ converges if $0 < \alpha < 1$ and diverges if $\alpha \geq 1$.

11. Suppose that X, the interarrival time between two customers entering a certain post office, satisfies

$$P(X > t) = \alpha e^{-\lambda t} + \beta e^{-\mu t}, \qquad t \geq 0,$$

    where $\alpha + \beta = 1$, $\alpha \geq 0$, $\beta \geq 0$, $\lambda > 0$, $\mu > 0$. Calculate the expected value of X.

HINT: For a fast calculation, use Remark 3 following Theorem 6.2.

12. For $n \geq 1$, let X_n be a continuous random variable with the probability density function

$$f_n(x) = \begin{cases} \dfrac{c_n}{x^{n+1}} & \text{if } x \geq c_n \\ 0 & \text{otherwise.} \end{cases}$$

X_n's are called *Pareto* random variables and are used to study income distributions.

(a) Calculate c_n, $n \geq 1$.
(b) Find $E(X_n)$, $n \geq 1$.
(c) Determine the density function of $Z_n = \ln X_n$, $n \geq 1$.
(d) For what values of m does $E(X_n^{m+1})$ exist?

13. Let X be a continuous random variable with the probability density function

$$f(x) = \begin{cases} \dfrac{1}{\pi} x \sin x & \text{if } 0 < x < \pi \\ 0 & \text{otherwise.} \end{cases}$$

Prove that

$$E(X^{n+1}) + (n+1)(n+2)E(X^{n-1}) = \pi^{n+1}.$$

14. Let X be a nonnegative random variable with distribution function F. Define

$$I(t) = \begin{cases} 1 & \text{if } X > t \\ 0 & \text{otherwise.} \end{cases}$$

(a) Prove that $\int_0^\infty I(t)dt = X$.
(b) By calculating the expected value of both sides of part (a), prove that

$$E(X) = \int_0^\infty [1 - F(t)]\, dt.$$

This is a special case of Theorem 6.2.

(c) For $r > 0$, use part (b) to prove that

$$E(X^r) = r \int_0^\infty t^{r-1}[1 - F(t)]\, dt.$$

15. Let X be a continuous random variable. Prove that

$$\sum_{n=1}^{\infty} P(|X| \geq n) \leq E(|X|) \leq 1 + \sum_{n=1}^{\infty} P(|X| \geq n).$$

These important inequalities show that $E(|X|) < \infty$ if and only if the series $\sum_{n=1}^{\infty} P(|X| \geq n)$ converges.

HINT: By Exercise 14,

$$E(|X|) = \int_{0}^{\infty} P(|X| > t)\, dt = \sum_{n=0}^{\infty} \int_{n}^{n+1} P(|X| > t)\, dt.$$

Note that on the interval $[n, n+1)$,

$$P(|X| \geq n+1) < P(|X| > t) \leq P(|X| \geq n).$$

16. Let X be the random variable introduced in Exercise 11. Applying the results of Exercise 14, calculate Var(X).

Review Problems

1. Let X be a random number from $(0, 1)$. Find the probability density function of $Y = 1/X$.

2. Let X be a continuous random variable with the probability density function

$$f(x) = \begin{cases} \dfrac{2}{x^3} & \text{if } x > 1 \\ 0 & \text{otherwise.} \end{cases}$$

Find $E(X)$ and Var(X) if they exist.

3. Let X be a continuous random variable with density function $f(x) = 6x(1 - x)$, $0 < x < 1$. What is the probability that X is within 2 standard deviations of the mean?

4. Let X be a random variable with density function

$$f(x) = \frac{e^{-|x|}}{2}, \qquad -\infty < x < \infty.$$

Find $P(-2 < X < 1)$.

5. Does there exist a constant c for which the following is a density function?

$$f(x) = \begin{cases} \dfrac{c}{1+x} & \text{if } x > 0 \\ 0 & \text{otherwise.} \end{cases}$$

6. Let X be a random variable with density function

$$f(x) = \begin{cases} \dfrac{4x^3}{15} & 1 \le x \le 2 \\ 0 & \text{otherwise.} \end{cases}$$

Find the density functions of $Y = e^X$, $Z = X^2$, and $W = (X - 1)^2$.

7. The probability density function of a continuous random variable X is

$$f(x) = \begin{cases} 30x^2(1 - x)^2 & \text{if } 0 < x < 1 \\ 0 & \text{otherwise.} \end{cases}$$

Find the probability density function of $Y = X^4$.

8. Let F, the distribution of a random variable X be defined by

$$F(x) = \begin{cases} 0 & x < -1 \\ \dfrac{1}{2} + \dfrac{\arcsin x}{\pi} & -1 \le x < 1 \\ 1 & x \ge 1, \end{cases}$$

where $\arcsin x$ lies between $-\pi/2$ and $\pi/2$. Find f, the probability density function of X and $E(X)$.

9. Prove or disprove: If $\sum_{i=1}^{n} \alpha_i = 1$, $\alpha_i \ge 0$, $\forall i$, and $\{f_i\}_{i=1}^{n}$ is a sequence of density functions, then $\sum_{i=1}^{n} \alpha_i f_i$ is a probability density function.

10. Let X be a continuous random variable with set of possible values $\{x : 0 < x < \alpha\}$ (where $\alpha < \infty$), distribution function F, and density function f. Using integration by parts, prove the following special case of Theorem 6.2.

$$E(X) = \int_0^{\alpha} [1 - F(t)] \, dt.$$

11. The lifetime (in hours) of a light bulb manufactured by a certain company is a random variable with probability density function

$$f(x) = \begin{cases} 0 & \text{if } x \leq 500 \\ \dfrac{5 \times 10^5}{x^3} & \text{if } x > 500. \end{cases}$$

Suppose that for all nonnegative real numbers a and b, the event that any light bulb lasts at least a hours is independent of the event that any other light bulb lasts at least b hours. Find the probability that of six such light bulbs selected at random, exactly two last over 1000 hours.

12. Let X be a continuous random variable with distribution function F and density function f. Find the distribution function and the density function of $Y = |X|$.

13. The density function of a continuous random variable X is symmetric about some point α. Show that if $E(X)$ exists, then $E(X) = \alpha$.

Chapter 7

SPECIAL CONTINUOUS DISTRIBUTIONS

In this chapter we study some examples of continuous random variables. These random variables appear frequently in theory and applications of probability, statistics, and branches of science and engineering.

7.1 Uniform Random Variable

In Sections 1.6 and 1.7 we explained that in random selection of a point from an interval (a, b), the probability of the occurrence of any particular point is zero. As a result, we stated that if $[\alpha, \beta] \subseteq (a, b)$, the events that the point falls in $[\alpha, \beta]$, (α, β), $[\alpha, \beta)$, and $(\alpha, \beta]$ are all equiprobable. Moreover, we said that a point is randomly selected from an interval (a, b) if any two of its subintervals which have the same length are equally likely to include the point. We also mentioned that the probability associated with the event that the subinterval (α, β) includes the point is defined to be $(\beta - \alpha)/(b - a)$. Applications of these facts have been discussed throughout the book. Therefore, their significance should be clear by now. In particular, in Chapter 12 we show that the core of computer simulations is selection of random points from intervals. In this section we introduce the concept of a uniform random variable. Then we study its properties and applications. As we will see now, uniform random variables are directly related to random selection of points from intervals.

Suppose that X is the value of the random point selected from an interval (a, b). Then X is called a *uniform random variable over* (a, b). Let F and f be probability distribution and density functions of X, respectively. Clearly,

$$F(t) = \begin{cases} 0 & t \leq a \\ \dfrac{t - a}{b - a} & a \leq t < b \\ 1 & t \geq b. \end{cases}$$

Therefore,

$$f(t) = F'(t) = \begin{cases} \dfrac{1}{b - a} & \text{if } a < t < b \\ 0 & \text{otherwise.} \end{cases} \tag{7.1}$$

DEFINITION   *A random variable X is said to be uniformly distributed over an interval (a, b) if its probability density function is given by (7.1).*

Another way of reaching this definition is to note that $f(x)$ is a measure which determines how likely it is for X to be close to x. Since for all $x \in (a, b)$, the probability that X is close to x is the same, f should be a nonzero constant on (a, b); zero, elsewhere. Therefore,

$$f(x) = \begin{cases} c & \text{if } a < x < b \\ 0 & \text{elsewhere.} \end{cases}$$

Now $\int_a^b f(x)\,dx = 1$ implies that $\int_a^b c\,dx = c(b - a) = 1$. Thus $c = 1/(b - a)$.

Figure 7.1 represents the graphs of f and F, the density function and the distribution function of a uniform random variable over the interval (a, b).

In random selections of a large number of points from (a, b), we expect that the average of the values of the points will be approximately $(a + b)/2$, the midpoint of (a, b). This can be shown by calculating $E(X)$, the expected value of a uniform random variable X over (a, b).

$$\begin{aligned} E(X) &= \int_a^b x\, \frac{1}{b - a}\,dx = \frac{1}{b - a}\left[\frac{1}{2}x^2\right]_a^b = \frac{1}{b - a}\left(\frac{1}{2}b^2 - \frac{1}{2}a^2\right) \\ &= \frac{(b - a)(b + a)}{2(b - a)} = \frac{a + b}{2}. \end{aligned}$$

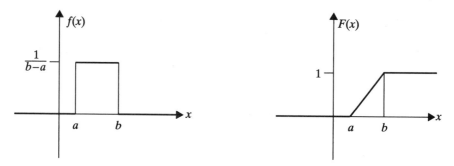

Figure 7.1   Density and distribution functions of a uniform random variable.

To find $\text{Var}(X)$, note that

$$E(X^2) = \int_a^b x^2 \frac{1}{b-a}\,dx = \frac{1}{3}\frac{b^3 - a^3}{b-a} = \frac{1}{3}(a^2 + ab + b^2).$$

Hence

$$\text{Var}(X) = E(X^2) - [E(X)]^2 = \frac{1}{3}(a^2 + ab + b^2) - \left(\frac{a+b}{2}\right)^2 = \frac{(b-a)^2}{12}.$$

It is interesting to note that the expected value and the variance of a randomly selected integer Y from the set $\{1, 2, 3, \ldots, N\}$ are very similar to $E(X)$ and $\text{Var}(X)$ obtained above. By Exercise 5 of Section 4.5, $E(Y) = (1 + N)/2$ and $\text{Var}(Y) = (N^2 - 1)/12$.

Example 7.1   Starting at 5:00 A.M., every half hour there is a flight from San Francisco airport to Los Angeles International airport. Suppose that none of these planes is completely sold out and that they always have room for passengers. A person who wants to fly to L.A. arrives at the airport at a random time between 8:45 A.M. and 9:45 A.M. Find the probability that she waits (a) at most 10 minutes; (b) at least 15 minutes.

SOLUTION: Let the passenger arrive at the airport X minutes past 8:45. Then X is a uniform random variable over the interval $(0, 60)$. Hence the density function of X is given by

$$f(x) = \begin{cases} \dfrac{1}{60} & \text{if } 0 < x < 60 \\ 0 & \text{elsewhere.} \end{cases}$$

Now, the passenger waits at most 10 minutes if she arrives between 8:50 and
9:00 or 9:20 and 9:30. That is, if $5 < X < 15$ or $35 < X < 45$. So the
answer to (a) is

$$P(5 < X < 15) + P(35 < X < 45) = \int_5^{15} \frac{1}{60} dx + \int_{35}^{45} \frac{1}{60} dx = \frac{1}{3}.$$

The passenger waits at least 15 minutes, if she arrives between 9:00 and 9:15
or 9:30 and 9:45. That is, if $15 < X < 30$ or $45 < X < 60$. Thus the answer
to (b) is

$$P(15 < X < 30) + P(45 < X < 60) = \int_{15}^{30} \frac{1}{60} dx + \int_{45}^{60} \frac{1}{60} dx = \frac{1}{2}. \quad \blacklozenge$$

Example 7.2   A person arrives at a bus station every day at 7:00 A.M. If a
bus arrives at a random time between 7:00 A.M. and 7:30 A.M., what is the
average time spent waiting?

SOLUTION: If the bus arrives X minutes past 7, then X is a uniform random
variable over the interval $(0, 30)$. Hence the average waiting time is

$$E(X) = \frac{0 + 30}{2} = 15 \text{ minutes.} \quad \blacklozenge$$

The uniform distribution is often used to solve elementary problems in
geometric probability. In Chapter 8 we solve several such problems. Here
we discuss only a famous problem introduced by the French mathematician
Joseph Bertrand (1822–1900) in 1889. It is called *Bertrand's paradox*.
Bertrand seriously doubted that probability could be defined on infinite
sample spaces. To make his point, he posed this problem:

Example 7.3   What is the probability that a *random chord* of a circle is
longer than a side of an equilateral triangle inscribed into the circle?

SOLUTION: Since the exact meaning of a *random chord* is not given, we
cannot solve this problem as stated. We interpret the expression *random
chord* in three different ways and solve the problem in each case. Let the
center of the circle be at C and its radius be r.
   First interpretation: To draw a random chord, by considerations of
symmetry, first choose a random point A on the circle and connect it to C,
the center, then choose a random number d from $(0, r)$ and place M on AC
so that $\overline{CM} = d$. Finally, from M draw a chord perpendicular to the radius
AC (see Figure 7.2). Since $d < r/2$ if and only if the chord is longer than

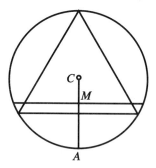

Figure 7.2   First interpretation.

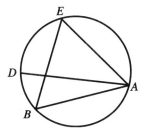

Figure 7.3   Second interpretation.

a side of an equilateral triangle inscribed into the circle and d is uniformly distributed over $(0, r)$, the desired probability is

$$P\left(d < \frac{r}{2}\right) = \frac{r/2}{r} = \frac{1}{2}.$$

Second interpretation: To draw a random chord, by considerations of symmetry, first choose a random point A and then another random point D on the circle and connect AD. Let B and E be the points on the circle that make ABE an equilateral triangle. The random chord AD is longer than a side of ABE if and only if D lies on the arc BE. Since the length of the arc BE is one-third of the length of the whole circle and D is a random point on the circle, D lies on the arc BE with probability $1/3$. Thus the desired probability is $1/3$ as well (see Figure 7.3).

Third interpretation: Since a chord is perpendicular to the radius connecting its midpoint to the center of the circle, every chord is uniquely determined by its midpoint. To draw a random chord, choose a random point M inside the circle, connect it to C, and draw a chord perpendicular to MC from M. It is clear that the chord is longer than a side of the equilateral

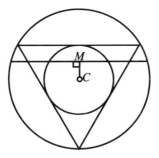

Figure 7.4   Third interpretation.

triangle inscribed into the circle if and only if its midpoint M lies inside
the circle centered at C with radius $r/2$ (see Figure 7.4). Since each choice
of M uniquely determines one choice of a chord, the desired probability is
the area of the small circle divided by the area of the original circle (a fact
discussed in detail in Section 8.1). It is equal to

$$\frac{\pi(r/2)^2}{\pi r^2} = \frac{1}{4}.$$

As we showed, three different interpretations of random chord resulted
in three different answers. Because of this, Bertrand's problem was once
considered a paradox. At that time, one did not pay attention to the fact that
the three interpretations correspond to three different experiments concerning
the selection of a random chord. In this process we are dealing with three
different probability functions defined on the same set of events.   ◆

EXERCISES

A

1. It takes a professor a random time between 20 and 27 minutes to walk
   from his home to school every day. If he has a class at 9:00 A.M. and he
   leaves home at 8:37 A.M., find the probability that he reaches his class
   on time.

2. Suppose that 15 points are selected at random and independently from
   the interval $(0, 1)$. How many of them can be expected to be greater
   than $3/4$?

3. The time at which a bus arrives at a station is uniform over an interval
   (a, b) with mean 2:00 P.M. and variance 12 minutes. Determine the
   values of a and b.

4. Suppose that b is a random number from the interval $(-3, 3)$. What is the probability that the quadratic equation $x^2 + bx + 1 = 0$ has at least one real root?

5. The radius of a sphere is a random number between 2 and 4. What is the expected value of its volume? What is the probability that its volume is at most 36π?

6. A point is selected at random on a line segment of length ℓ. What is the probability that the longer segment is at least twice as long as the shorter segment?

7. A point is selected at random on a line segment of length ℓ. What is the probability that none of the two segments is smaller than $\ell/3$?

8. From the class of all triangles one is selected at random. What is the probability that it is obtuse?
   HINT: The largest angle of a triangle is less than 180 degrees but greater than or equal to 60 degrees.

9. A farmer who has two pieces of lumber of lengths a and b $(a < b)$ decides to build a pen for his chickens in the shape of a triangle. He sends his foolish son out to cut the longer piece and the boy, without taking any thought as to the ultimate purpose, makes a cut on the lumber of length b, at a point selected randomly. What are the chances that these two resulting pieces and the piece of length a can be used to form a triangular pen?
   HINT: Three segments form a triangle if and only if the length of any one of them is less than the sum of the lengths of the remaining two.

10. Let θ be a random number between $-\pi/2$ and $\pi/2$. Find the probability density function of $X = \tan \theta$.

11. Let X be a random number from $[0, 1]$. Find the probability function of $[nX]$, the greatest integer less than or equal to nX.

B

12. Let X be a random number from $(0, 1)$. Find the density functions of (a) $Y = -\ln(1 - X)$, and (b) $Z = X^n$.

13. Let X be a continuous random variable with distribution function F. Prove that $F(X)$ is uniformly distributed over $(0, 1)$.

14. Let g be a nonnegative real-valued function on $\mathbf{R}$ that satisfies the relation $\int_{-\infty}^{\infty} g(t)\, dt = 1$. Show that if for a random variable X, the random variable $Y = \int_{-\infty}^{X} g(t)\, dt$ is uniform, then g is the density function of X.

15. The sample space of an experiment is $S = (0, 1)$, and for every subset A of S, $P(A) = \int_A dx$. Let X be a random variable defined on S by $X(\omega) = 5\omega - 1$. Prove that X is a uniform random variable over the interval $(-1, 4)$.

16. Let Y be a random number from $(0, 1)$. Let X be the second digit of $\sqrt{Y}$. Prove that for $n = 0, 1, 2, \ldots, 9$, $P(X = n)$ increases as n increases. This is remarkable because it shows that $P(X = n)$, $n = 1, 2, 3, \ldots$ is not constant. That is, Y is uniform but X is not.

7.2 Normal Random Variable

In search of formulas to approximate binomial probabilities, Poisson was not alone. Other mathematicians had also realized the importance of such investigations. In 1718, before Poisson, De Moivre had discovered the following approximation, which is completely different from Poisson's.

De Moivre's Theorem *Let X be a binomial random variable with parameters n and $1/2$. Then for any numbers a and b, $a < b$,*

$$\lim_{n \to \infty} P\left(a < \frac{X - (1/2)n}{(1/2)\sqrt{n}} < b\right) = \frac{1}{\sqrt{2\pi}} \int_a^b e^{-t^2/2}\, dt.$$

Note that in this formula $(1/2)n = E(X)$ and $(1/2)\sqrt{n} = \sigma_X$.

De Moivre's theorem was appreciated by Laplace, who recognized its importance. In 1812 he generalized it to binomial random variables with parameters n and p. He showed the following theorem, now called the De Moivre–Laplace theorem.

THEOREM 7.1 (De Moivre–Laplace Theorem) *Let X be a binomial random variable with parameters n and p. Then for any numbers a and b, $a < b$,*

$$\lim_{n \to \infty} P\left(a < \frac{X - np}{\sqrt{np(1 - p)}} < b\right) = \frac{1}{\sqrt{2\pi}} \int_a^b e^{-t^2/2}\, dt.$$

Note that $E(X) = np$ and $\sigma_X = \sqrt{np(1 - p)}$.

Poisson's approximation, as discussed, is good when n is large, p is relatively small, and np is appreciable. The De Moivre–Laplace formula yields excellent approximations for values of n and p for which $np(1 - p) \geq 10$.

To find his approximation, Poisson first considered a binomial random variable X with parameters (n, p). Then he showed that if $\lambda = np$, $P(X = i)$ is approximately $(e^{-\lambda}\lambda^i)/i!$ for large n. Finally, he proved that $(e^{-\lambda}\lambda^i)/i!$ itself is a probability function. The same procedure can be followed for the De Moivre–Laplace approximation as well. By this theorem, if X is a binomial random variable with parameters (n, p), the sequence of probabilities $P\left(\dfrac{X - np}{\sqrt{np(1 - p)}} \leq t\right)$, $n = 1, 2, 3, 4, \ldots$, converges to $(1/\sqrt{2\pi}) \int_{-\infty}^{t} e^{-x^2/2}\, dx$, where the function $\Phi(t) = (1\sqrt{2\pi}) \int_{-\infty}^{t} e^{-x^2/2}\, dx$ is a distribution function itself. To prove that Φ is a distribution function, note that it is increasing, continuous, and $\Phi(-\infty) = 0$. The proof of $\Phi(\infty) = 1$ is tricky. We use the following ingenious technique introduced by Gauss to show it. Let

$$I = \int_{-\infty}^{\infty} e^{-x^2/2}\, dx.$$

Then

$$I^2 = \left(\int_{-\infty}^{\infty} e^{-x^2/2}\, dx\right)\left(\int_{-\infty}^{\infty} e^{-y^2/2}\, dy\right)$$

$$= \int_{-\infty}^{\infty} \int_{-\infty}^{\infty} e^{-(x^2+y^2)/2}\, dx\, dy.$$

To evaluate this double integral, we change the variables to polar coordinates. That is, we let $x = r\cos\theta$, $y = r\sin\theta$. We get $dx\, dy = r\, d\theta\, dr$ and

$$I^2 = \int_{0}^{\infty} \int_{0}^{2\pi} e^{-r^2/2} r\, d\theta\, dr = \int_{0}^{\infty} e^{-r^2/2} r \left(\int_{0}^{2\pi} d\theta\right) dr$$

$$= 2\pi \int_{0}^{\infty} r e^{-r^2/2}\, dr = 2\pi \left[-e^{-r^2/2}\right]_{0}^{\infty} = 2\pi.$$

Thus

$$I = \int_{-\infty}^{\infty} e^{-x^2/2}\, dx = \sqrt{2\pi},$$

and hence

$$\Phi(\infty) = \frac{1}{\sqrt{2\pi}} \int_{-\infty}^{\infty} e^{-x^2/2}\, dx = 1.$$

Therefore, Φ is a distribution function.

DEFINITION *A random variable X is called standard normal if its distribution function is* Φ, *that is, if*

$$P(X \leq t) = \Phi(t) \equiv \frac{1}{\sqrt{2\pi}} \int_{-\infty}^{t} e^{-x^2/2} \, dx.$$

By the fundamental theorem of calculus, f, the density function of a standard normal random variable, is given by

$$f(x) = \Phi'(x) = \frac{1}{\sqrt{2\pi}} e^{-x^2/2}.$$

The standard normal density function is a bell-shaped curve that is symmetric about the y-axis (see Figure 7.5).

Since Φ is the distribution function of the standard normal random variable, $\Phi(t)$ is the area under this curve from $-\infty$ to t. Because $\Phi(\infty) = 1$ and the curve is symmetric about the y-axis, $\Phi(0) = 1/2$. Moreover, $\Phi(-t) = 1 - \Phi(t)$. To see this, note that

$$\Phi(-t) = \frac{1}{\sqrt{2\pi}} \int_{-\infty}^{-t} e^{-x^2/2} \, dx.$$

Substituting $u = -x$, we obtain

$$\Phi(-t) = \frac{-1}{\sqrt{2\pi}} \int_{\infty}^{t} e^{-u^2/2} \, du = \frac{1}{\sqrt{2\pi}} \int_{t}^{\infty} e^{-u^2/2} \, du$$

$$= \frac{1}{\sqrt{2\pi}} \int_{-\infty}^{\infty} e^{-u^2/2} \, du - \frac{1}{\sqrt{2\pi}} \int_{-\infty}^{t} e^{-u^2/2} \, du$$

$$= 1 - \Phi(t).$$

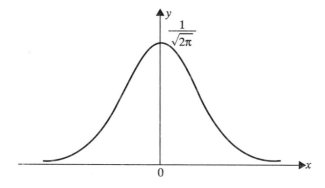

Figure 7.5 Graph of the standard normal density function.

Correction for Continuity

So far we have shown that the De Moivre–Laplace theorem approximates the distribution of a discrete random variable by that of a continuous one. But we have not demonstrated how this approximation works in practice. To do so, first we explain how, in general, a probability based on a discrete random variable is approximated by a probability based on a continuous one. Let X be a discrete random variable with probability function $p(x)$ and suppose that we want to find $P(i \leq X \leq j)$, $i < j$. Consider the *histogram* of X, as sketched in Figure 7.6 from i to j. In this figure the base of each rectangle equals 1 and the height (and therefore the area) of the rectangle with the base midpoint k is $p(k)$, $i \leq k \leq j$. Thus the sum of the areas of all rectangles is $\sum_{k=i}^{j} p(k)$, which is the exact value of $P(i \leq X \leq j)$. Now suppose that $f(x)$, the density function of a continuous random variable, sketched in Figure 7.7, is a good approximation to $p(x)$. Then as this figure shows, $P(i \leq X \leq j)$, the sum of the areas of all rectangles of the figure, is approximately the area under $f(x)$ from $i - 1/2$ to $j + 1/2$ rather than from i to j. That is,

$$P(i \leq X \leq j) \approx \int_{i-1/2}^{j+1/2} f(x)\,dx.$$

This adjustment is called *correction for continuity* and is necessary for approximation of the distribution of a discrete random variable with that of a continuous one. Similarly, the following corrections for continuity are made to calculate the given probabilities.

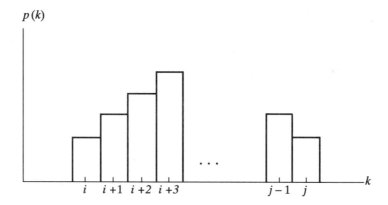

Figure 7.6 Histogram of X from i to j.

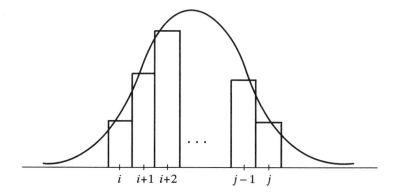

Figure 7.7   Histogram of X and the density function f.

$$P(X = k) \approx \int_{k-1/2}^{k+1/2} f(x)\, dx,$$

$$P(X \geq i) \approx \int_{i-1/2}^{\infty} f(x)\, dx,$$

$$P(X \leq j) \approx \int_{-\infty}^{j+1/2} f(x)\, dx.$$

In real-world problems, or even sometimes in theoretical ones, to apply the De Moivre–Laplace theorem, we need to calculate the numerical values of $\int_a^b e^{-x^2/2}\, dx$ for some real numbers a and b. Since $e^{-x^2/2}$ has no antiderivative in terms of elementary functions, such integrals are approximated by numerical techniques. Tables for such approximations have been available since 1799. Nowadays, most of the scientific calculators give excellent approximate values for these integrals. Table 1 of the Appendix gives the values of $\Phi(x) = (1/\sqrt{2\pi}) \int_{-\infty}^{x} e^{-y^2/2}$ for $x = 0$ to 3.69. We use $\Phi(x) = 1 - \Phi(-x)$ to find $\Phi(x)$ when $x < 0$. The following example shows how the De Moivre–Laplace theorem is applied.

Example 7.4   Suppose that of all the clouds that are seeded with silver iodide, 58% show splendid growth. If 60 clouds are seeded with silver iodide, what is the probability that exactly 35 show splendid growth?

SOLUTION: Let X be the number of clouds that show splendid growth. Then $E(X) = 60(0.58) = 34.80$ and $\sigma_X = \sqrt{60(0.58)(1 - 0.58)} = 3.82$. By correction for continuity and De Moivre–Laplace theorem,

$$P(X = 35) \approx P(34.5 < X < 35.5)$$

$$= P\left(\frac{34.5 - 34.80}{3.82} < \frac{X - 34.80}{3.82} < \frac{35.5 - 34.80}{3.82}\right)$$

$$= P\left(-0.08 < \frac{X - 34.8}{3.82} < 0.18\right) \approx \frac{1}{\sqrt{2\pi}} \int_{-0.08}^{0.18} e^{-x^2/2}\, dx$$

$$= \frac{1}{\sqrt{2\pi}} \int_{-\infty}^{0.18} e^{-x^2/2}\, dx - \frac{1}{\sqrt{2\pi}} \int_{-\infty}^{-0.08} e^{-x^2/2}\, dx$$

$$= \Phi(0.18) - \Phi(-0.08) = \Phi(0.18) - [1 - \Phi(0.08)]$$

$$= \Phi(0.18) + \Phi(0.08) - 1 \approx 0.5714 + 0.5319 - 1 = 0.1033.$$

The exact value of $P(X = 35)$ is $\binom{60}{35}(0.58)^{35}(0.42)^{25}$, which up to four decimal points equals 0.1039. Therefore, the answer obtained by the De Moivre–Laplace approximation is very close to the actual probability. ◆

Now that the importance of the standard normal random variable is clarified, we will calculate its expectation and variance. The graph of the standard normal density function (Figure 7.5) suggests that its expectation is zero. To prove this, let X be a standard normal random variable. Then

$$E(X) = \frac{1}{\sqrt{2\pi}} \int_{-\infty}^{\infty} x e^{-x^2/2}\, dx = 0,$$

because the integrand, $xe^{-x^2/2}$, is a finite odd function, and the integral is taken from $-\infty$ to $+\infty$. To calculate $\mathrm{Var}(X)$, note that

$$E(X^2) = \frac{1}{\sqrt{2\pi}} \int_{-\infty}^{\infty} x^2 e^{-x^2/2}\, dx.$$

Using integration by parts, we get (let $u = x$, $dv = xe^{-x^2/2}\, dx$)

$$\int_{-\infty}^{\infty} x x e^{-x^2/2}\, dx = \left[-x e^{-x^2/2}\right]_{-\infty}^{\infty} + \int_{-\infty}^{\infty} e^{-x^2/2}\, dx = 0 + \sqrt{2\pi} = \sqrt{2\pi}.$$

Therefore, $E(X^2) = 1$, $\mathrm{Var}(X) = E(X^2) - [E(X)]^2 = 1$, and $\sigma_X = \sqrt{\mathrm{Var}(X)} = 1$.

Although some natural phenomena obey a standard normal distribution, when it comes to the analysis of data, due to the lack of parameters in the standard normal distribution, it cannot be used. To overcome this difficulty, mathematicians generalized the standard normal distribution by introducing the following density function.

DEFINITION   *A random variable X is called normal, with parameters μ and σ, if its density function is given by*

$$f(x) = \frac{1}{\sigma\sqrt{2\pi}} \exp\left[\frac{-(x-\mu)^2}{2\sigma^2}\right], \qquad -\infty < x < \infty.$$

f is a density function because by the change of variable $y = (x-\mu)/\sigma$:

$$\int_{-\infty}^{\infty} \frac{1}{\sigma\sqrt{2\pi}} \exp\left[\frac{-(x-\mu)^2}{2\sigma^2}\right] dx = \frac{1}{\sqrt{2\pi}} \int_{-\infty}^{\infty} e^{-y^2/2}\, dy = \Phi(\infty) = 1.$$

If X is a normal random variable with parameters μ and σ, we write $X \sim N(\mu, \sigma^2)$.

One of the first applications of $N(\mu, \sigma^2)$ was given by Gauss in 1809. Gauss used $N(\mu, \sigma^2)$ to model the errors of observations in astronomy. For this reason, the normal distribution is sometimes called *the Gaussian distribution.*   Larsen and Marx, in their *An Introduction to Mathematical Statistics and Its Applications* (Prentice Hall, Englewood Cliffs, N.J., 1986), explain that $N(\mu, \sigma^2)$ was popularized by the Belgian scholar Lambert Quetelet (1796–1874), who used it successfully in data analysis in many situations.[†] The chest measurement of Scottish soldiers is one of the studies of $N(\mu, \sigma^2)$ by Quetelet.   He measured X_i, $i = 1, 2, 3, \ldots, 5738$, the respective sizes of the chests of 5738 Scottish soldiers and using statistical methods found that their average and standard deviation were 39.8 and 2.05 inches, respectively. Then he counted the number of soldiers who had a chest of size i, $i = 33, 34, \ldots, 48$, and calculated the relative frequencies of sizes 33 through 48. He showed that for $i = 33, \ldots, 48$ the relative frequency of the size i is very close to $P(i - 1/2 < X < i + 1/2)$, where X is $N(\mu, \sigma^2)$ with $\mu = 39.8$ and $\sigma = 2.05$. Hence he concluded that the measure of the chest of a Scottish soldier is approximately a normal random variable with parameters 39.8 and 2.05.

The following lemma shows that by a simple change of variable, $N(\mu, \sigma^2)$ can be transformed to $N(0, 1)$.

LEMMA   *If $X \sim N(\mu, \sigma^2)$, then $Z = (X - \mu)/\sigma$ is $N(0, 1)$. That is, if $X \sim N(\mu, \sigma^2)$, the standardized X is $N(0, 1)$.*

[†] An excellent treatise on the work of Quetelet concerning the applications of probability to the measurement of uncertainty in the social sciences can be found in Chapter 5 of "The History of Statistics" by Stephen M. Stigler, *The Belknap Press of Harvard University Press*, 1986.

PROOF: We show that the distribution function of Z is $(1/\sqrt{2\pi}) \int_{-\infty}^{x} e^{-y^2/2} \, dy$. Note that

$$P(Z \leq x) = P\left(\frac{X - \mu}{\sigma} \leq x\right) = P(X \leq \sigma x + \mu)$$

$$= \frac{1}{\sigma\sqrt{2\pi}} \int_{-\infty}^{\sigma x + \mu} \exp\left[\frac{-(t - \mu)^2}{2\sigma^2}\right] dt.$$

Let $y = (t - \mu)/\sigma$; then $dt = \sigma \, dy$ and we get

$$P(Z \leq x) = \frac{1}{\sigma\sqrt{2\pi}} \int_{-\infty}^{x} e^{-y^2/2} \sigma \, dy = \frac{1}{\sqrt{2\pi}} \int_{-\infty}^{x} e^{-y^2/2} \, dy. \quad \blacklozenge$$

Let $X \sim N(\mu, \sigma^2)$, Quetelet's analysis of the sizes of the chests of Scottish soldiers is an indication that μ and σ are the expected value and the standard deviation of X, respectively. To prove this, note that the random variable $Z = (X - \mu)/\sigma$ is $N(0, 1)$ and $X = \sigma Z + \mu$. Hence $E(X) = E(\sigma Z + \mu) = \sigma E(Z) + \mu = \mu$ and $\text{Var}(X) = \text{Var}(\sigma Z + \mu) = \sigma^2 \text{Var}(Z) = \sigma^2$. Therefore, the parameters μ and σ which appear in the formula of the density function of X are its expected value and standard deviation, respectively.

The graph of $f(x) = \dfrac{1}{\sigma\sqrt{2\pi}} \exp\left[\dfrac{-(t - \mu)^2}{2\sigma^2}\right]$ is a bell-shaped curve symmetric about $x = \mu$ with the maximum at $(\mu, \, 1/\sigma\sqrt{2\pi})$ and inflection points at $\mu \pm \sigma$ (see Figures 7.8 and 7.9).

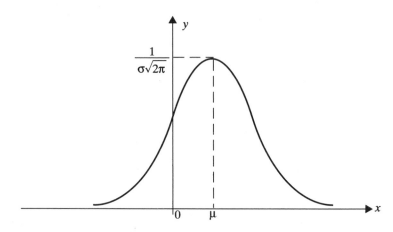

Figure 7.8   Density of $N(\mu, \sigma^2)$.

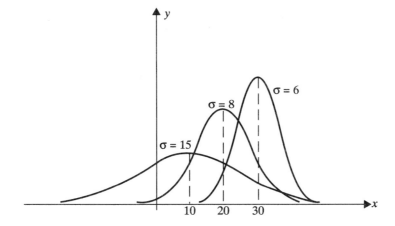

Figure 7.9 Different normal densities with specified parameters.

The transformation $Z = (X - \mu)/\sigma$ enables us to use Table 1 of the Appendix to calculate the probabilities concerning X. Some examples follow.

Example 7.5 Suppose that a Scottish soldier's chest size is normally distributed with mean 39.8 and standard deviation 2.05 inches, respectively. What is the probability that of 20 randomly selected Scottish soldiers, five have a chest of at least 40 inches?

SOLUTION: Let p be the probability that a randomly selected Scottish soldier has a chest of 40 or more inches. If X is the normal random variable with mean 39.8 and standard deviation 2.05, then

$$p = P(X \geq 40) = P\left(\frac{X - 39.8}{2.05} \geq \frac{40 - 39.8}{2.05}\right) = P\left(\frac{X - 39.8}{2.05} \geq 0.10\right)$$

$$= P(Z \geq 0.10) = 1 - \Phi(0.1) \approx 1 - 0.5398 \approx 0.46.$$

Therefore, the probability that of 20 randomly selected Scottish soldiers, five have a chest of at least 40 inches is

$$\binom{20}{5}(0.46)^5(0.54)^{15} \approx 0.03. \quad \blacklozenge$$

Example 7.6 Let X, the grade of a randomly selected student in a test of a probability course, be a normal random variable. A professor is said to grade such a test *on the curve* if he finds the average μ and the standard deviation

σ of the grades and then assigns letter grades according to the following table.

Range of the grade	$X \geq \mu + \sigma$	$\mu \leq X < \mu + \sigma$	$\mu - \sigma \leq X < \mu$	$\mu - 2\sigma \leq X < \mu - \sigma$	$X < \mu - 2\sigma$
Letter grade	A	B	C	D	F

Suppose that the professor of the probability course grades the test on the curve. Determine the percentage of the students who will get A, B, C, D, and F, respectively.

SOLUTION: By the fact that $(X - \mu)/\sigma$ is standard normal,

$$P(X \geq \mu + \sigma) = P\left(\frac{X - \mu}{\sigma} \geq 1\right) = 1 - \Phi(1) \approx 0.1587,$$

$$P(\mu \leq X < \mu + \sigma) = P\left(0 \leq \frac{X - \mu}{\sigma} < 1\right)$$
$$= \Phi(1) - \Phi(0) \approx 0.3413,$$

$$P(\mu - \sigma \leq X < \mu) = P\left(-1 \leq \frac{X - \mu}{\sigma} < 0\right) = \Phi(0) - \Phi(-1)$$
$$= \Phi(0) - 1 + \Phi(1) \approx 0.3413,$$

$$P(\mu - 2\sigma \leq X < \mu - \sigma) = P\left(-2 \leq \frac{X - \mu}{\sigma} < -1\right) = \Phi(-1) - \Phi(-2)$$
$$= 1 - \Phi(1) - 1 + \Phi(2)$$
$$= \Phi(2) - \Phi(1) \approx 0.1359,$$

$$P(X < \mu - 2\sigma) = P\left(\frac{X - \mu}{\sigma} < -2\right) = \Phi(-2) \approx 0.0228.$$

Therefore, approximately 16% should get A, 34% B, 34% C, 14% D, and 2% F. If an instructor grades a test on the curve, instead of calculating μ and σ, he or she may assign A to the top 16%, B to the next 34%, and so on. $\blacklozenge$

Example 7.7   The scores on an achievement test given to 100,000 students are normally distributed with mean 500 and standard deviation 100. What should the score of a student be to place him among the top 10% of all students?

SOLUTION: Letting X be a normal random variable with mean 500 and standard deviation 100, we must find x so that $P(X \geq x) = 0.10$ or $P(X < x) = 0.90$. This gives

$$P\left(\frac{X - 500}{100} < \frac{x - 500}{100}\right) = 0.90.$$

Thus

$$\Phi\left(\frac{x - 500}{100}\right) = 0.90.$$

From Table 1 of the Appendix, we have that $\Phi(1.28) \approx 0.8997$, implying that $(x - 500)/100 \approx 1.28$. This gives $x \approx 628$; therefore, a student should earn 628 or more to be among the top 10% of the students. ♦

EXERCISES

A

1. Suppose that 90% of the patients with a certain disease can be cured with a certain drug. What is the approximate probability that of 50 such patients, at least 45 can be cured with the drug?

2. A small college has 1095 students. What is the approximate probability that more than five students were born on Christmas day? Assume that the birth rates are constant throughout the year and that each year has 365 days.

3. Let $\Psi(x) = 2\Phi(x) - 1$. The function Ψ is called the positive normal distribution. Prove that if Z is standard normal, then $|Z|$ is positive normal.

4. Let Z be a standard normal random variable and α be a given constant. Find the real number x that maximizes $P(x < Z < x + \alpha)$.

5. Let X be a standard normal random variable. Calculate $E(X \cos X)$, $E(\sin X)$, and $E\left(\dfrac{X}{1 + X^2}\right)$.

6. The ages of subscribers to a certain newspaper are normally distributed with mean 35.5 years and standard deviation 4.8. What is the probability that the age of a random subscriber is (a) more than 35.5 years; (b) between 30 and 40 years?

7. The grades of a certain exam are normally distributed with mean 67 and variance 64. What percent of students get A($\geq$ 90), B(80 − 90), C(70 − 80), D(60 − 70), and F(< 60)?

8. Suppose that the distribution of the diastolic blood pressure of a randomly selected person in a certain population is normal with mean 80 mmHg and standard deviation 7 mmHg. If people with diastolic blood pressures 95 or above are considered hypertensive and people with diastolic blood pressures above 89 and below 95 are considered to have mild hypertension, what percent of this population have mild hypertension and what percent are hypertensive? Assume that in this population no one has abnormal systolic blood pressure.

9. The length of an aluminum-coated steel sheet manufactured by a certain factory is approximately normal with mean 75 centimeters and standard deviation 1 centimeter. Find the probability that a randomly selected sheet manufactured by this factory is between 74.5 and 75.8 centimeters.

10. Suppose that the IQ of a randomly selected student from a university is normal with mean 110 and standard deviation 20. Determine the interval of values that is centered at the mean and includes 50% of the IQs of the students of this university.

11. The amount of cereal in a box is normal with mean 16.5 ounces. If the packager is required to fill at least 90% of the cereal boxes with 16 or more ounces of cereal, what is the largest standard deviation for the amount of cereal in a box?

12. Suppose that the scores on a certain manual dexterity test are normal with mean 12 and standard deviation 3. If eight randomly selected individuals take the test, what is the probability that none will make a score less than 14?

13. A number t is said to be the *median* of a continuous random variable X if $P(X \leq t) = P(X \geq t) = 1/2$. Calculate the median of the normal random variable with parameters μ and σ^2.

14. Let $X \sim N(\mu, \sigma^2)$. Prove that $P(|X - \mu| > k\sigma)$ does not depend on μ or σ.

15. Suppose that lifetimes of light bulbs produced by a certain company are normal random variables with mean 1000 hours and standard deviation 100 hours. Is this company correct when it claims that 95% of its light bulbs last at least 900 hours?

16. Suppose that lifetimes of light bulbs produced by a certain company are normal random variables with mean 1000 hours and standard deviation 100 hours. Suppose that lifetimes of light bulbs produced by a second company are normal random variables with mean 900 hours and standard

deviation 150 hours. Howard buys one light bulb manufactured by the first company and one by the second company. What is the probability that at least one of them lasts 980 or more hours?

17. Find the expected value and the variance of a random variable with the probability density function

$$f(x) = \sqrt{\frac{2}{\pi}} e^{-2(x-1)^2}.$$

18. Let $X \sim N(\mu, \sigma^2)$. Find the probability distribution function of $|X - \mu|$ and its expected value.

19. Determine the value(s) of k for which the following is the probability density function of a normal random variable.

$$f(x) = \sqrt{k} e^{-k^2 x^2 - 2kx - 1}, \qquad -\infty < x < \infty.$$

20. The viscosity of a brand of motor oil is normal with mean 37 and standard deviation 10. What is the lowest possible viscosity for a specimen that has viscosity higher than at least 90% of this brand of motor oil?

21. In a certain town the length of residence of a family in a home is normal with mean 80 months and variance 900. What is the probability that of 12 independent families living on a certain street of this town, at least three will have lived there more than eight years?

22. Let $X \sim N(0, \sigma^2)$. Calculate the density function of $Y = X^2$.

23. Let $X \sim N(\mu, \sigma^2)$. Calculate the density function of $Y = e^X$.

24. Let $X \sim N(0, 1)$. Calculate the density function of $Y = \sqrt{|X|}$.

B

25. Suppose that the odds in favor of a customer of a particular bookstore buying a certain fiction bestseller are one in 5000. If 800 customers enter the store every day, how many copies of this bestseller should the store carry every month so that, with a probability of more than 98%, it does not run out of this book? For simplicity, assume that a month is 30 days.

26. Every day a factory produces 5000 light bulbs, of which 2500 are type I and 2500 are type II. If a sample of 40 light bulbs is selected at random to be examined for defects, what is the approximate probability that this sample contains at least 18 light bulbs of each type?

27. To examine the accuracy of an algorithm that selects random numbers from the set $\{1, 2, \ldots, 40\}$, 100,000 numbers are selected and there are 3500 ones. Given that the expected number of ones is 2500, is it fair to say that this algorithm is not accurate?

28. Prove that for some constant k, $f(x) = ka^{-x^2}$, $a \in (0, \infty)$, is a normal probability density function.

29. (a) Prove that for all $x > 0$,

$$\frac{1}{x\sqrt{2\pi}} \left(1 - \frac{1}{x^2} \right) e^{-x^2/2} < 1 - \Phi(x) < \frac{1}{x\sqrt{2\pi}} e^{-x^2/2}.$$

HINT: Integrate the following inequalities:

$$(1 - 3y^{-4})e^{-y^2/2} < e^{-y^2/2} < (1 + y^{-2})e^{-y^2/2}.$$

(b) Use part (a) to prove that

$$1 - \Phi(x) \sim \frac{1}{x\sqrt{2\pi}} e^{-x^2/2}.$$

That is, as $x \to \infty$, the ratio of the two sides approaches 1.

30. Let Z be a standard normal random variable. Show that for $x > 0$,

$$\lim_{t \to \infty} P\left(Z > t + \frac{x}{t} \mid Z \ge t \right) = e^{-x}.$$

HINT: Use part (b) of Exercise 29.

31. The amount of soft drink in a bottle is a normal random variable. Suppose that in 7% of the bottles containing this soft drink there is less than 15.5 ounces, and in 10% of them there are more than 16.3 ounces. What are the mean and standard deviation of the amount of soft drink in a randomly selected bottle?

32. At an archaeological site 130 skeletons are found and their heights are measured and found to be approximately normal with mean 172 centimeters and variance 81 centimeters. At a nearby site, five skeletons are discovered and it is found that the heights of exactly three of them are above 185 centimeters. Based on this information is it reasonable to assume that the second group of skeletons belongs to the same family as the first group of skeletons?

33. In a forest, the number of trees that grow in a region of area R has a Poisson distribution with mean λR, where λ is a positive real number.

Find the expected value of the distance from a certain tree to its nearest neighbor.

34. Let $I = \int_0^\infty e^{-x^2/2}\,dx$; then

$$I^2 = \int_0^\infty \left[\int_0^\infty e^{-(x^2+y^2)} dy \right] dx.$$

Let $y/x = s$ and change the order of integration to show that $I^2 = \pi/4$. This gives an alternative proof of the fact that Φ is a distribution function. The advantage of this method is that it avoids polar coordinates.

7.3   Exponential Random Variable

Let $\{N(t), t \geq 0\}$ be a Poisson process. Then, as discussed in Section 5.2, $N(t)$ is the number of "events" that have occurred at or prior to time t. Let X_1 be the time of the first event, X_2 be the elapsed time between the first and the second events, X_3 be the elapsed time between the second and third events, and so on. The sequence of random variables $\{X_1, X_2, X_3, \ldots\}$ is called *the sequence of interarrival times of the Poisson process* $\{N(t), t \geq 0\}$. Let $\lambda = E[N(1)]$; then

$$P(N(t) = n) = \frac{e^{-\lambda t}(\lambda t)^n}{n!}.$$

This enables us to calculate the probability distribution functions of the random variables X_i, $i \geq 1$. For $t \geq 0$,

$$P(X_1 > t) = P(N(t) = 0) = e^{-\lambda t}.$$

Therefore,

$$P(X_1 \leq t) = 1 - P(X_1 > t) = 1 - e^{-\lambda t}.$$

Since a Poisson process is stationary and possesses independent increments, at any time t, the process probabilistically starts all over again. Hence the interarrival time of any two consecutive events has the same distribution as X_1; that is, the sequence $\{X_1, X_2, X_3, \ldots\}$ is identically distributed. Therefore for, all $n \geq 1$,

$$P(X_n \leq t) = P(X_1 \leq t) = \begin{cases} 1 - e^{-\lambda t} & t \geq 0 \\ 0 & t < 0. \end{cases}$$

Let

$$F(t) = \begin{cases} 1 - e^{-\lambda t} & t \geq 0 \\ 0 & t < 0 \end{cases}$$

for some $\lambda > 0$. Then F is the distribution function of X_n for all $n \geq 1$. It is called *exponential distribution* and is one of the most important distributions of pure and applied probability. Since

$$f(t) = F'(t) = \begin{cases} \lambda e^{-\lambda t} & t \geq 0 \\ 0 & t < 0 \end{cases} \tag{7.2}$$

and

$$\int_0^\infty \lambda e^{-\lambda t} \, dt = \lim_{b \to \infty} \int_0^b \lambda e^{-\lambda t} \, dt = \lim_{b \to \infty} \left[-e^{-\lambda t} \right]_0^b$$

$$= \lim_{b \to \infty} (1 - e^{-\lambda b}) = 1,$$

f is a density function.

DEFINITION   *A continuous random variable X is called exponential with parameter $\lambda > 0$ if its density function is given by (7.2).*

Because the interarrival times of a Poisson process are exponential, the following are examples of random variables that might be exponential.

1. The interarrival time between two customers at a post office.
2. The duration of Jim's next telephone call.
3. The time between two consecutive earthquakes in California.
4. The time between two accidents at an intersection.
5. The time until the next baby is born in a hospital.
6. The time until the next crime in a certain town.
7. The time to failure of the next fiber segment in a large group of such segments when all of them are initially fault-free.
8. The time interval between the observation of two consecutive shooting stars on a summer evening.
9. The time between two consecutive fish caught by a fisherman from a large lake with lots of fish.

From Section 5.2 we know that λ is the average number of the events in one time unit; that is $E[N(1)] = \lambda$. Therefore, we should expect an average time $1/\lambda$ between two consecutive events. To prove this, let X be an exponential random variable with parameter λ; then

$$E(X) = \int_{-\infty}^{\infty} xf(x)\,dx = \int_{0}^{\infty} x(\lambda e^{-\lambda x})\,dx.$$

Using integration by parts with $u = x$ and $dv = \lambda e^{-\lambda x}\,dx$, we obtain

$$E(X) = \left[-xe^{-\lambda x}\right]_0^{\infty} + \int_{0}^{\infty} e^{-\lambda x}\,dx = 0 - \left[\frac{1}{\lambda}e^{-\lambda x}\right]_0^{\infty} = \frac{1}{\lambda}.$$

A similar calculation shows that

$$E(X^2) = \int_{-\infty}^{\infty} x^2 f(x)\,dx = \int_{0}^{\infty} x^2(\lambda e^{-\lambda x})\,dx = \frac{2}{\lambda^2}.$$

Hence

$$\mathrm{Var}(X) = E(X^2) - [E(X)]^2 = \frac{2}{\lambda^2} - \frac{1}{\lambda^2} = \frac{1}{\lambda^2},$$

and therefore,

$$\sigma_X = \frac{1}{\lambda}.$$

Figures 7.10 and 7.11 represent the graphs of the exponential density and exponential distribution functions, respectively.

Example 7.8 Suppose that, on average, every three months an earthquake occurs in California. What is the probability that the next earthquake occurs after three but before seven months?

SOLUTION: Let X be the time (in months) until the next earthquake, it can be assumed that X is an exponential random variable with $1/\lambda = 3$ or $\lambda = 1/3$. To calculate $P(3 < X < 7)$, note that since F, the distribution function of X, is given by

$$F(t) = P(X \le t) = 1 - e^{-t/3} \quad \text{for } t > 0,$$

we can write

$$P(3 < X < 7) = F(7) - F(3) = (1 - e^{-7/3}) - (1 - e^{-1})$$
$$= e^{-1} - e^{-7/3} \approx 0.27. \quad \blacklozenge$$

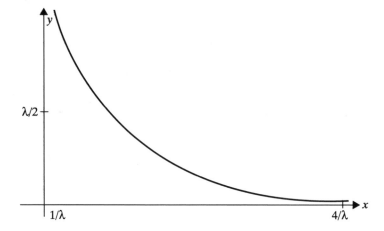

Figure 7.10 Exponential density function with parameter λ.

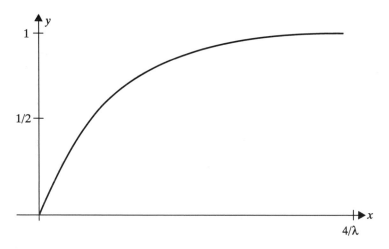

Figure 7.11 Exponential distribution function.

Example 7.9 At an intersection, on average, there are two accidents per day. What is the probability that after the next accident there will be no accidents at all for the next two days?

SOLUTION: Let X be the time (in days) between the next two accidents. It can be assumed that X is exponential with parameter λ, satisfying $1/\lambda = 1/2$, so that $\lambda = 2$. To find $P(X > 2)$, note that F, the distribution function of X, is given by

$$F(t) = 1 - e^{-2t} \quad \text{for } t > 0.$$

Hence

$$P(X > 2) = 1 - P(X \le 2) = 1 - F(2) = e^{-4} \approx 0.02. \quad \blacklozenge$$

An important feature of exponential distribution is its *memoryless property*. A nonnegative random variable X is called *memoryless* if for all $s, t \ge 0$,

$$P(X > s + t \mid X > t) = P(X > s). \tag{7.3}$$

If, for example, X is the lifetime of some type of instrument, then (7.3) means that there is no deterioration with the age of the instrument. The probability that a new instrument lasts more than s years is the same as the probability that a used instrument that has lasted more than t years lasts at least another s years. In other words, the probability that such an instrument deteriorates in the next s years does not depend on the age of the instrument.

To show that an exponential distribution is memoryless, note that (7.3) is equivalent to

$$\frac{P(X > s + t, X > t)}{P(X > t)} = P(X > s)$$

and

$$P(X > s + t) = P(X > s)P(X > t). \tag{7.4}$$

Now since

$$P(X > s + t) = 1 - \left[1 - e^{-\lambda(s+t)}\right] = e^{-\lambda(s+t)},$$
$$P(X > s) = 1 - (1 - e^{-\lambda s}) = e^{-\lambda s},$$

and

$$P(X > t) = 1 - (1 - e^{-\lambda t}) = e^{-\lambda t},$$

we have that (7.4) follows. Hence X is memoryless. It can be shown that exponential is the only continuous distribution which possesses memoryless property (see Exercise 13).

Example 7.10   The lifetime of a TV tube (in years) is an exponential random variable with mean 10. If Jim bought his TV set 10 years ago, what is the probability that its tube will last another 10 years?

SOLUTION: Let X be the lifetime of the tube. Since X is an exponential random variable, there is no deterioration with the age of the tube. Hence

$$P(X > 20 \mid X > 10) = P(X > 10) = 1 - \left[1 - e^{(-1/10)10}\right] \approx 0.37. \quad \blacklozenge$$

Example 7.11   Suppose that, on average, two earthquakes occur in San Francisco and two in Los Angeles every year. If the last earthquake in San Francisco occurred 10 months ago and the last earthquake in Los Angeles occurred two months ago, what is the probability that the next earthquake in San Francisco occurs after the next earthquake in Los Angeles?

SOLUTION: It can be assumed that the number of earthquakes in San Francisco and Los Angeles are both Poisson processes with common rate $\lambda = 2$. Hence the times between two consecutive earthquakes in Los Angeles and two consecutive earthquakes in San Francisco are both exponentially distributed with the common mean $1/\lambda = 1/2$. Because of the memoryless property of the exponential distribution, it does not matter when the last earthquakes in San Francisco and Los Angeles have occurred. The times between now and the next earthquake in San Francisco and the next earthquake in Los Angeles both have the same distribution. Since these time periods are exponentially distributed with the same parameter, by symmetry, the probability that the next earthquake in San Francisco occurs after that in Los Angeles is 1/2. $\blacklozenge$

EXERCISES

A

1. Customers arrive at a post office at a Poisson rate of three per minute. What is the probability that the next customer does not arrive during the next 3 minutes?

2. Find the median of an exponential random variable with rate λ. Recall that for a continuous distribution F, the median $Q_{0.5}$ is the point at which $F(Q_{0.5}) = 1/2$.

3. Let X be an exponential random variable with mean 1. Find the probability density function of $Y = -\ln X$.

4. The time between the first and second heart attacks in a certain group of people is an exponential random variable. If 50% of those who have had a heart attack will have another one within the next five years, what is the probability that a person who had one heart attack five years ago will not have another one in the next five years?

5. Customers arrive at a hotel in accordance with a Poisson process at a rate of five per hour. Suppose that for the last 10 minutes no customer has arrived. What is the probability that (a) the next one will arrive in less than 2 minutes; (b) from the arrival of the tenth to the arrival of the eleventh customer takes no more than 2 minutes?

6. Let X be an exponential random variable with parameter λ. Find

$$P(|X - E(X)| \geq 2\sigma_X).$$

7. Suppose that at an Italian restaurant, the time, in minutes, between two customers ordering pizza is exponential with parameter λ. What is the probability that (a) no customer orders pizza during the next t minutes; (b) the next pizza order is placed in at least t minutes but no later than s minutes $(t < s)$?

8. Suppose that the time it takes for a novice secretary to type a document is exponential with mean 1 hour. If at the beginning of a certain eight-hour working day the secretary receives 12 documents to type, what is the probability that she will finish them all by the end of the day?

9. The profit on each computer assembled by a certain person is $350. Suppose that the assembler guarantees his computers for one year and the time between two failures of a computer is exponential with mean 18 months. If it costs the assembler $40 to repair a failed computer, what is the expected profit per computer?
   HINT: Let $N(t)$ be the number of times that the computer fails in $[0, t]$. Then $\{N(t) : t \geq 0\}$ is a Poisson process with parameter $\lambda = 1/18$.

10. Mr. Jones is at a train station waiting to make a phone call. There are two public telephone booths next to each other and occupied by two persons, say A and B. If the duration of each telephone call is an exponential random variable with $\lambda = 1/8$, what is the probability that among Mr. Jones, A, and B, Mr. Jones will not be the last to finish his call?

B

11. The random variable X is called *double exponentially distributed* if its density function is given by

$$f(x) = ce^{-|x|}, \quad -\infty < x < +\infty.$$

   (a) Find the value of c.
   (b) Prove that $E(X^{2n}) = (2n)!$ and $E(X^{2n+1}) = 0$.

12. Let X, the lifetime (in years) of a radio tube, be exponentially distributed with mean $1/\lambda$. Prove that the integer part of X, $[X]$, which is the complete number of years that the tube works, is a geometric random variable.

13. Prove that if X is a positive, continuous, memoryless random variable with distribution function F, then $F(t) = 1 - e^{-\lambda t}$ for some $\lambda > 0$. This shows that the exponential is the only distribution on $(0, \infty)$ with the memoryless property.

7.4   Gamma Distribution

Let $\{N(t), t \geq 0\}$ be a Poisson process, X_1 be the time of the first event, and for $n \geq 2$, let X_n be the time between the $(n-1)$st and nth events. As we explained in Section 7.3, $\{X_1, X_2, X_3, \ldots\}$ is a sequence of identically distributed exponential random variables with mean $1/\lambda$, where λ is the rate of $\{N(t), t \geq 0\}$. For this Poisson process let X be the time of the nth event. Then X is said to have a *gamma distribution with parameters* (n, λ). To find f, the density function of X, note that $\{X \leq t\}$ occurs if the time of the nth event is in $[0, t]$, that is, if the number of events occurring in $[0, t]$ is at least n. Hence F, the distribution function of X is given by

$$F(t) = P(X \leq t) = P\{N(t) \geq n\} = \sum_{i=n}^{\infty} \frac{e^{-\lambda t}(\lambda t)^i}{i!}.$$

Differentiating F, the density function f is obtained:

$$f(t) = \sum_{i=n}^{\infty} \left[-\lambda e^{-\lambda t} \frac{(\lambda t)^i}{i!} + \lambda e^{-\lambda t} \frac{(\lambda t)^{i-1}}{(i-1)!} \right]$$

$$= \sum_{i=n}^{\infty} -\lambda e^{-\lambda t} \frac{(\lambda t)^i}{i!} + \left[\lambda e^{-\lambda t} \frac{(\lambda t)^{n-1}}{(n-1)!} + \sum_{i=n+1}^{\infty} \lambda e^{-\lambda t} \frac{(\lambda t)^{i-1}}{(i-1)!} \right]$$

$$= \left[-\sum_{i=n}^{\infty} \lambda e^{-\lambda t} \frac{(\lambda t)^i}{i!} \right] + \lambda e^{-\lambda t} \frac{(\lambda t)^{n-1}}{(n-1)!} + \left[\sum_{i=n}^{\infty} \lambda e^{-\lambda t} \frac{(\lambda t)^i}{i!} \right]$$

$$= \lambda e^{-\lambda t} \frac{(\lambda t)^{n-1}}{(n-1)!}.$$

The density function

$$f(x) = \begin{cases} \lambda e^{-\lambda x} \dfrac{(\lambda x)^{n-1}}{(n-1)!} & \text{if } x \geq 0 \\ 0 & \text{elsewhere} \end{cases}$$

is called *the gamma (or n-Erlang) density with parameters* (n, λ).

Now we extend the definition of the gamma density from parameters (n, λ) to (r, λ), where $r > 0$ is not necessarily a positive integer. As we shall see later, this extension has useful applications in probability and statistics. In the formula of the gamma density function, the term $(n-1)!$ is defined only for positive integers. So the only obstacle in such an extension is to find a function of r that has the basic property of the factorial function, namely $n! = n \cdot (n-1)!$, and coincides with $(n-1)!$ when n is a positive integer. The function with these properties is $\Gamma\colon (0, \infty) \to \mathbf{R}$ defined by

$$\Gamma(r) = \int_0^\infty t^{r-1} e^{-t} \, dt.$$

The property analogous to $n! = n \cdot (n-1)!$ is

$$\Gamma(r+1) = r\Gamma(r), \quad r > 1,$$

which is obtained by an integration by parts applied to $\Gamma(r+1)$ with $u = t^r$ and $dv = e^{-t} \, dt$:

$$\Gamma(r+1) = \int_0^\infty t^r e^{-t} \, dt = \left[-t^r e^{-t}\right]_0^\infty + r \int_0^\infty t^{r-1} e^{-t} \, dt$$

$$= r \int_0^\infty t^{r-1} e^{-t} \, dt = r\Gamma(r).$$

To show that $\Gamma(n)$ coincides with $(n-1)!$ when n is a positive integer, note that

$$\Gamma(1) = \int_0^\infty e^{-t} \, dt = 1.$$

Therefore,

$$\Gamma(2) = (2-1)\Gamma(2-1) = 1 = 1!,$$
$$\Gamma(3) = (3-1)\Gamma(3-1) = 2 \cdot 1 = 2!,$$
$$\Gamma(4) = (4-1)\Gamma(4-1) = 3 \cdot 2 \cdot 1 = 3!.$$

Repetition of this process or a simple induction implies that

$$\Gamma(n + 1) = n!.$$

Hence $\Gamma(r + 1)$ is the natural generalization of $n!$ for a noninteger $r > 0$. This motivates the following definition.

DEFINITION *A random variable X with probability density function*

$$f(x) = \begin{cases} \dfrac{\lambda e^{-\lambda x}(\lambda x)^{r-1}}{\Gamma(r)} & \textit{if } x \geq 0 \\ 0 & \textit{elsewhere} \end{cases}$$

*is said to have a **gamma distribution** with parameters (r, λ), $\lambda > 0$, $r > 0$.*

Figures 7.12 and 7.13 demonstrate the shape of the gamma density function for several values of r and λ.

$$X \sim G(3, 4)$$

Example 7.12 Suppose that, on average, the number of β-particles emitted from a radioactive substance is four every second. What is the probability that it takes at least 2 seconds before the next two β-particles are emitted?

Solution: Let $N(t)$ denote the number of β-particles emitted from a radioactive substance in $[0, t]$. It is reasonable to assume that $\{N(t) : t \geq 0\}$ is a Poisson process. Let 1 second be the time unit; then $\lambda = E[N(1)] = 4$. X, the time between now and when the second β-particle is emitted, has a gamma distribution with parameters $(2, 4)$. Therefore,

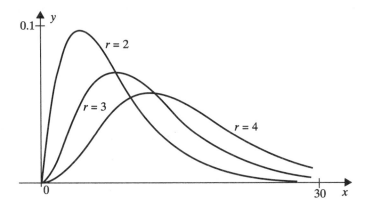

Figure 7.12 Gamma densities for $\lambda = 1/4$.

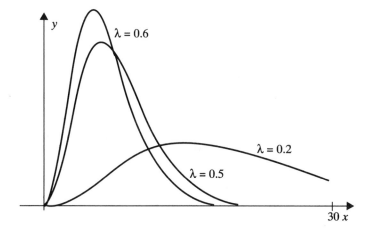

Figure 7.13   Gamma densities for $r = 4$.

$$P(X \geq 2) = \int_2^\infty \frac{4e^{-4x}(4x)^{2-1}}{\Gamma(2)}\,dx = \int_2^\infty 16xe^{-4x}\,dx$$

$$= \left[-4xe^{-4x}\right]_2^\infty - \int_2^\infty -4e^{-4x}\,dx$$

$$= 8e^{-8} + e^{-8} \approx 0.003.$$

Note that an alternative solution to this problem is

$$P(X \geq 2) = P\{N(2) \leq 1\} = P\{N(2) = 0\} + P\{N(2) = 1\}$$

$$= \frac{e^{-8}(8)^0}{0!} + \frac{e^{-8}(8)^1}{1!} = 9e^{-8} \approx 0.003. \quad \blacklozenge$$

Let X be a gamma random variable with parameters (r, λ). To find $E(X)$ and $\mathrm{Var}(X)$, note that for all $n \geq 0$,

$$E(X^n) = \int_0^\infty x^n \frac{\lambda e^{-\lambda x}(\lambda x)^{r-1}}{\Gamma(r)}\,dx = \frac{\lambda^r}{\Gamma(r)}\int_0^\infty x^{n+r-1}e^{-\lambda x}\,dx.$$

Let $t = \lambda x$; then $dt = \lambda\,dx$, so

$$E(X^n) = \frac{\lambda^r}{\Gamma(r)}\int_0^\infty \frac{t^{n+r-1}}{\lambda^{n+r-1}}e^{-t}\frac{1}{\lambda}\,dt$$

$$= \frac{\lambda^r}{\Gamma(r)\lambda^{n+r}}\int_0^\infty t^{n+r-1}e^{-t}\,dt = \frac{\Gamma(n+r)}{\Gamma(r)\lambda^n}.$$

For $n = 1$, this gives

$$E(X) = \frac{\Gamma(r+1)}{\Gamma(r)\lambda} = \frac{r\Gamma(r)}{\lambda\Gamma(r)} = \frac{r}{\lambda}.$$

For $n = 2$,

$$E(X^2) = \frac{\Gamma(r+2)}{\Gamma(r)\lambda^2} = \frac{(r+1)\Gamma(r+1)}{\lambda^2\Gamma(r)} = \frac{(r+1)r\Gamma(r)}{\lambda^2\Gamma(r)} = \frac{r^2+r}{\lambda^2}.$$

Thus

$$\text{Var}(X) = \frac{r^2+r}{\lambda^2} - \left(\frac{r}{\lambda}\right)^2 = \frac{r}{\lambda^2}.$$

Example 7.13   There are 100 questions in a test. Suppose that for all $s > 0$ and $t > 0$ the event that it takes t minutes to answer one question is independent of the event that it takes s minutes to answer another one. If the time that it takes to answer a question is exponential with mean 1/2, find the distribution, the average time, and the standard deviation of the time it takes to do the entire test.

SOLUTION: Let X be the time to answer a question and $N(t)$ be the number of questions answered by time t. Then $\{N(t) : t \geq 0\}$ is a Poisson process at the rate of $\lambda = 1/E(X) = 2$ per minute. Therefore the time that it takes to do all the questions is gamma with parameters $(100, 2)$. The average time to finish the test is $r/\lambda = 100/2 = 50$ minutes with standard deviation $\sqrt{r/\lambda^2} = \sqrt{100/4} = 5$.   ◆

EXERCISES

A

1. Show that the gamma density function with parameters (r, λ) has a unique maximum at $(r - 1)/\lambda$.

2. Let X be a gamma random variable with parameters (r, λ). Find the distribution function of cX, where c is a positive constant.

3. In a hospital, babies are born at a Poisson rate of 12 per day. What is the probability that it takes at least seven hours before the next three babies are born?

4. Let f be the density function of a gamma random variable X, with parameters (r, λ). Prove that $\int_{-\infty}^{\infty} f(x)\,dx = 1$.

5. Customers arrive at a restaurant at a Poisson rate of 12 per hour. If the restaurant makes a profit only after 30 customers have arrived, what is the expected length of time until the restaurant starts to make profit?

B

6. For $n = 0, 1, 2, 3, \ldots$, calculate $\Gamma(n + 1/2)$.

7. Howard enters a bank that has n tellers. All the tellers are busy serving customers and there is exactly one queue feeding all tellers, with one customer ahead of Howard waiting to be served. If the service time of a customer is exponential with parameter λ, find the distribution of the waiting time of Howard until his service time.

7.5 Beta Distribution *(DON'T WORRY ABOUT)*

A random variable X is called *beta with parameters* (α, β), $\alpha > 0$, $\beta > 0$, if f, its density function, is given by

$$
f(x) = \begin{cases} \dfrac{1}{B(\alpha, \beta)} x^{\alpha-1}(1-x)^{\beta-1} & \text{if } 0 < x < 1 \\ 0 & \text{otherwise,} \end{cases}
$$

where

$$
B(\alpha, \beta) = \int_0^1 x^{\alpha-1}(1-x)^{\beta-1}\,dx.
$$

$B(\alpha, \beta)$ is related to the gamma function by the relation

$$
B(\alpha, \beta) = \frac{\Gamma(\alpha)\Gamma(\beta)}{\Gamma(\alpha + \beta)}.
$$

Beta density occurs in a natural way in the study of the median of a sample of random points from $(0, 1)$. Let $X_{(1)}$ be the smallest of these numbers, $X_{(2)}$ be the second smallest, $\ldots$, $X_{(i)}$ be the ith smallest, $\ldots$, and $X_{(n)}$ be the largest of these numbers. If $n = 2k + 1$ is odd, $X_{(k)}$ is called the *median* of these n random numbers, whereas if $n = 2k$ is even, $[X_{(k)} + X_{(k+1)}]/2$ is called the *median*. It can be shown that the median of $(2n + 1)$ random numbers from the interval $(0, 1)$ is a *beta* random variable with parameters $(n + 1, n + 1)$. In statistics, beta distributions are used as prior distributions in the Bayesian estimation of parameters. As Figures 7.14 and 7.15 show, by changing the values of parameters α and β, beta densities cover a wide

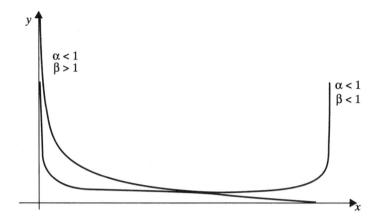

Figure 7.14   Beta densities for $\alpha < 1$, $\beta < 1$ and $\alpha < 1$, $\beta > 1$.

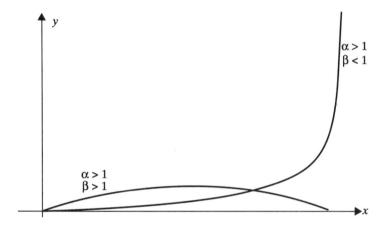

Figure 7.15   Beta densities for $\alpha > 1$, $\beta > 1$ and $\alpha > 1$, $\beta < 1$.

range of different shapes. If $\alpha = \beta$, the median is $x = 1/2$ and the density of the beta random variable is symmetric about the median. In particular, if $\alpha = \beta = 1$, the uniform density over the interval $(0, 1)$ is obtained.

Beta distributions are often appropriate models for random variables that vary between two finite limits—an upper and a lower, such as the fraction of people in a community who use a certain product in a given period of time, the percentage of total farm acreage that produces healthy watermelons, and the distance from one end of a tree to the point at which it breaks in a severe storm. In these three instances the random variables are restricted between 0 and 1, 0 and 100, and 0 and the length of the tree, respectively.

To find the expected value and the variance of a beta random variable X, with parameters (α, β), note that for $n \geq 1$,

$$E(X^n) = \frac{1}{B(\alpha, \beta)} \int_0^1 x^{\alpha+n-1}(1-x)^{\beta-1}\, dx = \frac{B(\alpha+n, \beta)}{B(\alpha, \beta)}$$

$$= \frac{\Gamma(\alpha+n)\Gamma(\alpha+\beta)}{\Gamma(\alpha)\Gamma(\alpha+\beta+n)}.$$

Letting $n = 1$ and $n = 2$ in this relation, we find that

$$E(X) = \frac{\Gamma(\alpha+1)\Gamma(\alpha+\beta)}{\Gamma(\alpha)\Gamma(\alpha+\beta+1)} = \frac{\alpha\Gamma(\alpha)\Gamma(\alpha+\beta)}{\Gamma(\alpha)\,(\alpha+\beta)\Gamma(\alpha+\beta)} = \frac{\alpha}{\alpha+\beta},$$

$$E(X^2) = \frac{\Gamma(\alpha+2)\Gamma(\alpha+\beta)}{\Gamma(\alpha)\Gamma(\alpha+\beta+2)} = \frac{(\alpha+1)\alpha}{(\alpha+\beta+1)(\alpha+\beta)}.$$

Thus

$$\text{Var}(X) = E(X^2) - [E(X)]^2 = \frac{\alpha\beta}{(\alpha+\beta+1)(\alpha+\beta)^2}.$$

There is also an interesting relation between beta and binomial distributions that the following theorem explains.

THEOREM 7.2 *Let α and β be positive integers, X be a beta random variable with parameters α and β, and Y be a binomial random variable with parameters $\alpha + \beta - 1$ and p, $0 < p < 1$. Then*

$$P(X \leq p) = P(Y \geq \alpha).$$

PROOF: From the definition,

$$P(X \leq p) = \frac{(\alpha+\beta-1)!}{(\alpha-1)!\,(\beta-1)!} \int_0^p x^{\alpha-1}(1-x)^{\beta-1}\, dx$$

$$= (\alpha+\beta-1)\binom{\alpha+\beta-2}{\alpha-1} \int_0^p x^{\alpha-1}(1-x)^{\beta-1}\, dx$$

$$= \sum_{i=\alpha}^{\alpha+\beta-1} \binom{\alpha+\beta-1}{i} p^i (1-p)^{\alpha+\beta-1-i}$$

$$= P(Y \geq \alpha),$$

where the third equality follows from using integration by parts $(\alpha - 1)$ times. ◆

Example 7.14   Suppose that all 25 passengers of a commuter plane that departs at 2:00 P.M. arrive at random times between 12:45 and 1:45 P.M. Find the probability that the median of the arrival times of the passengers is at or prior to 1:12 P.M.

SOLUTION: Let the arrival time of a randomly selected passenger be W hours past 12:45. Then W is a uniform random variable over the interval $(0, 1)$. Let X be the median of the arrival times of the 25 passengers. Since the median of $(2n + 1)$ random numbers from $(0, 1)$ is beta with parameters $(n + 1, n + 1)$, we have that X is beta with parameters $(13, 13)$. Since the length of the time interval from 12:45 to 1:12 is 27 minutes, the desired quantity is $P(X \leq 27/60)$. To calculate this, let Y be a binomial random variable with parameters 25 and $p = 27/60 = 0.45$. Then by Theorem 7.2,

$$P(X \leq 0.45) = P(Y \geq 13) = \sum_{i=13}^{25} \binom{25}{i}(0.45)^i (0.55)^{25-i} \approx 0.18. \quad ◆$$

EXERCISES

A

1. At a certain university, the fraction of students who get a C in any section of a certain course is uniform over $(0, 1)$. Find the probability that the median of these fractions for the 13 sections of the course that are offered next semester is at least 0.40.

2. Suppose that while in fantasy, the fraction X of the time that one commits brave deeds is beta with parameters $(5, 21)$. What is the probability that next time Jeff is in fantasy, he commits brave deeds at least 1/4 of the time?

B

3. Prove that

$$B(\alpha, \beta) = \frac{\Gamma(\alpha)\Gamma(\beta)}{\Gamma(\alpha + \beta)}.$$

Review Problems

1. For a restaurant, the time it takes to deliver pizza (in minutes) is uniform over the interval (25, 37). Determine the proportion of deliveries that are made in less than half an hour.

2. It is known that the weight of a random woman from a community is normal with mean 130 pounds and standard deviation 20. From the women in this community who weigh above 140 pounds, what percent weigh over 170 pounds?

3. One thousand random digits are generated. What is the probability that digit 5 is generated at most 93 times?

4. Let X, the lifetime of a light bulb, be an exponential random variable with parameter λ. Is it possible that X satisfies the following relation?

$$P(X \le 2) = 2P(2 < X \le 3).$$

If so, for what value of λ?

5. The time that it takes for a computer system to fail is exponential with mean 1700 hours. If a lab has 20 such computer systems, what is the probability that at least two fail before working 1700 hours?

6. Let X be a uniform random variable over the interval (0, 1). Calculate $E(-\ln X)$.

7. Suppose that the diameter of a randomly selected disk produced by a certain manufacturer is normal with mean 4 inches and standard deviation 1 inch. Find the distribution function of the diameter of a randomly chosen disk in centimeters.

8. Let X be an exponential random variable with parameter λ. Prove that

$$P(\alpha \le X \le \alpha + \beta) \le P(0 \le X \le \beta).$$

9. The time that it takes for a calculus student to answer all the questions on a certain exam is an exponential random variable with mean 1 hour and 15 minutes. If all 10 students of a calculus class are taking this exam, what is the probability that at least one of them completes it in less than one hour?

10. Determine the value(s) of k for which the following is a density function.

$$f(x) = ke^{-x^2+3x+2}, \qquad -\infty < x < \infty.$$

11. The grades of students in a calculus-based probability course are normal with mean 72 and standard deviation 7. If 90, 80, 70, and 60 are the respective lowest A, B, C, and D, what percent of students in this course get A's, B's, C's, D's, and F's?

12. The number of minutes that a train from Milan to Rome is late is an exponential random variable X with parameter λ. Find $P\{X > E(X)\}$.

13. In a measurement, a number is rounded off to the nearest k decimal places. Let X be the rounding error. Determine the probability distribution function of X and its parameters.

14. Suppose that the weights of passengers taking an elevator in a certain building are normal with mean 175 pounds and standard deviation 22. What is the minimum weight for a passenger who outweighs at least 90% of the other passengers?

15. The breaking strength of a certain type of yarn produced by a certain vendor is normal with mean 95 and standard deviation 11. What is the probability that in a random sample of size 10 from the stock of this vendor, the breaking strengths of at least two are over 100?

Chapter 8

JOINT DISTRIBUTIONS

8.1 Bivariate Distributions

Joint Probability Functions

So far we have only studied probability functions of single discrete random variables and probability density functions of single continuous random variables. Sometimes it is desirable to study two or more random variables that are defined on the same sample space simultaneously. In this section we consider such cases with two variables. Cases of three or more variables are studied in Section 8.4.

DEFINITION *Let X and Y be two discrete random variables defined on the same sample space. Let the sets of possible values of X and Y be A and B, respectively. The function*

$$p(x, y) = P(X = x, Y = y)$$

is called the joint probability function of X and Y.

Note that $p(x, y) \geq 0$. If $x \notin A$ or $y \notin B$, then $p(x, y) = 0$. Also,

$$\sum_{x\in A, y\in B} p(x, y) = 1. \tag{8.1}$$

Let X and Y have joint probability function $p(x, y)$. Let p_X be the probability function of X. Then

$$p_X(x) = P(X = x) = P(X = x, Y \in B) = \sum_{y\in B} P(X = x, Y = y)$$

$$= \sum_{y\in B} p(x, y).$$

Similarly, p_Y, the probability function of Y, is given by

$$p_Y(y) = \sum_{x\in A} p(x, y).$$

These relations motivate the following definition.

DEFINITION   *Let X and Y have joint probability function $p(x, y)$. Let A be the set of possible values of X and B be the set of possible values of Y. Then the functions $p_X(x) = \sum_{y\in B} p(x, y)$ and $p_Y(y) = \sum_{x\in A} p(x, y)$ are called, respectively, the marginal probability functions of X and Y.*

Example 8.1   The College of Natural and Mathematical Sciences of a university has 90 male and 30 female professors. An ad hoc committee of five is selected at random to write the vision and mission of the college. Let X and Y be the number of men and women in this committee, respectively.

(a) Find the joint probability function of X and Y.

(b) Find p_X and p_Y, the marginal probability functions of X and Y.

SOLUTION: (a) The set of possible values for both X and Y is $\{0, 1, 2, 3, 4, 5\}$. The joint probability function of X and Y, $p(x, y)$, is given by

$$p(x, y) = \begin{cases} \dfrac{\dbinom{90}{x}\dbinom{30}{y}}{\dbinom{120}{5}} & \text{if } x, y \in \{0, 1, 2, 3, 4, 5\}, \quad x + y = 5 \\ \\ 0 & \text{otherwise.} \end{cases}$$

(b) To find p_X and p_Y, the respective marginal probability functions of X and Y, note that $p_X(x) = \sum_{y=0}^{5} p(x, y)$, $p_Y(y) = \sum_{x=0}^{5} p(x, y)$. Since

$p(x, y) = 0$ if $x + y \neq 5$, $\sum_{y=0}^{5} p(x, y) = p(x, 5 - x)$ and $\sum_{x=0}^{5} p(x, y) = p(5 - y, y)$. Therefore

$$p_X(x) = \frac{\binom{90}{x}\binom{30}{5-x}}{\binom{120}{5}}, \qquad x \in \{0, 1, 2, 3, 4, 5\},$$

$$p_Y(y) = \frac{\binom{90}{5-y}\binom{30}{y}}{\binom{120}{5}}, \qquad y \in \{0, 1, 2, 3, 4, 5\}.$$

Note that, as expected, p_X and p_Y are hypergeometric. ◆

Example 8.2 Roll a balanced die and let the outcome be X. Then toss a fair coin X times and let Y denote the number of tails. What is the joint probability function of X and Y and the marginal probability functions of X and Y?

SOLUTION: Let $p(x, y)$ be the joint probability function of X and Y. Clearly, $X \in \{1, 2, 3, 4, 5, 6\}$ and $Y \in \{0, 1, 2, 3, 4, 5, 6\}$. Now if $X = 1$, then $Y = 0$ or 1; we have

$$p(1, 0) = P(X = 1, Y = 0) = P(X = 1)P(Y = 0 \mid X = 1)$$
$$= \frac{1}{6} \cdot \frac{1}{2} = \frac{1}{12},$$
$$p(1, 1) = P(X = 1, Y = 1) = P(X = 1)P(Y = 1 \mid X = 1)$$
$$= \frac{1}{6} \cdot \frac{1}{2} = \frac{1}{12}.$$

If $X = 2$, then $y = 0, 1$, or 2, where

$$p(2, 0) = P(X = 2, Y = 0) = P(X = 2)P(Y = 0 \mid X = 2) = \frac{1}{6} \cdot \frac{1}{4} = \frac{1}{24}.$$

Similarly, $p(2, 1) = 1/12$, $p(2, 2) = 1/24$. If $X = 3$, then $y = 0, 1, 2$, or 3, and

$$p(3, 0) = P(X = 3, Y = 0) = P(X = 3)P(Y = 0 \mid X = 3)$$
$$= \frac{1}{6}\binom{3}{0}\left(\frac{1}{2}\right)^0\left(\frac{1}{2}\right)^3 = \frac{1}{48},$$

$$p(3, 1) = P(X = 3, Y = 1) = P(X = 3)P(Y = 1 \mid X = 3)$$

$$= \frac{1}{6}\binom{3}{1}\left(\frac{1}{2}\right)^2\left(\frac{1}{2}\right)^1 = \frac{3}{48}.$$

Similarly, $p(3, 2) = 3/48$, $p(3, 3) = 1/48$. Similar calculations will yield the following table for $p(x, y)$.

	\multicolumn{7}{c}{y}	
x	0	1	2	3	4	5	6	$p_X(x)$
1	1/12	1/12	0	0	0	0	0	1/6
2	1/24	2/24	1/24	0	0	0	0	1/6
3	1/48	3/48	3/48	1/48	0	0	0	1/6
4	1/96	4/96	6/96	4/96	1/96	0	0	1/6
5	1/192	5/192	10/192	10/192	5/192	1/192	0	1/6
6	1/384	6/384	15/384	20/384	15/384	6/384	1/384	1/6
$p_Y(y)$	63/384	120/384	99/384	64/384	29/384	8/384	1/384	

Note that $p_X(x) = P(X = x)$ and $p_Y(y) = P(Y = y)$, the probability functions of X and Y, are obtained by summing up the rows and the columns of this table, respectively. ◆

EXERCISES

A

1. Let the joint probability function of discrete random variables X and Y be given by

$$p(x, y) = \begin{cases} k\left(\dfrac{x}{y}\right) & \text{if } x = 1, 2, \quad y = 1, 2 \\ 0 & \text{otherwise.} \end{cases}$$

Determine (a) the value of the constant k, (b) the marginal probability functions of X and Y, and (c) $P(X > 1 \mid Y = 1)$.

2. Let the joint probability function of discrete random variables X and Y be given by

$$p(x, y) = \begin{cases} c(x + y) & \text{if } x = 1, 2, 3, \quad y = 1, 2 \\ 0 & \text{otherwise.} \end{cases}$$

Determine (a) the value of the constant c, (b) the marginal probability functions of X and Y, and (c) $P(X \geq 2 \mid Y = 1)$.

3. Let the joint probability function of discrete random variables X and Y be given by

$$p(x, y) = \begin{cases} k(x^2 + y^2) & \text{if } (x, y) = (1, 1), (1, 3), (2, 3) \\ 0 & \text{otherwise.} \end{cases}$$

Determine (a) the value of the constant c, (b) the marginal probability functions of X and Y, and (c) $P(X \geq 2 \mid Y = 1)$.

4. Let the joint probability function of discrete random variables X and Y be given by

$$p(x, y) = \begin{cases} \dfrac{1}{25}(x^2 + y^2) & \text{if } x = 1, 2, \quad y = 0, 1, 2 \\ 0 & \text{otherwise.} \end{cases}$$

Find $P(X > Y)$, $P(X + Y \leq 2)$, and $P(X + Y = 2)$.

5. Thieves took four animals at random from a farm that had seven sheep, eight goats, and five burros. Calculate the joint probability function of the number of sheep and goats stolen.

6. Two dice are rolled. The sum of the outcomes is denoted by X and the absolute value of their difference by Y. Calculate the joint probability function of X and Y, and the marginal probability functions of X and Y.

7. In a community 30% of the adults are Republicans, 50% are Democrats, and the rest are independent. For a randomly selected person, let

$$X = \begin{cases} 1 & \text{if he or she is a Republican} \\ 0 & \text{otherwise,} \end{cases}$$

$$Y = \begin{cases} 1 & \text{if he or she is a Democrat} \\ 0 & \text{otherwise.} \end{cases}$$

Calculate the joint probability function of X and Y.

8. From an ordinary deck of 52 cards, seven cards are drawn at random and without replacement. Let X and Y be the number of hearts and the number of spades drawn, respectively.
(a) Find the joint probability function of X and Y.
(b) Calculate $P(X \geq Y)$.

Joint Probability Density Functions

To define the concept of *joint probability density function* of two continuous
random variables X and Y, recall from Section 6.1 that a single random
variable X is called continuous if there exists a nonnegative real-valued
function $f : \mathbf{R} \to [0, \infty)$ such that for any subset A of real numbers which
can be constructed from intervals by a countable number of set operations,

$$P(X \in A) = \int_A f(x)\, dx.$$

This definition is generalized in the following obvious way:

DEFINITION   *Two random variables X and Y, defined on the same sample space, have a continuous joint distribution if there exists a nonnega-
tive function of two variables, $f(x, y)$ on $\mathbf{R} \times \mathbf{R}$, such that for any region R in the xy-plane that can be formed from rectangles by a count-
able number of set operations,*

$$P\{(X, Y) \in R\} = \iint\limits_R f(x, y)\, dx\, dy. \tag{8.2}$$

*The function $f(x, y)$ is called the joint probability density function of X
and Y.*

Note that if in (8.2) the region R is a plane curve, then $P\{(X, Y) \in
R\} = 0$. That is, the probability that (X, Y) lies on any curve (in particular,
a circle, an ellipse, a straight line, etc.) is 0.
Let $R = \{(x, y) : x \in A, y \in B\}$, where A and B are *any* subsets of
real numbers which can be constructed from intervals by a countable number
of set operations. Then (8.2) gives

$$P(X \in A, Y \in B) = \int_B \int_A f(x, y)\, dx\, dy. \tag{8.3}$$

Letting $A = (-\infty, \infty)$, $B = (-\infty, \infty)$, (8.3) implies the relation

$$\int_{-\infty}^{\infty} \int_{-\infty}^{\infty} f(x, y)\, dx\, dy = 1,$$

which is the continuous analog of (8.1). The relation (8.3) also implies that

$$P(X = a, Y = b) = \int_b^b \int_a^a f(x, y)\, dx\, dy = 0.$$

Hence for $a < b$ and $c < d$,

$$P(a < X \le b, c \le Y \le d) = P(a < X < b, c < Y < d)$$
$$= P(a \le X < b, c \le Y < d) = \cdots$$
$$= \int_c^d \left(\int_a^b f(x, y)\, dx \right) dy.$$

Because for real numbers a and b, $P(X = a, Y = b) = 0$, in general, $f(a, b)$ is not equal to $P(X = a, Y = b)$. At no point do the values of f represent probabilities. Intuitively, $f(a, b)$ is a measure that determines how likely it is that X is close to a and Y is close to b. To see this, note that if ε and δ are very small positive numbers, then $P(a - \varepsilon < X < a + \varepsilon, b - \delta < Y < b + \delta)$ is the probability that X is close to a and Y is close to b. Now

$$P(a - \varepsilon < X < a + \varepsilon, b - \delta < Y < b + \delta) = \int_{b-\delta}^{b+\delta} \int_{a-\varepsilon}^{a+\varepsilon} f(x, y)\, dx\, dy$$

is the volume under the surface $z = f(x, y)$ above the region $(a - \varepsilon, a + \varepsilon) \times (b - \delta, b + \delta)$. For infinitesimal ε and δ, this volume is approximately equal to the volume of a rectangular parallelepiped with sides of lengths 2ε, 2δ, and height $f(a, b)$ [i.e., $(2\varepsilon)(2\delta) f(a, b) = 4\varepsilon\delta f(a, b)$]. Therefore,

$$P(a - \varepsilon < X < a + \varepsilon, b - \delta < Y < b + \delta) \approx 4\varepsilon\delta f(a, b).$$

Hence for fixed small values of ε and δ, we observe that a larger value of $f(a, b)$ makes $P(a - \varepsilon < X < a + \varepsilon, b - \delta < Y < b + \delta)$ larger. That is, the larger the value of $f(a, b)$, the higher the probability that X and Y are close to a and b, respectively.

Let X and Y have joint probability density function $f(x, y)$. Let f_Y be the probability density function of Y. To find f_Y in terms of f, note that on the one hand, for any subset B of $\mathbf{R}$,

$$P(Y \in B) = \int_B f_Y(y)\, dy, \tag{8.4}$$

and on the other hand, using (8.3),

$$P(Y \in B) = P\{X \in (-\infty, \infty), Y \in B\} = \int_B \left(\int_{-\infty}^{\infty} f(x, y)\, dx \right) dy.$$

Comparing this with (8.4), we can write

$$f_Y(y) = \int_{-\infty}^{\infty} f(x, y)\, dx. \tag{8.5}$$

Similarly,

$$f_X(x) = \int_{-\infty}^{\infty} f(x, y)\, dy. \tag{8.6}$$

Therefore, it is reasonable to make the following definition:

DEFINITION *Let X and Y have joint probability density function f(x, y); then the functions f_X and f_Y, given by (8.6) and (8.5), are called, respectively, the marginal probability density functions of X and Y.*

Note that while from the joint probability density function of X and Y we can find the marginal probability density functions of X and Y, it is not possible, in general, to find the joint probability density function of two random variables from their marginals. This is because the marginal probability density function of a random variable X gives information about X without looking at the possible values of other random variables. However, sometimes with more information the marginal probability density functions enable us to find the joint probability density function of two random variables. An example of such a case, studied in Section 8.2, is when X and Y are independent random variables.

Let X and Y be two random variables (discrete, continuous, or mixed). *The joint probability distribution function, or joint cumulative probability distribution function, or simply the joint distribution of X and Y, is defined by*

$$F(t, u) = P(X \le t, Y \le u)$$

for all $-\infty < t, u < \infty$. The marginal probability distribution function of X, F_X, can be found from F as follows:

$$F_X(t) = P(X \le t) = P(X \le t, Y < \infty) = P(\lim_{n \to \infty} \{X \le t, Y \le n\})$$
$$= \lim_{n \to \infty} P\{X \le t, Y \le n\} = \lim_{n \to \infty} F(t, n) \equiv F(t, \infty).$$

To justify the fourth equality, note that the sequence of events $\{X \le t, Y \le n\}$, $n \ge 1$, is an increasing sequence. Therefore, by continuity of probability function (Theorem 1.7),

$$P(\lim_{n \to \infty} \{X \le t, Y \le n\}) = \lim_{n \to \infty} P\{X \le t, Y \le n\}.$$

Similarly, F_Y, the *marginal probability distribution function of Y*, is

$$F_Y(u) = P(Y \le u) = \lim_{n \to \infty} F(n, u) \equiv F(\infty, u).$$

Now suppose that the joint probability density function of X and Y is $f(x, y)$. Then

$$F(x, y) = P(X \le x, Y \le y) = P\{X \in (-\infty, x], Y \in (-\infty, y]\}$$

$$= \int_{-\infty}^{y} \int_{-\infty}^{x} f(x, y) \, dx \, dy. \tag{8.7}$$

Assuming that the partial derivatives of F exist, by differentiation of (8.7), we get

$$f(x, y) = \frac{\partial^2}{\partial x \, \partial y} F(x, y).$$

Moreover, from (8.7), we obtain that

$$F_X(x) = F(x, \infty) = \int_{-\infty}^{\infty} \left(\int_{-\infty}^{x} f(x, y) \, dx \right) dy$$

$$= \int_{-\infty}^{x} \left(\int_{-\infty}^{\infty} f(x, y) \, dy \right) dx = \int_{-\infty}^{x} f_X(x) \, dx,$$

and similarly,

$$F_Y(y) = \int_{-\infty}^{y} f_Y(y) \, dy.$$

These relations show that if X and Y have joint probability density function $f(x, y)$, then X and Y are continuous random variables with density functions f_X and f_Y, and distribution functions F_X and F_Y, respectively. Therefore, $F_X'(x) = f_X(x)$ and $F_Y'(y) = f_Y(y)$.

Example 8.3   The joint probability density function of random variables X and Y is given by

$$f(x, y) = \begin{cases} \lambda x y^2 & 0 \le x \le y \le 1 \\ 0 & \text{otherwise.} \end{cases}$$

(a) Determine the value of λ.

(b) Find the marginal probability density functions of X and Y.

SOLUTION: (a) To find λ, note that

$$\int_{-\infty}^{\infty} \int_{-\infty}^{\infty} f(x, y)\, dx\, dy = 1.$$

This gives

$$\int_{0}^{1} \left(\int_{x}^{1} \lambda xy^2\, dy \right) dx = 1.$$

Therefore,

$$1 = \int_{0}^{1} \left(\int_{x}^{1} y^2\, dy \right) \lambda x\, dx = \int_{0}^{1} \left[\frac{1}{3}y^3 \right]_{x}^{1} \lambda x\, dx$$

$$= \lambda \int_{0}^{1} \left(\frac{1}{3} - \frac{1}{3}x^3 \right) x\, dx = \frac{\lambda}{3} \int_{0}^{1} (1 - x^3)x\, dx = \frac{\lambda}{3} \int_{0}^{1} (x - x^4)\, dx$$

$$= \frac{\lambda}{3} \left[\frac{1}{2}x^2 - \frac{1}{5}x^5 \right]_{0}^{1} = \frac{\lambda}{10},$$

and hence $\lambda = 10$.

(b) To find f_X and f_Y, the respective marginal probability density functions of X and Y, we use (8.6) and (8.5):

$$f_X(x) = \int_{-\infty}^{\infty} f(x, y)\, dy = \int_{x}^{1} 10xy^2\, dy = \left[\frac{10}{3}xy^3 \right]_{x}^{1}$$

$$= \frac{10}{3}x(1 - x^3), \qquad 0 \le x \le 1;$$

$$f_Y(y) = \int_{-\infty}^{\infty} f(x, y)\, dx = \int_{0}^{y} 10xy^2\, dx = \left[5x^2y^2 \right]_{0}^{y}$$

$$= 5y^4, \qquad 0 \le y \le 1. \quad \blacklozenge$$

Example 8.4   For $\lambda > 0$, let

$$F(x, y) = \begin{cases} 1 - \lambda e^{-\lambda(x+y)} & \text{if } x > 0, \quad y > 0 \\ 0 & \text{otherwise.} \end{cases}$$

Determine if F is the joint probability distribution function of two random variables X and Y.

SOLUTION: If F is the joint probability distribution function of two random variables X and Y, then $\dfrac{\partial^2}{\partial x \, \partial y} F(x, y)$ is the joint probability density function of X and Y. But

$$\frac{\partial^2}{\partial x \, \partial y} F(x, y) = \begin{cases} -\lambda^3 e^{-\lambda(x+y)} & \text{if } x > 0, \quad y > 0 \\ 0 & \text{otherwise.} \end{cases}$$

Since $\dfrac{\partial^2}{\partial x \, \partial y} F(x, y) < 0$, it cannot be a joint probability density function. Therefore, F is not a joint probability distribution function. ◆

Example 8.5   A circle of radius 1 is inscribed in a square with sides of length 2.   A point is selected at random from the square.   What is the probability that it is inside the circle?   Note that by a point being selected at random from the square we mean that the point is selected in a way that all the subsets of equal areas of the square are equally likely to contain the point.

SOLUTION: Let the square and the circle be situated in the coordinate system as shown in Figure 8.1. Let the coordinates of the point selected at random be (X, Y); then X and Y are random variables. By definition, regions inside the square with equal areas are equally likely to contain (X, Y). Hence all the points of the square have the same probability of being at a certain distance from (X, Y). Let $f(x, y)$ be the joint probability density function

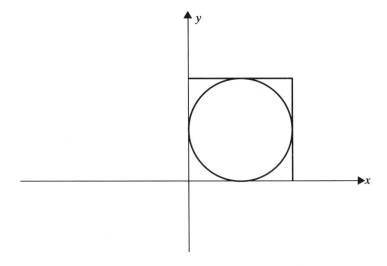

Figure 8.1   Geometric model of Example 8.5.

of X and Y. Since $f(x, y)$ is a measure that determines how likely it is that X is close to x and Y is close to y, $f(x, y)$ must be constant for the points inside the square, and 0 elsewhere. Therefore, for some $c > 0$,

$$f(x, y) = \begin{cases} c & \text{if } 0 < x < 2, \quad 0 < y < 2 \\ 0 & \text{otherwise,} \end{cases}$$

where

$$\int_{-\infty}^{\infty} \int_{-\infty}^{\infty} f(x, y) \, dx \, dy = 1.$$

This gives

$$\int_{0}^{2} \int_{0}^{2} c \, dx \, dy = 1,$$

implying that $c = 1/4$. Now let R be the region inside the circle. Then the desired probability is

$$\iint_R f(x, y) \, dx \, dy = \iint_R \frac{1}{4} \, dx \, dy = \frac{\displaystyle\iint_R dx \, dy}{4}.$$

Note that $\displaystyle\iint_R dx \, dy$ is the area of the circle, and 4 is the area of the square. Thus the desired probability is

$$\frac{\text{area of the circle}}{\text{area of the square}} = \frac{\pi(1)^2}{4} = \frac{\pi}{4}. \quad \blacklozenge$$

What we showed in Example 8.5 is true in general. Let S be a bounded region in the Euclidean plane and suppose that R is a region inside S. Fix a coordinate system, and let the coordinates of a point selected at random from S be (X, Y). By an argument similar to that of Example 8.5, we have that for some $c > 0$, the joint probability density function of X and Y, $f(x, y)$, is given by

$$f(x, y) = \begin{cases} c & \text{if } (x, y) \in S \\ 0 & \text{otherwise,} \end{cases}$$

where

$$\iint_S f(x, y)\, dx\, dy = 1.$$

This gives $c \iint_S dx\, dy = 1$, or equivalently, $c \times \text{area}(S) = 1$. Therefore, $c = 1/\text{area}(S)$ and hence

$$f(x, y) = \begin{cases} \dfrac{1}{\text{area}(S)} & \text{if } (x, y) \in S \\[2mm] 0 & \text{otherwise.} \end{cases}$$

Thus

$$P\{(X, Y) \in R\} = \iint_R f(x, y)dxdy = \frac{1}{\text{area}(S)} \iint_R dx\, dy = \frac{\text{area}(R)}{\text{area}(S)}.$$

Based on these observations, we make the following definition.

DEFINITION   *Let S be a subset of the plane with area $A(S)$. A point is said to be randomly selected from S if for any subset R of S with area $A(R)$, the probability that R contains the point is $A(R)/A(S)$.*

This definition is essential in the field of *geometric probability*. By the following examples, we will show how it can help to solve problems easily. An additional interesting example called *Buffon's needle problem* is illustrated in Section 12.5.

Example 8.6   A man invites his fiancée to a gorgeous hotel for a fancy Sunday brunch. They decide to meet in the lobby of the hotel between 11:30 A.M. and 12 noon. If they arrive at random times during this period, what is the probability that they meet within 10 minutes?

SOLUTION: Let X and Y be the minutes past 11:30 A.M. that the man and his fiancée arrive at the lobby, respectively. Let $S = \{(x, y) : 0 \le x \le 30,\ 0 \le y \le 30\}$, and $R = \{(x, y) \in S : |x - y| \le 10\}$. Then the desired probability, $P(|X - Y| \le 10)$, is given by

$$P(|X - Y| \le 10) = \frac{\text{area of } R}{\text{area of } S} = \frac{\text{area of } R}{30 \times 30} = \frac{\text{area}(R)}{900}.$$

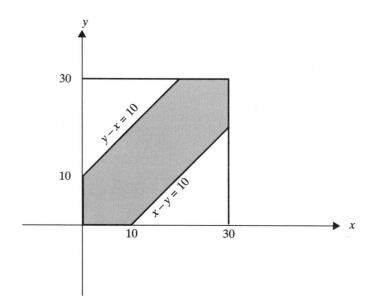

Figure 8.2   Geometric model of Example 8.6.

Now $R = \{(x, y) \in S : x - y \le 10$ and $y - x \le 10\}$ is the shaded region of Figure 8.2, and its area is the area of the square minus the areas of the two unshaded triangles: $(30)(30) - 2(1/2 \times 20 \times 20) = 500$. Hence the desired probability is $500/900 = 5/9$. ♦

Example 8.7   A farmer decides to build a pen for his chickens in the shape of a triangle. He sends his foolish son out to cut the lumber and the boy, without taking any thought as to the ultimate purpose, makes two cuts at two points selected at random. What are the chances that the resulting three pieces of lumber can be used to form a triangular pen?

SOLUTION: Suppose that the length of the lumber is ℓ. Let A and B be the random points placed on the lumber; let the distances of A and B from the left end of the lumber be denoted by X and Y, respectively. If $X < Y$, the lumber is divided into three parts of lengths X, $Y - X$, and $\ell - Y$; otherwise, it is divided into three parts of lengths Y, $X - Y$, and $\ell - X$. Because of symmetry, we calculate the probability that $X < Y$, and X, $Y - X$, and $\ell - Y$ form a triangle. Then we multiply the result by 2 to obtain the desired probability. We know that three segments form a triangle if and only if the length of any one of them is less than the sum of the lengths of the remaining two. Therefore, we must have

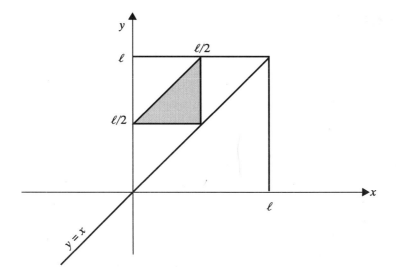

Figure 8.3 Geometric model of Example 8.7.

$$\begin{cases} X < Y \\ X < (Y - X) + (\ell - Y) \\ (Y - X) < X + (\ell - Y) \\ \ell - Y < X + (Y - X), \end{cases}$$

or equivalently,

$$\begin{cases} X < Y \\ X < \dfrac{\ell}{2} \\ Y < X + \dfrac{\ell}{2} \\ Y > \dfrac{\ell}{2}. \end{cases}$$

Since (X, Y) is a random point from the square $(0, \ell) \times (0, \ell)$, the probability that it satisfies these inequalities is the area of R, where

$$R = \left\{ (x, y) : x < y, x < \frac{\ell}{2}, y < x + \frac{\ell}{2}, y > \frac{\ell}{2} \right\},$$

divided by ℓ^2, the area of the square. Note that R is the shaded region of Figure 8.3 and its area is 1/8 of the area of the square. Thus the probability that (X, Y) lies in R is 1/8 and hence the desired probability is 1/4. ◆

EXERCISES

A

1. Let the joint probability density function of random variables X and Y be given by

$$f(x, y) = \begin{cases} 2 & \text{if } 0 \leq y \leq x \leq 1 \\ 0 & \text{elsewhere.} \end{cases}$$

   (a) Calculate the marginal probability density functions of X and Y, respectively.
   (b) Calculate $P(X < 1/2)$, $P(X < 2Y)$, and $P(X = Y)$.

2. Let the joint probability density function of random variables X and Y be given by

$$f(x, y) = \begin{cases} 8xy & \text{if } 0 \leq y \leq x \leq 1 \\ 0 & \text{elsewhere.} \end{cases}$$

   Calculate the marginal probability density functions of X and Y, respectively.

3. Let the joint probability density function of random variables X and Y be given by

$$f(x, y) = \begin{cases} \dfrac{1}{2}ye^{-x} & \text{if } x > 0, \quad 0 < y < 2 \\ 0 & \text{elsewhere.} \end{cases}$$

   Find the marginal probability density functions of X and Y.

4. Let X and Y have the joint probability density function

$$f(x, y) = \begin{cases} 1 & \text{if } 0 \leq x \leq 1, \quad 0 \leq y \leq 1 \\ 0 & \text{elsewhere.} \end{cases}$$

   Calculate $P(X + Y \leq 1/2)$, $P(X - Y \leq 1/2)$, $P(XY \leq 1/4)$, and $P(X^2 + Y^2 \leq 1)$.

5. Let R be the bounded region between $y = x$ and $y = x^2$. A random point (X, Y) is selected from R. Find the joint probability density function of X and Y.

6. A man invites his fiancée to a gorgeous hotel for a fancy Sunday brunch. They decide to meet in the lobby of the hotel between 11:30 A.M. and

12 noon. If they arrive at random times during this period, what is the probability that the first to arrive has to wait at least 12 minutes?

7. On a line segment AB of length ℓ, two points C and D are placed at random and independently. What is the probability that C is closer to D than to A?

8. Two points X and Y are selected at random and independently from the interval $(0, 1)$. Calculate $P(Y \leq X \text{ and } X^2 + Y^2 \leq 1)$.

B

9. Suppose that h is the probability density function of a continuous random variable. Let the joint probability density function of two random variables X and Y be given by

$$f(x, y) = h(x)h(y), \qquad x \in \mathbb{R}, \quad y \in \mathbb{R}.$$

Prove that $P(X \geq Y) = 1/2$.

10. Let g and h be two probability density functions with probability distribution functions G and H, respectively. Show that for $-1 \leq \alpha \leq 1$, the function

$$f(x, y) = g(x)h(y)\{1 + \alpha[2G(x) - 1][2H(y) - 1]\}$$

is a joint probability density function of two random variables. Moreover, prove that g and h are the marginal probability density functions of f. NOTE: This exercise gives an infinite family of joint probability density functions all with the same marginal probability density function of X and of Y.

11. Three points M, N, and L are placed on a circle at random and independently. What is the probability that MNL is an acute angle?

12. Two numbers x and y are selected at random from the interval $(0, 1)$. For $i = 0, 1, 2$, determine the probability that the integer nearest to $x + y$ is i. NOTE: This problem was given by Hilton and Pedersen in the paper "A Role for Untraditional Geometry in the Curriculum," published in the December 1989 issue of *Kolloquium Mathematik-Didaktik der Universität Bayreuth*. It is not hard to solve, but it has interesting consequences in arithmetic.

13. A farmer who has two pieces of lumber of lengths a and b ($a < b$) decides to build a pen for his chickens in the shape of a triangle. He sends his foolish son out to cut the lumber and the boy, without taking

any thought as to the ultimate purpose, makes two cuts, one on each piece, at randomly selected points. He then chooses three of the resulting pieces at random and take them to his father. What are the chances that they can be used to form a triangular pen?

14. Two points are placed on a segment of length ℓ independently and at random to divide the line into three parts. What is the probability that the length of none of the three parts exceeds a given value α, $\ell/3 \le \alpha \le \ell$?

15. A point is selected at random and uniformly from the region

$$R = \{(x, y) : |x| + |y| \le 1\}.$$

Find the probability density function of the x-coordinate of the point selected at random.

16. For $\alpha > 0$, $\beta > 0$, and $\gamma > 0$, the following function is called the *bivariate Dirichlet probability density function.*

$$f(x, y) = \frac{\Gamma(\alpha + \beta + \gamma)}{\Gamma(\alpha)\Gamma(\beta)\Gamma(\gamma)} x^{\alpha-1} y^{\beta-1} (1 - x - y)^{\gamma-1}$$

if $x \ge 0$, $y \ge 0$, and $x + y \le 1$; $f(x, y) = 0$, otherwise. Prove that f_X, the marginal probability density function of X, is beta with parameters $(\alpha, \beta + \gamma)$; and f_Y is beta with the parameters $(\beta, \alpha + \gamma)$.
HINT: Note that

$$\frac{\Gamma(\alpha + \beta + \gamma)}{\Gamma(\alpha)\Gamma(\beta)\Gamma(\gamma)} = \left[\frac{\Gamma(\alpha + \beta + \gamma)}{\Gamma(\alpha)\Gamma(\beta + \gamma)}\right]\left[\frac{\Gamma(\beta + \gamma)}{\Gamma(\beta)\Gamma(\gamma)}\right]$$

$$= \frac{1}{B(\alpha, \beta + \gamma)} \frac{1}{B(\beta, \gamma)}.$$

8.2   Independent Random Variables

Two random variables X and Y are called *independent* if for arbitrary subsets A and B of real numbers, the events $\{X \in A\}$ and $\{Y \in B\}$ are independent, that is, if

$$P(X \in A, Y \in B) = P(X \in A)P(Y \in B). \tag{8.8}$$

Using the axioms of probability, one can prove that X and Y are independent if and only if for any two real numbers a and b,

$$P(X \le a, Y \le b) = P(X \le a)P(Y \le b). \tag{8.9}$$

Hence (8.8) and (8.9) are equivalent.

If X and Y are discrete random variables with sets of possible values E and F, respectively, and joint probability function $p(x, y)$, then the definition of independence is satisfied if for all $x \in E$ and $y \in F$, the events $\{X = x\}$ and $\{Y = y\}$ are independent; that is,

$$P(X = x, Y = y) = P(X = x)P(Y = y). \tag{8.10}$$

Relation (8.9) states that X and Y are independent random variables, if and only if their joint probability distribution function is the product of their marginal distribution functions. If X and Y are discrete, relation (8.10) shows that their joint probability function is the product of their marginal probability functions as well. The following theorems state these facts.

THEOREM 8.1 *Let X and Y be two random variables defined on the same sample space. If F is the joint probability distribution function of X and Y, then X and Y are independent if and only if for all real numbers t and u,*

$$F(t, u) = F_X(t)F_Y(u).$$

THEOREM 8.2 *Let X and Y be two discrete random variables defined on the same sample space. If p is the joint probability function of X and Y, then X and Y are independent if and only if for all real numbers x and y,*

$$p(x, y) = p_X(x)p_Y(y).$$

Let X and Y be discrete independent random variables with sets of possible values E and F, respectively. Then (8.10) implies that for all $x \in E$ and $y \in F$,

$$P(X = x \mid Y = y) = P(X = x)$$

and

$$P(Y = y \mid X = x) = P(Y = y).$$

Hence X and Y are *independent if knowing the value of one of them does not change the probability function of the other.*

For jointly continuous random variables X and Y having joint probability distribution function $F(x, y)$ and joint probability density function $f(x, y)$, Theorem 8.1 implies that for any two real numbers x and y,

$$F(x, y) = F_X(x)F_Y(y). \tag{8.11}$$

Differentiating (8.11) with respect to x, we get

$$\frac{\partial F}{\partial x} = f_X(x)F_Y(y),$$

which upon differentiation with respect to y yields

$$\frac{\partial^2 F}{\partial y \partial x} = f_X(x)f_Y(y),$$

or equivalently,

$$f(x, y) = f_X(x)f_Y(y). \tag{8.12}$$

We leave it as an exercise that if (8.12) is valid, X and Y are independent. We have the following theorem:

THEOREM 8.3   *Let X and Y be jointly continuous random variables with joint probability density function $f(x, y)$. Then X and Y are independent if and only if $f(x, y)$ is the product of their marginal densities $f_X(x)$ and $f_Y(y)$.*

Example 8.8   Suppose that 4% of the bicycle fenders cut by a stamping machine from the strips of steel need smoothing. What is the probability that of the next 13 bicycle fenders cut by this stamping machine, two need smoothing, and of the next 20, three need smoothing?

SOLUTION: Let X be the number of bicycle fenders among the first 13 that need smoothing. Let Y be the number of those among the next 7 that need smoothing. We want to calculate $P(X = 2, Y = 1)$. Since X and Y are independent binomial random variables with parameters $(13, 0.04)$ and $(7, 0.04)$, respectively, we can write

$$P(X = 2, Y = 1) = P(X = 2)P(Y = 1)$$

$$= \binom{13}{2}(0.04)^2(0.96)^{11}\binom{7}{1}(0.04)^1(0.96)^6 \approx 0.0175. \quad \blacklozenge$$

The idea of Example 8.8 can be stated in general terms: Suppose that $n + m$ independent Bernoulli trials each with parameter p are performed. If X and Y are the number of successes in the first n and in the last m trials, then X and Y are binomial random variables with parameters (n, p) and

(m, p), respectively. Furthermore, they are independent because knowing the number of successes in the first n trials does not change the probability function of the number of successes in the last m trials. Hence

$$P(X = i, Y = j) = P(X = i)P(Y = j)$$

$$= \binom{n}{i} p^i (1 - p)^{n-i} \binom{m}{j} p^j (1 - p)^{m-j}$$

$$= \binom{n}{i} \binom{m}{j} p^{i+j} (1 - p)^{(n+m)-(i+j)}.$$

Example 8.9 Stores A and B, which belong to the same owner, are located in two different towns. If the probability density function of the weekly profit of each store, in thousands of dollars, is given by

$$f(x) = \begin{cases} \dfrac{x}{4} & \text{if } 1 < x < 3 \\ 0 & \text{otherwise,} \end{cases}$$

and the profit of one store is independent of the other, what is the probability that next week one store makes at least $500 more than the other store?

SOLUTION: Let X and Y denote next week's profits of A and B, respectively. The desired probability is $P(X > Y + 1/2) + P(Y > X + 1/2)$. Since X and Y have the same probability density function, by symmetry, this sum equals $2P(X > Y + 1/2)$. To calculate this, we need to know $f(x, y)$, the joint probability density function of X and Y. Since X and Y are independent,

$$f(x, y) = f_X(x) f_Y(y),$$

where

$$f_X(x) = \begin{cases} \dfrac{x}{4} & \text{if } 1 < x < 3 \\ 0 & \text{otherwise} \end{cases}$$

$$f_Y(y) = \begin{cases} \dfrac{y}{4} & \text{if } 1 < y < 3 \\ 0 & \text{otherwise.} \end{cases}$$

Thus

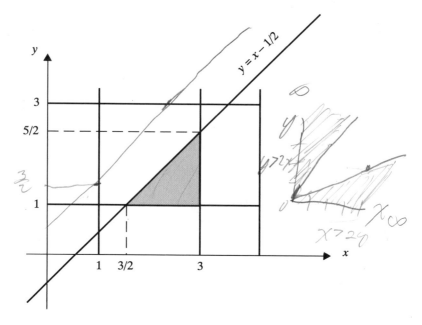

Figure 8.4   Figure of Example 8.9.

$$f(x, y) = \begin{cases} \dfrac{xy}{16} & \text{if } 1 < x < 3, \ 1 < y < 3 \\ 0 & \text{otherwise.} \end{cases}$$

The desired probability, as seen from Figure 8.4, is therefore equal to

$$2P\left(X > Y + \frac{1}{2}\right) = 2P\left\{(X, Y) \in \left\{(x, y) : \frac{3}{2} < x < 3, 1 < y < x - \frac{1}{2}\right\}\right\}$$

$$= 2\int_{3/2}^{3}\left(\int_{1}^{x-1/2} \frac{xy}{16}dy\right)dx = \frac{1}{8}\int_{3/2}^{3}\left[\frac{xy^2}{2}\right]_{1}^{x-1/2}dx$$

$$= \frac{1}{16}\int_{3/2}^{3} x\left[(x - \frac{1}{2})^2 - 1\right]dx$$

$$= \frac{1}{16}\int_{3/2}^{3}\left(x^3 - x^2 - \frac{3}{4}x\right)dx$$

$$= \frac{1}{16}\left[\frac{1}{4}x^4 - \frac{1}{3}x^3 - \frac{3}{8}x^2\right]_{3/2}^{3} = \frac{549}{1024} \approx 0.54. \quad \blacklozenge$$

Example 8.10   A point is selected at random from the rectangle

$$R = \{(x, y) \in \mathbf{R}^2 : 0 < x < a, 0 < y < b\}.$$

Let X be the x-coordinate and Y be the y-coordinate of the point selected. Determine if X and Y are independent random variables.

SOLUTION: From Section 8.1 we know that $f(x, y)$, the joint probability density function of X and Y, is given by

$$f(x, y) = \begin{cases} \dfrac{1}{\text{area}(R)} = \dfrac{1}{ab} & \text{if } (x, y) \in R \\ 0 & \text{elsewhere.} \end{cases}$$

Now the marginal density functions of X and Y, f_X and f_Y, are given by

$$f_X(x) = \int_0^b \frac{1}{ab} dy = \frac{1}{a}, \qquad x \in (0, a),$$

$$f_Y(y) = \int_0^a \frac{1}{ab} dx = \frac{1}{b}, \qquad y \in (0, b).$$

Therefore, $f(x, y) = f_X(x) f_Y(y)$, $\forall x, y \in \mathbf{R}$, and hence X and Y are independent. ◆

Example 8.11   Prove that two random variables X and Y with the following joint probability density function are not independent.

$$f(x, y) = \begin{cases} 8xy & 0 \le x \le y \le 1 \\ 0 & \text{otherwise.} \end{cases}$$

SOLUTION: To check the validity of (8.12), we first calculate f_X and f_Y.

$$f_X(x) = \int_x^1 8xy \, dy = 4x(1 - x^2), \qquad 0 \le x \le 1,$$

$$f_Y(y) = \int_0^y 8xy \, dx = 4y^3, \qquad 0 \le y \le 1.$$

Now since $f(x, y) \ne f_X(x) f_Y(y)$, X and Y are dependent. This is expected because of the range $0 \le x \le y \le 1$, which is not a Cartesian product of one-dimensional regions. ◆

We now prove that functions of independent random variables are also independent.

THEOREM 8.4 *Let X and Y be independent random variables and $f : \mathbf{R} \rightarrow \mathbf{R}$ and $g : \mathbf{R} \rightarrow \mathbf{R}$ be real-valued functions, then $f(X)$ and $g(Y)$ are also independent random variables.*

PROOF: $f(X)$ and $g(Y)$ are independent since for arbitrary subsets A and B of $\mathbf{R}$ we can write

$$P\{f(X) \in A, g(Y) \in B\} = P\{X \in f^{-1}(A). \; Y \in g^{-1}(B)\}$$
$$= P\{X \in f^{-1}(A)\}P\{Y \in g^{-1}(B)\}$$
$$= P\{f(X) \in A\}P\{f(Y) \in B\}. \quad \blacklozenge$$

By this theorem, if X and Y are independent random variables, then sets such as $\{X^2, Y\}$, $\{\sin X, e^Y\}$, $\{X^2 - 2X, Y^3 + 3Y\}$ are sets of independent random variables.

Another important property of independent random variables is that the expected value of their product is equal to the product of their expected values. To prove this, we need to state the following generalization of Theorem 6.3 from 1 to dimension 2.

THEOREM 8.5 *Let $f(x, y)$ be the joint probability density function of random variables X and Y. If h is a function of two variables from $\mathbf{R}^2$ to $\mathbf{R}$, then $Z = h(X, Y)$ is a random variable with the expected value given by*

$$E(Z) = \int_{-\infty}^{\infty} \int_{-\infty}^{\infty} h(x, y) f(x, y) \, dx \, dy,$$

provided that the integral is absolutely convergent.

COROLLARY *For random variables X and Y,*

$$E(X + Y) = E(X) + E(Y).$$

PROOF: We prove this for continuous random variables. For discrete random variables the proof is similar. In Theorem 8.5 let $h(x, y) = x + y$. Then

$$E(X+Y) = \int_{-\infty}^{\infty}\int_{-\infty}^{\infty}(x+y)f(x,y)\,dx\,dy$$

$$= \int_{-\infty}^{\infty}\int_{-\infty}^{\infty}xf(x,y)\,dx\,dy + \int_{-\infty}^{\infty}\int_{-\infty}^{\infty}yf(x,y)\,dx\,dy$$

$$= E(X)+E(Y). \quad \blacklozenge$$

In Section 9.3 we study some applications of a generalization of this basic and important corollary.

Example 8.12   Let X and Y have joint probability density function

$$f(x,y) = \begin{cases} \dfrac{3}{2}(x^2+y^2) & \text{if } 0 < x < 1, \quad 0 < y < 1 \\ 0 & \text{otherwise.} \end{cases}$$

Find $E(X^2+Y^2)$.

SOLUTION: By Theorem 8.5,

$$E(X^2+Y^2) = \int_{-\infty}^{\infty}\int_{-\infty}^{\infty}(x^2+y^2)f(x,y)\,dx\,dy$$

$$= \int_0^1\int_0^1 \frac{3}{2}(x^2+y^2)^2\,dx\,dy$$

$$= \frac{3}{2}\int_0^1\int_0^1(x^4+2x^2y^2+y^4)\,dx\,dy = \frac{14}{15}. \quad \blacklozenge$$

Note that the following discrete version of Theorem 8.5 is also valid: If X and Y are discrete random variables with joint probability function $p(x,y)$, and Z and h are as before, then

$$E(Z) = \sum_x\sum_y h(x,y)p(x,y),$$

provided that the sum is absolutely convergent.

THEOREM 8.6   *Let X and Y be independent random variables. Then for all real-valued functions $g: \mathbf{R} \to \mathbf{R}$ and $h: \mathbf{R} \to \mathbf{R}$,*

$$E[g(X)h(Y)] = E[g(X)]E[h(Y)],$$

where, as usual, we assume that $E[g(X)]$ and $E[h(Y)]$ are finite.

PROOF: Again, for convenience, we prove this for continuous random variables. For discrete random variables the proof is similar. Let $f(x, y)$ be the joint probability density function of X and Y. Then

$$
\begin{aligned}
E[g(X)h(Y)] &= \int_{-\infty}^{\infty} \int_{-\infty}^{\infty} g(x)h(y)f(x, y)\,dx\,dy \\
&= \int_{-\infty}^{\infty} \int_{-\infty}^{\infty} g(x)h(y)f_X(x)f_Y(y)\,dx\,dy \\
&= \int_{-\infty}^{\infty} h(y)f_Y(y) \left(\int_{-\infty}^{\infty} g(x)f_X(x)dx \right) dy \\
&= \left(\int_{-\infty}^{\infty} g(x)f_X(x)\,dx \right) \left(\int_{-\infty}^{\infty} h(y)f_Y(y)\,dy \right) \\
&= E[g(X)]E[h(Y)]. \quad \blacklozenge
\end{aligned}
$$

By this theorem, if X and Y are independent, then $E(XY) = E(X)E(Y)$, and relations such as the following are valid.

$$
E\left(X^2|Y|\right) = E\left(X^2\right) E(|Y|),
$$
$$
E\left[(\sin X)e^Y\right] = E(\sin X)E\left(e^Y\right),
$$
$$
E\left[\left(X^2 - 2X\right)\left(Y^3 + 3Y\right)\right] = E\left(X^2 - 2X\right) E\left(Y^3 + 3Y\right).
$$

The converse of Theorem 8.6 is not necessarily true. That is, two random variables X and Y might be dependent while $E(XY) = E(X)E(Y)$. Here is an example:

Example 8.13 Let X be a random variable with the set of possible values $\{-1, 0, 1\}$ and probability function $p(-1) = p(0) = p(1) = 1/3$. Letting $Y = X^2$, we have

$$
E(X) = -1 \cdot \frac{1}{3} + 0 \cdot \frac{1}{3} + 1 \cdot \frac{1}{3} = 0,
$$
$$
E(Y) = E(X^2) = (-1)^2 \cdot \frac{1}{3} + 0^2 \cdot \frac{1}{3} + (1)^2 \cdot \frac{1}{3} = \frac{2}{3},
$$
$$
E(XY) = E(X^3) = (-1)^3 \cdot \frac{1}{3} + 0^3 \cdot \frac{1}{3} + (1)^3 \cdot \frac{1}{3} = 0.
$$

Thus $E(XY) = E(X)E(Y)$ while, clearly, X and Y are dependent. $\blacklozenge$

EXERCISES

A

1. Let the joint probability function of random variables X and Y be given by

$$p(x, y) = \begin{cases} \dfrac{1}{25}(x^2 + y^2) & \text{if } x = 1, 2, \quad y = 0, 1, 2 \\ 0 & \text{elsewhere.} \end{cases}$$

Are X and Y independent? Why or why not?

2. Let the joint probability function of random variables X and Y be given by

$$p(x, y) = \begin{cases} \dfrac{1}{7}x^2 y & \text{if } (x, y) = (1, 1), (1, 2), (2, 1) \\ 0 & \text{elsewhere.} \end{cases}$$

Are X and Y independent? Why or why not?

3. Let X and Y be independent random variables each having the probability function

$$p(x) = \frac{1}{2}\left(\frac{2}{3}\right)^x, \qquad x = 1, 2, 3, \ldots.$$

Find $P(X = 1, Y = 3)$ and $P(X + Y = 3)$.

4. From an ordinary deck of 52 cards, eight cards are drawn at random and without replacement. Let X and Y be the number of clubs and spades, respectively. Are X and Y independent?

5. What is the probability that there are exactly two girls among the first seven and exactly four girls among the first 15 babies born in a hospital in a given week? Assume that the events that a child born is a girl and is a boy are equiprobable.

6. Let X and Y be two independent random variables with distribution functions F and G, respectively. Find the distribution functions of $\max(X, Y)$ and $\min(X, Y)$.

7. A fair coin is tossed n times by Adam and n times by Andrew. What is the probability that they get the same number of heads?

8. The joint probability function $p(x, y)$ of the random variables X and Y is given by the following table. Determine if X and Y are independent.

	y			
x	0	1	2	3
0	0.1681	0.1804	0.0574	0.0041
1	0.1804	0.1936	0.0616	0.0044
2	0.0574	0.0616	0.0196	0.0014
3	0.0041	0.0044	0.0014	0.0001

9. Let the joint probability density function of random variables X and Y be given by

$$f(x, y) = \begin{cases} 2 & \text{if } 0 \le y \le x \le 1 \\ 0 & \text{elsewhere.} \end{cases}$$

Are X and Y independent? Why or why not?

10. Suppose that the amount of cholesterol in a certain type of sandwich is $100X$ milligrams, where X is a random variable with the following density function:

$$f(x) = \begin{cases} \dfrac{2x + 3}{18} & \text{if } 2 < x < 4 \\ 0 & \text{otherwise.} \end{cases}$$

Find the probability that two such sandwiches made independently have the same amount of cholesterol.

11. Let the joint probability density function of random variables X and Y be given by

$$f(x, y) = \begin{cases} x^2 e^{-x(y+1)} & \text{if } x \ge 0, \quad y \ge 0 \\ 0 & \text{elsewhere.} \end{cases}$$

Are X and Y independent? Why or why not?

12. Let the joint probability density function of X and Y be given by

$$f(x, y) = \begin{cases} 8xy & \text{if } 0 \le x < y \le 1 \\ 0 & \text{otherwise.} \end{cases}$$

Determine if $E(XY) = E(X)E(Y)$.

13. Let the joint probability density function of X and Y be given by

$$f(x, y) = \begin{cases} 2e^{-(x+2y)} & \text{if } x \geq 0, \quad y \geq 0 \\ 0 & \text{otherwise.} \end{cases}$$

Find $E(X^2 Y)$.

14. Let X and Y be two independent random variables with the same probability density function given by

$$f(x) = \begin{cases} e^{-x} & \text{if } 0 < x < \infty \\ 0 & \text{elsewhere.} \end{cases}$$

Show that g, the probability density function of X/Y, is given by

$$g(t) = \begin{cases} \dfrac{1}{(1+t)^2} & \text{if } 0 < t < \infty \\ 0 & t \leq 0. \end{cases}$$

15. A farmer makes cuts at two points selected at random on a piece of lumber of length ℓ. What is the expected value of the length of the middle piece?

16. Let X and Y be independent exponential random variables both with mean 1. Find $E[\max(X, Y)]$.

17. Let X and Y be independent random points from the interval $(-1, 1)$. Find $E[\max(X, Y)]$.

18. Let X and Y be independent random points from the interval $(0, 1)$. Find the probability density function of the random variable XY.

19. A point is selected at random from the disk

$$R = \{(x, y) \in \mathbf{R}^2 : x^2 + y^2 \leq 1\}.$$

Let X be the x-coordinate and Y be the y-coordinate of the point selected. Determine if X and Y are independent random variables.

20. Six brothers and sisters who are all either under 10 or in their early teens, are having dinner with their parents and four grandparents. Their mother unintentionally feeds the entire family (including herself) a type of poisonous mushrooms that makes 20% of the adults and 30% of the children sick. What is the probability that more adults get sick than children?

21. The lifetimes of mufflers manufactured by company A are random with the following density function:

$$f(x) = \begin{cases} \dfrac{1}{6}e^{-x/6} & \text{if } x > 0 \\ 0 & \text{elsewhere.} \end{cases}$$

The lifetimes of mufflers manufactured by company B are random with the following density function:

$$g(y) = \begin{cases} \dfrac{2}{11}e^{-2y/11} & \text{if } y > 0 \\ 0 & \text{elsewhere.} \end{cases}$$

Elizabeth buys two mufflers, one from company A and the other one from company B and installs them on her cars at the same time. What is the probability that the muffler of company B outlasts that of company A?

B

22. Let E be an event; the random variable I_E, defined as follows is called the *indicator of E*:

$$I_E = \begin{cases} 1 & \text{if } E \text{ occurs} \\ 0 & \text{otherwise.} \end{cases}$$

Show that A and B are independent events if and only if I_A and I_B are independent random variables.

23. Let B and C be two independent random variables both having the following density function:

$$f(x) = \begin{cases} \dfrac{3x^2}{26} & \text{if } 1 < x < 3 \\ 0 & \text{otherwise.} \end{cases}$$

What is the probability that the quadratic equation $X^2 + BX + C = 0$ has two real roots?

24. Let the joint probability density function of two random variables X and Y satisfy

$$f(x, y) = g(x)h(y), \qquad -\infty < x < \infty, \quad -\infty < y < \infty,$$

where g and h are two functions from $\mathbf{R}$ to $\mathbf{R}$. Show that X and Y are independent.

25. Let X and Y be two independent random points from $(0, 1)$. Calculate the probability distribution function and the probability density function of $\max(X, Y)/ \min(X, Y)$.

26. Suppose that X and Y are independent, identically distributed exponential random variables with mean $1/\lambda$. Prove that $X/(X + Y)$ is uniform over $(0, 1)$.

27. Let X and Y be continuous random variables with joint probability density function $f(x, y)$. Let $Z = Y/X$, $X \neq 0$. Prove that the probability density function of Z is given by

$$f_Z(z) = \int_{-\infty}^{\infty} |x| f(x, xz) \, dx.$$

28. Let $f(x, y)$ be the joint probability density function of two continuous random variables. f is called *circularly symmetrical* if it is a function of $\sqrt{x^2 + y^2}$, the distance of (x, y) from the origin, that is, if there exists a function φ so that $f(x, y) = \varphi(\sqrt{x^2 + y^2})$. Prove that if X and Y are independent random variables, their joint probability density function is circularly symmetrical if and only if they are both normal with mean 0 and equal variance.

HINT: Suppose that f is circularly symmetrical; then

$$f_X(x) f_Y(y) = \varphi\left(\sqrt{x^2 + y^2}\right).$$

Differentiating this relation with respect to x yields

$$\frac{\varphi'(\sqrt{x^2 + y^2})}{\varphi(\sqrt{x^2 + y^2})\sqrt{x^2 + y^2}} = \frac{f_X'(x)}{x f_X(x)}.$$

This implies that both sides are constants, so that

$$\frac{f_X'(x)}{x f_X(x)} = k$$

for some constant k. Solve this and use the fact that f_X is a probability density function to show that f_X is normal. Repeat the same procedure for f_Y.

8.3   Conditional Distributions

Let X and Y be two discrete or two continuous random variables. In this chapter we study the distribution function and the expected value of the random variable X given that $Y = y$.

Let X be a discrete random variable with set of possible values A, and let Y be a discrete random variable with set of possible values B. Let $p(x, y)$ be the joint probability function of X and Y, and let p_X and p_Y be the marginal probability functions of X and Y. When no information is given about the value of Y,

$$p_X(x) = P(X = x) = \sum_{y \in B} P(X = x, Y = y) = \sum_{y \in B} p(x, y)$$

is used to calculate the probabilities of events concerning X. However, if the value of Y is known, then instead of $p_X(x)$ *the conditional probability function of X given that $Y = y$ is used.* This function, denoted by $p_{X|Y}(x|y)$, is defined as follows:

$$p_{X|Y}(x|y) = P(X = x \mid Y = y) = \frac{P(X = x, Y = y)}{P(Y = y)} = \frac{p(x, y)}{p_Y(y)},$$

where $x \in A$, $y \in B$, and $p_Y(y) > 0$. Note that

$$\sum_{x \in A} p_{X|Y}(x|y) = \sum_{x \in A} \frac{p(x, y)}{p_Y(y)} = \frac{1}{p_Y(y)} \sum_{x \in A} p(x, y) = \frac{1}{p_Y(y)} p_Y(y) = 1.$$

Hence for any fixed $y \in B$, $p_{X|Y}(x|y)$ is itself a probability function with the set of possible values A. If X and Y are independent, $p_{X|Y}$ coincides with p_X because

$$p_{X|Y}(x|y) = \frac{p(x, y)}{p_Y(y)} = \frac{P(X = x, Y = y)}{P(Y = y)} = \frac{P(X = x)P(Y = y)}{P(Y = y)}$$
$$= P(X = x) = p_X(x).$$

Similar to $p_{X|Y}(x|y)$, *the conditional distribution function of X given that $Y = y$ is defined as follows:*

$$F_{X|Y}(x|y) = P(X \leq x \mid Y = y) = \sum_{t \leq x} P(X = t \mid Y = y) = \sum_{t \leq x} p_{X|Y}(t|y).$$

Example 8.14   Let the joint probability function of X and Y be given by

$$p(x, y) = \begin{cases} \dfrac{1}{15}(x + y) & \text{if } x = 0, 1, 2, \; y = 1, 2 \\ 0 & \text{otherwise.} \end{cases}$$

Find $p_{X|Y}(x|y)$ and $P(X = 0|Y = 2)$.

SOLUTION: To use $p_{X|Y}(x|y) = p(x, y)/p_Y(y)$, we must first calculate $p_Y(y)$.

$$p_Y(y) = \sum_{x=0}^{2} p(x, y) = \frac{1 + y}{5}.$$

Therefore,

$$p_{X|Y}(x|y) = \frac{(x + y)/15}{(1 + y)/5} = \frac{x + y}{3(1 + y)}, \qquad x = 0, 1, 2 \text{ when } y = 1 \text{ or } 2.$$

In particular, $P(X = 0|Y = 2) = \dfrac{0 + 2}{3(1 + 2)} = \dfrac{2}{9}$.   ◆

Example 8.15   Let $N(t)$ be the number of males who enter a certain post office at or prior to time t. Let $M(t)$ be the number of females who enter a certain post office at or prior to t. Suppose that $\{N(t) : t \geq 0\}$ and $\{M(t) : t \geq 0\}$ are independent Poisson processes with rates λ and μ, respectively. So for all $t > 0$ and $s > 0$, $N(t)$ is independent of $M(s)$. If at some instant t, $N(t) + M(t) = n$, what is the conditional probability function of $N(t)$?

SOLUTION: For simplicity, let $K(t) = N(t) + M(t)$ and $p(x, n)$ be the joint probability function of $N(t)$ and $K(t)$. Then $p_{N(t)|K(t)}(x|n)$, the desired probability function, is found as follows:

$$p_{N(t)|K(t)}(x|n) = \frac{p(x, n)}{p_{K(t)}(n)} = \frac{P\{N(t) = x, K(t) = n\}}{P\{K(t) = n\}}$$

$$= \frac{P\{N(t) = x, M(t) = n - x\}}{P\{K(t) = n\}}.$$

Since $N(t)$ and $M(t)$ are independent random variables and $K(t) = N(t) + M(t)$ is a Poisson random variable with rate $\lambda t + \mu t$ (accept this for now; we will prove it in Theorem 9.4):

$$p_{N(t)|K(t)}(x|n) = \frac{P\{N(t) = x\}P\{M(t) = n - x\}}{P\{K(t) = n\}}$$

$$= \frac{\dfrac{e^{-\lambda t}(\lambda t)^x}{x!} \dfrac{e^{-\mu t}(\mu t)^{n-x}}{(n-x)!}}{\dfrac{e^{-(\lambda t + \mu t)}(\lambda t + \mu t)^n}{n!}}$$

$$= \frac{n!}{x!\,(n-x)!} \frac{\lambda^x \mu^{n-x}}{(\lambda + \mu)^n}$$

$$= \binom{n}{x}\left(\frac{\lambda}{\lambda + \mu}\right)^x \left(\frac{\mu}{\lambda + \mu}\right)^{n-x}$$

$$= \binom{n}{x}\left(\frac{\lambda}{\lambda + \mu}\right)^x \left(1 - \frac{\lambda}{\lambda + \mu}\right)^{n-x}, \qquad x = 0, 1, \ldots, n.$$

This shows that the conditional probability function of the number of males who have entered the post office at or prior to t, given that altogether n persons have entered the post office during this period, is binomial with parameters n and $\lambda/(\lambda + \mu)$. ◆

Now let X and Y be two continuous random variables with the joint probability density function $f(x, y)$. Again, when no information is given about the value of Y, $f_X(x) = \int_{-\infty}^{\infty} f(x, y)\,dy$, is used to calculate the probabilities of events concerning X. However, when the value of Y is known, to find such probabilities, $f_{X|Y}(x|y)$, *the conditional probability density function of X given that $Y = y$ is used*. Similar to the discrete case, $f_{X|Y}(x|y)$ is defined as follows:

$$f_{X|Y}(x|y) = \frac{f(x, y)}{f_Y(y)},$$

provided that $f_Y(y) > 0$. Note that

$$\int_{-\infty}^{\infty} f_{X|Y}(x|y)\,dx = \int_{-\infty}^{\infty} \frac{f(x, y)}{f_Y(y)}\,dx = \frac{1}{f_Y(y)}\int_{-\infty}^{\infty} f(x, y)\,dx$$

$$= \frac{1}{f_Y(y)} f_Y(y) = 1,$$

showing that for a fixed y, $f_{X|Y}(x|y)$ is itself a probability density function. If X and Y are independent, then $f_{X|Y}$ coincides with f_X because

$$f_{X|Y}(x|y) = \frac{f(x, y)}{f_Y(y)} = \frac{f_X(x) f_Y(y)}{f_Y(y)} = f_X(x).$$

Similarly, *the conditional probability density function of Y given that X = x is defined by*

$$f_{Y|X}(y|x) = \frac{f(x, y)}{f_X(x)},$$

provided that $f_X(x) > 0$. Also as one expects, $F_{X|Y}(x|y)$, *the conditional probability distribution function of X given that Y = y is defined as follows:*

$$F_{X|Y}(x|y) = P(X \le x \mid Y = y) = \int_{-\infty}^{x} f_{X|Y}(t|y)\, dt.$$

Therefore,

$$\frac{d}{dt} F_{X|Y}(x|y) = f_{X|Y}(x|y).$$

Example 8.16   Let X and Y be continuous random variables with joint probability density function

$$f(x, y) = \begin{cases} \dfrac{3}{2}(x^2 + y^2) & \text{if } 0 < x < 1, \quad 0 < y < 1 \\ 0 & \text{otherwise.} \end{cases}$$

Find $f_{X|Y}(x|y)$.

SOLUTION: By definition, $f_{X|Y}(x|y) = \dfrac{f(x, y)}{f_Y(y)}$, where

$$f_Y(y) = \int_{-\infty}^{\infty} f(x, y)\, dx = \int_{0}^{1} \frac{3}{2}(x^2 + y^2)\, dx = \frac{3}{2}y^2 + \frac{1}{2}.$$

Thus

$$f_{X|Y}(x|y) = \frac{3/2(x^2 + y^2)}{(3/2)y^2 + 1/2} = \frac{3(x^2 + y^2)}{3y^2 + 1}$$

for $0 < x < 1$ and $0 < y < 1$. Everywhere else, $f_{X|Y}(x|y) = 0$. ◆

Example 8.17   First a point Y is selected at random from the interval $(0, 1)$. Then another point X is chosen at random from the interval $(0, Y)$. Find the probability density function of X.

SOLUTION: Let $f(x, y)$ be the joint probability density function of X and Y. Then

$$f_X(x) = \int_{-\infty}^{\infty} f(x, y)\, dy,$$

where from

$$f_{X|Y}(x|y) = \frac{f(x, y)}{f_Y(y)},$$

we obtain

$$f(x, y) = f_{X|Y}(x|y) f_Y(y).$$

Therefore

$$f_X(x) = \int_{-\infty}^{\infty} f_{X|Y}(x|y) f_Y(y)\, dy.$$

Since Y is uniformly distributed over $(0, 1)$,

$$f_Y(y) = \begin{cases} 1 & \text{if } 0 < y < 1 \\ 0 & \text{elsewhere.} \end{cases}$$

Since given $Y = y$, X is uniformly distributed over $(0, y)$,

$$f_{X|Y}(x|y) = \begin{cases} \dfrac{1}{y} & \text{if } 0 < y < 1, \quad 0 < x < y \\ 0 & \text{elsewhere.} \end{cases}$$

Thus

$$f_X(x) = \int_{-\infty}^{\infty} f_{X|Y}(x|y) f_Y(y)\, dy = \int_{x}^{1} \frac{dy}{y} = \ln 1 - \ln x = -\ln x.$$

Therefore,

$$f_X(x) = \begin{cases} -\ln x & \text{if } 0 < x < 1 \\ 0 & \text{elsewhere.} \end{cases} \quad \blacklozenge$$

Example 8.18 Let the conditional probability density function of X given that $Y = y$ be given by

$$f_{X|Y}(x|y) = \frac{x+y}{1+y}e^{-x}, \qquad 0 < x < \infty, \quad 0 < y < \infty.$$

Find $P(X < 1|Y = 2)$.

SOLUTION: The probability density function of X given $Y = 2$ is

$$f_{X|Y}(x|2) = \frac{x+2}{3}e^{-x}, \qquad 0 < x < \infty.$$

Therefore,

$$
\begin{aligned}
P(X < 1|Y = 2) &= \int_0^1 \frac{x+2}{3}e^{-x}\,dx \\
&= \frac{1}{3}\left[\int_0^1 xe^{-x}\,dx + \int_0^1 2e^{-x}\,dx\right] \\
&= \frac{1}{3}\left[-xe^{-x} - e^{-x}\right]_0^1 - \frac{2}{3}\left[e^{-x}\right]_0^1 = 1 - \frac{4}{3}e^{-1}.
\end{aligned}
$$

Note that while we easily calculated $P(X < 1|Y = 2)$, it is not possible to calculate probabilities such as $P(X < 1|2 < Y < 3)$ using $f_{X|Y}(x|y)$. This is because the probability density function of X given that $2 < Y < 3$ cannot be found from the conditional probability density function of X given $Y = y$. ♦

We now generalize the concept of mathematical expectation to the conditional case. Let X and Y be discrete random variables, and let the set of possible values of X be A. *The conditional expectation of the random variable X given that $Y = y$ is as follows:*

$$E(X \mid Y = y) = \sum_{x \in A} xP(X = x \mid Y = y) = \sum_{x \in A} xp_{X|Y}(x|y),$$

where $p_Y(y) > 0$.

Similarly, for continuous random variables X and Y with joint probability density function $f(x, y)$, *the conditional expectation of X given that $Y = y$ is as follows:*

$$E(X \mid Y = y) = \int_{-\infty}^{\infty} xf_{X|Y}(x|y)\,dx,$$

where $f_Y(y) > 0$.

Note that conditional expectations are simply ordinary expectations computed relative to conditional distributions. For this reason, they satisfy the same properties that ordinary expectations do. For example, if h is an ordinary function from $\mathbf{R}$ to $\mathbf{R}$, then for the discrete random variables X and Y with set of possible values A for X, the expected value of $h(X)$ is obtained from

$$E[h(X) \mid Y = y] = \sum_{x \in A} h(x) p_{X|Y}(x|y).$$

If X and Y are continuous with joint probability density function $f(x, y)$, then

$$E[h(X) \mid Y = y] = \int_{-\infty}^{\infty} h(x) f_{X|Y}(x|y)\, dx.$$

In particular, this implies that the conditional variance of X given that $Y = y$ is given by

$$\sigma_{X|Y=y}^2 = \int_{-\infty}^{\infty} [x - E(X \mid Y = y)]^2\, f_{X|Y}(x|y)\, dx.$$

Example 8.19   Calculate the expected number of aces in a randomly selected poker hand that is found to have exactly two jacks.

SOLUTION: Let X and Y be the number of aces and jacks in a random poker hand, respectively. Then

$$E[X \mid Y = 2] = \sum_{x=0}^{3} x p_{X|Y}(x|2) = \sum_{x=0}^{3} x \frac{p(x, 2)}{p_Y(2)}$$

$$= \sum_{x=0}^{3} x \frac{\dfrac{\dbinom{4}{2}\dbinom{4}{x}\dbinom{44}{3-x}}{\dbinom{52}{5}}}{\dfrac{\dbinom{4}{2}\dbinom{48}{3}}{\dbinom{52}{5}}} = \sum_{x=0}^{3} x \frac{\dbinom{4}{x}\dbinom{44}{3-x}}{\dbinom{48}{3}} \approx 0.25. \quad \blacklozenge$$

Example 8.20   Let X and Y be continuous random variables with joint probability density function

$$f(x, y) = \begin{cases} e^{-y} & \text{if } y > 0 \quad 0 < x < 1 \\ 0 & \text{elsewhere.} \end{cases}$$

Find $E(X \mid Y = 2)$.

SOLUTION: From the definition,

$$E(X \mid Y = 2) = \int_{-\infty}^{\infty} x f_{X|Y}(x|2)\, dx = \int_{0}^{1} x \frac{f(x, 2)}{f_Y(2)}\, dx = \int_{0}^{1} x \frac{e^{-2}}{f_Y(2)}\, dx.$$

But $f_Y(2) = \int_0^1 f(x, 2)\, dx = \int_0^1 e^{-2}\, dx = e^{-2}$; therefore,

$$E(X \mid Y = 2) = \int_{0}^{1} x \frac{e^{-2}}{e^{-2}}\, dx = \frac{1}{2}. \quad \blacklozenge$$

Example 8.21   While rolling a balanced die successively, the first 6 occurred on the third roll. What is the expected number of rolls until the first 1?

SOLUTION: Let X and Y be the number of rolls until the first 1 and the first 6, respectively. The required quantity, $E(X \mid Y = 3)$, is calculated from

$$E(X \mid Y = 3) = \sum_{x=1}^{\infty} x p_{X|Y}(x|3) = \sum_{x=1}^{\infty} x \frac{p(x, 3)}{p_Y(3)}.$$

Clearly, $p_Y(3) = (5/6)^2(1/6) = 25/216$, $p(1, 3) = (1/6)(5/6)(1/6) = 5/216$, $p(2, 3) = (4/6)(1/6)(1/6) = 4/216$, $p(3, 3) = 0$, and for $x > 3$,

$$p(x, 3) = \left(\frac{4}{6}\right)^2 \left(\frac{1}{6}\right) \left(\frac{5}{6}\right)^{x-4} \left(\frac{1}{6}\right) = \frac{1}{81} \left(\frac{5}{6}\right)^{x-4}.$$

Therefore,

$$E(X \mid Y = 3) = \sum_{x=1}^{\infty} x \frac{p(x, 3)}{p_Y(3)} = \frac{216}{25} \sum_{x=1}^{\infty} x p(x, 3)$$

$$= \frac{216}{25} \left[1 \cdot \frac{5}{216} + 2 \cdot \frac{4}{216} + \sum_{x=4}^{\infty} x \cdot \frac{1}{81} \left(\frac{5}{6}\right)^{x-4} \right]$$

$$= \frac{13}{25} + \frac{8}{75} \sum_{x=4}^{\infty} x \left(\frac{5}{6}\right)^{x-4}$$

$$= \frac{13}{25} + \frac{8}{75} \sum_{x=4}^{\infty} (x - 4 + 4) \left(\frac{5}{6}\right)^{x-4}$$

$$= \frac{13}{25} + \frac{8}{75} \left[\sum_{x=4}^{\infty} (x - 4) \left(\frac{5}{6}\right)^{x-4} + 4 \sum_{x=4}^{\infty} \left(\frac{5}{6}\right)^{x-4} \right]$$

$$= \frac{13}{25} + \frac{8}{75} \left[\sum_{x=0}^{\infty} x \left(\frac{5}{6}\right)^{x} + 4 \sum_{x=0}^{\infty} \left(\frac{5}{6}\right)^{x} \right]$$

$$= \frac{13}{25} + \frac{8}{75} \left[\frac{5/6}{(1 - 5/6)^2} + 4 \left(\frac{1}{1 - 5/6}\right) \right] = 6.28. \quad \blacklozenge$$

EXERCISES

A

1. Let the joint probability function of discrete random variables X and Y be given by

$$p(x, y) = \begin{cases} \dfrac{1}{25}(x^2 + y^2) & \text{if } x = 1, 2, \quad y = 0, 1, 2 \\ 0 & \text{otherwise.} \end{cases}$$

Find $p_{X|Y}(x|y)$, $P(X = 2|Y = 1)$, and $E(X|Y = 1)$.

2. Let the joint probability density function of continuous random variables X and Y be given by

$$f(x, y) = \begin{cases} 2 & \text{if } 0 < x < y < 1 \\ 0 & \text{elsewhere.} \end{cases}$$

Find $f_{X|Y}(x|y)$.

3. An unbiased coin is flipped until the sixth head is obtained. If the third head occurs in the fifth flip, what is the probability function of the number of flips?

4. Let the conditional probability density function of X given that $Y = y$ be given by

$$f_{X|Y}(x|y) = \frac{3(x^2 + y^2)}{3y^2 + 1}, \qquad 0 < x < 1, \quad 0 < y < 1.$$

Find $P(1/4 < X < 1/2 | Y = 3/4)$.

5. Let X and Y be independent discrete random variables. Prove that for all y, $E(X \mid Y = y) = E(X)$. Do the same for continuous random variables X and Y.

6. Let X and Y be continuous random variables with joint probability density function

$$f(x, y) = \begin{cases} x + y & \text{if } 0 \leq x \leq 1, \quad 0 \leq y \leq 1 \\ 0 & \text{elsewhere.} \end{cases}$$

Calculate $f_{X|Y}(x|y)$.

7. Let X and Y be continuous random variables with joint probability density function given by

$$f(x, y) = \begin{cases} e^{-x(y+1)} & \text{if } x \geq 0, \quad 0 \leq y \leq e - 1 \\ 0 & \text{elsewhere.} \end{cases}$$

Calculate $E(X \mid Y = y)$.

8. First a point Y is selected at random from the interval $(0, 1)$. Then another point X is selected at random from the interval $(Y, 1)$. Find the probability density function of X.

9. Let (X, Y) be a random point from the unit disk centered at the origin. Find $P(0 \leq X \leq 4/11 \mid Y = 4/5)$.

10. The joint probability density function of X and Y is given by

$$f(x, y) = \begin{cases} ce^{-x} & \text{if } x \geq 0, \quad |y| < x \\ 0 & \text{otherwise.} \end{cases}$$

   (a) Determine the constant c.
   (b) Find $f_{X|Y}(x|y)$ and $f_{Y|X}(y|x)$.
   (c) Calculate $E(Y \mid X = x)$ and $\text{Var}(Y|X = x)$.

11. Leon leaves his office every day at a random time between 4:30 P.M. and 5:00 P.M. If he leaves t minutes past 4:30, the time it will take him to reach home is a random number between 20 and $20 + (2t)/3$ minutes. Let Y be the number of minutes past 4:30 that Leon leaves his office tomorrow and X be the number of minutes it takes him to reach home. Find the joint probability density function of X and Y.

12. Show that if $\{N(t) : t \geq 0\}$ is a Poisson process, the conditional distribution of the first arrival time given $N(t) = 1$ is uniform on $(0, t)$.

13. In a sequence of independent Bernoulli trials, let X be the number of successes in the first m trials and Y be the number of successes in the first n trials, $m < n$. Show that the conditional distribution of X given $Y = y$ is hypergeometric. Also find the conditional distribution of Y given $X = x$.

B

14. A point is selected at random and uniformly from the region

$$R = \{(x, y) : |x| + |y| \leq 1\}.$$

Find the conditional probability density function of X given $Y = y$.

15. Let $\{N(t) : t \geq 0\}$ be a Poisson process. For $s < t$ show that the conditional distribution of $N(s)$ given $N(t) = n$ is binomial with parameters n and $p = s/t$. Also find the conditional distribution of $N(t)$ given $N(s) = k$.

16. Cards are drawn from an ordinary deck of 52, one at a time, randomly, and with replacement. Let X and Y denote the number of draws until the first ace and the first king, respectively. Find $E(X \mid Y = 5)$.

17. A box contains 10 red and 12 blue chips. Suppose that 18 chips are drawn, one by one, at random, and with replacement. If it is known that 10 of them are blue, show that the expected number of blue chips in the first nine draws is 5.

18. Let X and Y be continuous random variables with joint probability density function

$$f(x, y) = \begin{cases} n(n-1)(y-x)^{n-2} & \text{if } 0 \leq x \leq y \leq 1 \\ 0 & \text{otherwise.} \end{cases}$$

Find the conditional expectation of Y given that $X = x$.

19. A point (X, Y) is selected randomly from the triangle with vertices $(0, 0)$, $(0, 1)$, and $(1, 0)$.
  (a) Find the joint probability density function of X and Y.
  (b) Calculate $f_{X|Y}(x|y)$.
  (c) Evaluate $E(X \mid Y = y)$.

20. Let X and Y be discrete random variables with joint probability function

$$p(x, y) = \frac{1}{e^2 y! \, (x - y)!}, \qquad x = 0, 1, 2, \ldots, \qquad y = 0, 1, 2, \ldots, x,$$

$p(x, y) = 0$, elsewhere. Find $E(Y \mid X = x)$.

8.4   Multivariate Distributions

The following definition generalizes the concept of joint probability function of two discrete random variables to $n > 2$ discrete random variables.

DEFINITION   *Let $X_1, X_2, \ldots, X_n$ be discrete random variables defined on the same sample space with sets of possible values $A_1, A_2, \ldots, A_n$, respectively. The function*

$$p(x_1, x_2, \ldots, x_n) = P(X_1 = x_1, X_2 = x_2, \ldots, X_n = x_n)$$

is called the joint probability function of $X_1, X_2, \ldots, X_n$.

Note that

(a) $p(x_1, x_2, \ldots, x_n) \geq 0$.
(b) If for some i, $1 \leq i \leq n$, $x_i \notin A_i$, then $p(x_1, x_2, \ldots, x_n) = 0$.
(c) $\sum_{x_i \in A_i, \, 1 \leq i \leq n} p(x_1, x_2, \ldots, x_n) = 1$.

Moreover, if the joint probability function of random variables X_1, X_2, X_3, $\ldots$, X_n, $p(x_1, x_2, \ldots, x_n)$, is given, then for $1 \leq i \leq n$, the *marginal probability function* of X_i, p_{X_i}, can be found from $p(x_1, x_2, \ldots, x_n)$ by

$$p_{X_i}(x_i) = P(X_i = x_i) = P(X_i = x_i; \; X_j \in A_j, \, 1 \leq j \leq n, j \neq i)$$

$$= \sum_{x_j \in A_j, \, j \neq i} p(x_1, x_2, \ldots, x_n).$$

More generally, to find the *marginal joint probability function* of a given set of k of these random variables, we sum up $p(x_1, x_2, \ldots, x_n)$ over all possible values of the remaining $n - k$ random variables. For example, if $p(x, y, z)$ denotes the joint probability function of random variables X, Y, and Z, then

$$p_{X,Y}(x, y) = \sum_z p(x, y, z)$$

is the marginal joint probability function of X and Y, whereas

$$p_{Y,Z}(y, z) = \sum_{x} p(x, y, z)$$

is the marginal joint probability function of Y and Z.

Example 8.22   Dr. Shams has 23 hypertensive patients, of whom five do not use any medicine but try to lower their blood pressures by self-help: dieting, exercise, not smoking, relaxation, and so on. Of the remaining 18 patients, 10 use beta blockers and eight use diuretics. A random sample of seven of all these patients is selected. Let X, Y, and Z be the number of the patients in the sample trying to lower their blood pressures by self-help, beta blockers, and diuretics, respectively. Find the joint probability function and the marginal probability functions of X, Y, and Z.

SOLUTION: Let $p(x, y, z)$ be the joint probability function of X, Y, and Z. Then for $0 \leq x \leq 5$, $0 \leq y \leq 7$, $0 \leq z \leq 7$, $x + y + z = 7$,

$$p(x, y, z) = \frac{\binom{5}{x}\binom{10}{y}\binom{8}{z}}{\binom{23}{7}};$$

$p(x, y, z) = 0$, otherwise. p_X, the marginal probability function of X, is obtained as follows:

$$p_X(x) = \sum_{\substack{x+y+z=7 \\ 0\leq y\leq 7 \\ 0\leq z\leq 7}} p(x, y, z) = \sum_{y=0}^{7-x} \frac{\binom{5}{x}\binom{10}{y}\binom{8}{7-x-y}}{\binom{23}{7}}$$

$$= \frac{\binom{5}{x}}{\binom{23}{7}} \sum_{y=0}^{7-x} \binom{10}{y}\binom{8}{7-x-y}.$$

Now $\displaystyle\sum_{y=0}^{7-x} \binom{10}{y}\binom{8}{7-x-y}$ is the total number of the ways we can choose $7 - x$ patients from 18 (in Exercise 37 of Section 2.4, let $m = 10$, $n = 8$, and $r = 7 - x$), hence it is equal to $\binom{18}{7-x}$. Therefore, as expected,

$$p_X(x) = \frac{\binom{5}{x}\binom{18}{7-x}}{\binom{23}{7}}, \qquad x = 0, 1, 2, 3, 4, 5.$$

Similarly,

$$p_Y(y) = \frac{\binom{10}{y}\binom{13}{7-y}}{\binom{23}{7}}, \qquad y = 0, 1, 2, \ldots, 7,$$

$$p_Z(z) = \frac{\binom{8}{z}\binom{15}{7-z}}{\binom{23}{7}}, \qquad z = 0, 1, 2, \ldots, 7. \quad \blacklozenge$$

A joint probability function such as $p(x, y, z)$ of Example 8.22 is called *multivariate hypergeometric*. In general, suppose that a box contains n_1 marbles of type 1, n_2 marbles of type 2, ..., and n_r marbles of type r. If n marbles are drawn at random and X_i ($i = 1, 2, \ldots, r$) is the number of the marbles of type i drawn, the joint probability function of $X_1, X_2, \ldots, X_r$ is called *multivariate hypergeometric* and is given by

$$p(x_1, x_2, \ldots, x_r) = \frac{\binom{n_1}{x_1}\binom{n_2}{x_2}\cdots\binom{n_r}{x_r}}{\binom{n_1 + n_2 + \cdots + n_r}{n}},$$

where $x_1 + x_2 + \cdots + x_r = n$.

We now define the concept of jointly continuous density function for more than two random variables.

DEFINITION   *Let $X_1, X_2, \ldots, X_n$ be continuous random variables defined on the same sample space. We say that $X_1, X_2, \ldots, X_n$ have a continuous joint distribution if there exists a nonnegative function of n variables, $f(x_1, x_2, \ldots, x_n)$, on $\mathbf{R} \times \mathbf{R} \times \cdots \times \mathbf{R} \equiv \mathbf{R}^n$ such that for any region R in $\mathbf{R}^n$ that can be formed from n-dimensional rectangles by a countable number of set operations,*

$$P\{(X_1, X_2, \ldots, X_n) \in R\}$$

$$= \int \cdots \int_R f(x_1, x_2, \ldots, x_n) \, dx_1 \, dx_2 \cdots dx_n. \qquad (8.13)$$

The function $f(x_1, x_2, \ldots, x_n)$ is called the joint probability density function of $X_1, X_2, \ldots, X_n$.

Let $R = \{(x_1, x_2, \ldots, x_n) : x_i \in A_i, \ 1 \le i \le n\}$, where $A_i, \ 1 \le i \le n$, is any subset of real numbers which can be constructed from intervals by a countable number of set operations. Then (8.13) gives

$$P(X_1 \in A_1, X_2 \in A_2, \ldots, X_n \in A_n)$$

$$= \int_{A_n} \int_{A_{n-1}} \cdots \int_{A_1} f(x_1, x_2, \ldots, x_n) \, dx_1 \, dx_2 \cdots dx_n.$$

Letting $A_i = (-\infty, +\infty), \ 1 \le i \le n$, this implies that

$$\int_{-\infty}^{+\infty} \cdots \int_{-\infty}^{+\infty} f(x_1, x_2, \ldots, x_n) \, dx_1 \, dx_2 \cdots dx_n = 1.$$

Let f_{X_i} be the marginal probability density function of $X_i, \ 1 \le i \le n$. Then

$$f_{X_i}(x_i) = \underbrace{\int_{-\infty}^{+\infty} \cdots \int_{-\infty}^{+\infty}}_{n-1 \text{ terms}} f(x_1, x_2, \ldots, x_n) \, dx_1 \cdots dx_{i-1} \, dx_{i+1} \cdots dx_n.$$

Therefore, for instance,

$$f_{X_3}(x_3) = \underbrace{\int_{-\infty}^{+\infty} \cdots \int_{-\infty}^{+\infty}}_{n-1 \text{ terms}} f(x_1, x_2, \ldots, x_n) \, dx_1 \, dx_2 \, dx_4 \cdots dx_n.$$

More generally, to find the *marginal joint probability density function* of a given set of k of these random variables, we integrate $f(x_1, x_2, \ldots, x_n)$ over all possible values of the remaining $n - k$ random variables. For example, if $f(x, y, z, t)$ denotes the joint probability density function of random variables X, Y, Z, and T, then

$$f_{Y,T}(y, t) = \int_{-\infty}^{+\infty} \int_{-\infty}^{+\infty} f(x, y, z, t) \, dx \, dz$$

is the marginal joint probability density function of Y and T, whereas

$$f_{X,Z,T}(x, z, t) = \int_{-\infty}^{+\infty} f(x, y, z, t)\, dy$$

is the marginal joint probability density function of X, Z, and T.

Example 8.23

(a) Prove that the following is a joint probability density function:

$$f(x, y, z, t) = \begin{cases} \dfrac{1}{xyz} & \text{if } 0 < t \le z \le y \le x \le 1 \\ 0 & \text{elsewhere.} \end{cases}$$

(b) Suppose that f is the joint probability density function of random variables X, Y, Z, and T. Find $f_{Y,Z,T}(y, z, t)$, $f_{X,T}(x, t)$, and $f_Z(z)$.

SOLUTION: (a) Since $f(x, y, z, t) \ge 0$ and

$$\int_0^1 \int_0^x \int_0^y \int_0^z \frac{1}{xyz}\, dt\, dz\, dy\, dx = \int_0^1 \int_0^x \int_0^y \frac{1}{xy}\, dz\, dy\, dx$$

$$= \int_0^1 \int_0^x \frac{1}{x}\, dy\, dx = \int_0^1 dx = 1,$$

f is a joint probability density function.
    (b) For $0 < t \le z \le y \le 1$,

$$f_{Y,Z,T}(y, z, t) = \int_y^1 \frac{1}{xyz}\, dx = \frac{1}{yz} \ln x \Big|_y^1 = -\frac{\ln y}{yz}.$$

Therefore,

$$f_{Y,Z,T}(y, z, t) = \begin{cases} -\dfrac{\ln y}{yz} & \text{if } 0 < t \le z \le y \le 1 \\ 0 & \text{elsewhere.} \end{cases}$$

To find $f_{X,T}(x, t)$, we have that for $0 < t \le x \le 1$,

$$f_{X,T}(x, t) = \int_t^x \int_t^y \frac{1}{xyz}\, dz\, dy = \int_t^x \left[\frac{\ln z}{xy} \right]_t^y dy$$

$$= \int_t^x \left(\frac{\ln y}{xy} - \frac{\ln t}{xy} \right) dy = \left[\frac{1}{2x} (\ln y)^2 - \frac{\ln t}{x} \ln y \right]_t^x$$

$$= \frac{1}{2x}(\ln x)^2 - \frac{1}{x}(\ln t)(\ln x) + \frac{1}{2x}(\ln t)^2 = \frac{1}{2x}(\ln x - \ln t)^2$$

$$= \frac{1}{2x}\ln^2\frac{x}{t}.$$

Therefore,

$$f_{X,T}(x,t) = \begin{cases} \dfrac{1}{2x}\ln^2\dfrac{x}{t} & \text{if } 0 < t \le x \le 1 \\[2mm] 0 & \text{otherwise.} \end{cases}$$

To find $f_Z(z)$, we have that for $0 < z \le 1$,

$$f_Z(z) = \int_z^1 \int_z^x \int_0^z \frac{1}{xyz}\,dt\,dy\,dx = \int_z^1 \int_z^x \frac{1}{xy}\,dy\,dx$$

$$= \int_z^1 \left[\frac{1}{x}\ln y\right]_z^x dx = \int_z^1 \left(\frac{1}{x}\ln x - \frac{1}{x}\ln z\right) dx$$

$$= \left[\frac{1}{2}(\ln x)^2 - (\ln x)(\ln z)\right]_z^1 = \frac{1}{2}(\ln z)^2.$$

Thus

$$f_Z(z) = \begin{cases} \dfrac{1}{2}(\ln z)^2 & \text{if } 0 < z \le 1 \\[2mm] 0 & \text{otherwise.} \end{cases} \quad \blacklozenge$$

We now extend the definition of the joint probability distribution from 2 to $n > 2$ random variables. Let $X_1, X_2, \ldots, X_n$ be n random variables (discrete, continuous, or mixed). Then the *joint probability distribution function* of $X_1, X_2, \ldots, X_n$ is defined by

$$F(t_1, t_2, \ldots, t_n) = P(X_1 \le t_1, X_2 \le t_2, \ldots, X_n \le t_n)$$

for all $-\infty < t_i < +\infty$, $i = 1, 2, \ldots, n$. The *marginal probability distribution function* of X_i, $1 \le i \le n$, can be found from F as follows:

$$F_{X_i}(t_i) = P(X_i \le t_i)$$

$$= P(X_1 < \infty, \ldots, X_{i-1} < \infty, X_i \le t_i, X_{i+1} < \infty, \ldots, X_n < \infty)$$

$$= \lim_{\substack{t_j \to \infty \\ 1 \le j \le n,\ j \ne i}} F(t_1, t_2, \ldots, t_n).$$

More generally, to find the *marginal joint probability distribution function* of a given set of k of these random variables, we calculate the limit of $F(t_1, t_2, \ldots, t_n)$ as $t_j \to \infty$, for every j that belongs to one of the remaining $n - k$ variables. For example, if $F(x, y, z, t)$ denotes the joint probability density function of random variables X, Y, Z, and T, then

$$F_{Y,T}(y, t) = \lim_{x,z \to \infty} F(x, y, z, t).$$

Just as the joint probability distribution function of two random variables, we have that F, the joint probability distribution of n random variables, satisfies the following:

(a) F is nondecreasing in each argument.

(b) F is right continuous in each argument.

(c) $F(t_1, t_2, \ldots, t_{i-1}, -\infty, t_{i+1}, \ldots, t_n) = 0$ for $i = 1, 2, \ldots, n$.

(d) $F(\infty, \infty, \ldots, \infty) = 1$.

Moreover, if F is the joint probability distribution function of jointly continuous random variables $X_1, X_2, \ldots, X_n$ with the joint probability density function $f(x_1, x_2, \ldots, x_n)$, then

$$F(t_1, t_2, \ldots, t_n)$$
$$= \int_{-\infty}^{t_n} \int_{-\infty}^{t_{n-1}} \cdots \int_{-\infty}^{t_1} f(x_1, x_2, \ldots, x_n) \, dx_1 \cdots dx_n,$$

and

$$f(x_1, x_2, \ldots, x_n) = \frac{\partial^n F(x_1, x_2, \ldots, x_n)}{\partial x_1 \partial x_2 \cdots \partial x_n}.$$

We now generalize the concept of independence of random variables from two to any number of random variables.

Suppose that $X_1, X_2, \ldots, X_n$ are random variables on a sample space. We say that they are *independent* if for arbitrary subsets $A_1, A_2, \ldots, A_n$ of real numbers,

$$P(X_1 \in A_1, X_2 \in A_2, \ldots, X_n \in A_n)$$
$$= P(X_1 \in A_1)P(X_2 \in A_2) \cdots P(X_n \in A_n).$$

Similar to the case of two random variables, $X_1, X_2, \ldots, X_n$ are independent if and only if for any $x_i \in \mathbf{R}$, $i = 1, 2, 3, \ldots, n$,

$$P(X_1 \leq x_1, X_2 \leq x_2, \ldots, X_n \leq x_n)$$
$$= P(X_1 \leq x_1)P(X_2 \leq x_2) \cdots P(X_n \leq x_n).$$

That is,

$$F(x_1, x_2, \ldots, x_n) = F_{X_1}(x_1)F_{X_2}(x_2) \cdots F_{X_n}(x_n).$$

If $X_1, X_2, \ldots, X_n$ are discrete, the definition of independence reduces to the following condition:

$$P(X_1 = x_1, X_2 = x_2, \ldots, X_n = x_n)$$
$$= P(X_1 = x_1)P(X_2 = x_2) \cdots P(X_n = x_n)$$

for any set of points, x_i, $i = 1, 2, \ldots, n$.

Let X, Y, and Z be independent random variables. Then by definition, for arbitrary subsets A_1, A_2, and A_3 of **R**,

$$P(X \in A_1, Y \in A_2, Z \in A_3) = P(X \in A_1)P(Y \in A_2)P(Z \in A_3). \quad (8.14)$$

Now if in (8.14), we let $A_2 = \mathbf{R}$, then since the event $Y \in \mathbf{R}$ is certain and has probability 1, we get

$$P(X \in A_1, Z \in A_3) = P(X \in A_1)P(Z \in A_3).$$

This shows that X and Z are independent random variables. In the same way it can be shown that $\{X, Y\}$, and $\{Y, Z\}$ are also independent sets. Similarly, if $\{X_1, X_2, \ldots, X_n\}$ is a sequence of independent random variables, its subsets are also independent sets of random variables. This observation motivates the following definition for the concept of independence of any collection of random variables.

DEFINITION   *A collection of random variables is called independent if all of its finite subcollections are independent.*

Sometimes independence of a collection of random variables is self-evident and requires no checking. For example, let E be an experiment with the sample space S. Let X be a random variable defined on S. If the experiment E is repeated n times independently and on the ith experiment X is called X_i, the sequence $\{X_1, X_2, \ldots, X_n\}$ is an independent sequence of random variables.

Just as in the case of two random variables, if $\{X_1, X_2, \ldots\}$ is a sequence of independent random variables and for $i = 1, 2, 3, \ldots$, $f_i : \mathbf{R} \to \mathbf{R}$ is a real-valued function, then the sequence $\{f_1(X_1), f_2(X_2), \ldots\}$ is also an independent sequence of random variables.

The following theorem is the generalization of Theorem 8.3 and is proved the same way.

THEOREM 8.7 *Let $X_1, X_2, \ldots, X_n$ be jointly continuous random variables with the joint probability density function $f(x_1, x_2, \ldots, x_n)$. Then $X_1, X_2, X_3, \ldots, X_n$ are independent if and only if $f(x_1, x_2, \ldots, x_n)$ is the product of their marginal densities $f_{X_1}(x_1), f_{X_2}(x_2), f_{X_3}(x_3), \ldots, f_{X_n}(x_n)$.*

Example 8.24 A system has n components, the lifetime of each being an exponential random variable with parameter λ. Suppose that the lifetimes of the components are independent random variables, and the system fails as soon as any of its components fails. Find the probability density function of the time until the system fails.

SOLUTION: Let $X_1, X_2, \ldots, X_n$ be the lifetimes of the components. Then $X_1, X_2, \ldots, X_n$ are independent random variables and for $i = 1, 2, \ldots, n$,

$$P(X_i \le t) = 1 - e^{-\lambda t}.$$

Letting X be the time until the system fails, we have

$$X = \min(X_1, X_2, \ldots, X_n).$$

Therefore,

$$
\begin{aligned}
P(X > t) &= P\{\min(X_1, X_2, \ldots, X_n) > t\} \\
&= P(X_1 > t, X_2 > t, \ldots, X_n > t) \\
&= P(X_1 > t)P(X_2 > t) \cdots P(X_n > t) \\
&= \left(e^{-\lambda t}\right)\left(e^{-\lambda t}\right) \cdots \left(e^{-\lambda t}\right) = e^{-n\lambda t}.
\end{aligned}
$$

Let f be the probability density function of X, then

$$f(t) = \frac{d}{dt}P(X \le t) = \frac{d}{dt}(1 - e^{-n\lambda t}) = n\lambda e^{-n\lambda t}. \quad \blacklozenge$$

Note that the discrete version of Theorem 8.7 is also true and its proof is similar to that of this theorem. Generalization of Theorem 8.5 from dimension 2 to n is also valid in both continuous and discrete cases. We will state it in the continuous case; the statement is similar for the discrete case.

THEOREM 8.8 *Let* $f(x_1, x_2, \ldots, x_n)$ *be the joint probability density function of random variables* $X_1, X_2, \ldots, X_n$. *If h is a function of n variables from* $\mathbf{R}^n$ *to* $\mathbf{R}$, *then* $Y = h(X_1, X_2, \ldots, X_n)$ *is a random variable with expected value given by*

$$E(Y) = \int_{-\infty}^{\infty} \cdots \int_{-\infty}^{\infty} h(x_1, x_2, \ldots, x_n) f(x_1, x_2, \ldots, x_n) \, dx_1 \, dx_2 \cdots dx_n,$$

provided that the integral is absolutely convergent.

Using Theorems 8.7 and 8.8, an almost identical proof to that of Theorem 8.6 implies that *the expected value of the product of several independent random variables is equal to the product of their expected values.*

EXERCISES

A

1. From an ordinary deck of 52 cards, 13 cards are selected at random. Calculate the joint probability function of the numbers of hearts, diamonds, clubs, and spades selected.

2. A jury of 12 persons is randomly selected from a group of eight Afro-American, seven Hispanic, three Native American, and 20 white potential jurors. Let A, H, N, and W be the number of Afro-American, Hispanic, Native American, and white jurors selected, respectively. Calculate the joint probability function of A, H, N, W and the marginal probability function of A.

3. Let $p(x, y, z) = (xyz)/162$, $x = 4, 5$, $y = 1, 2, 3$, and $z = 1, 2$, be the joint probability function of the random variables X, Y, Z. Calculate the joint marginal probability functions of X, Y; Y, Z; and X, Z.

4. Let the joint probability density function of X, Y, and Z be given by

$$f(x, y, z) = \begin{cases} 6e^{-x-y-z} & \text{if } 0 < x < y < z < \infty \\ 0 & \text{elsewhere.} \end{cases}$$

Find the marginal joint probability density function of X, Y; X, Z; and Y, Z.

5. From the set of families with two children a *family* is selected at random. Let $X_1 = 1$ if the first child of the family is a girl; $X_2 = 1$ if the second child of the family is a girl; and $X_3 = 1$ if the family has exactly one

boy. For $i = 1, 2, 3$, let $X_i = 0$ in other cases. Determine if X_1, X_2, and X_3 are independent. Assume that in a family the probability that a child is a girl is independent of the gender of the other children and is 1/2.

6. Let X, Y, and Z be jointly continuous with the following joint probability density function

$$f(x, y, z) = \begin{cases} x^2 e^{-x(1+y+z)} & \text{if } x, y, z > 0 \\ 0 & \text{otherwise.} \end{cases}$$

Are X, Y, and Z independent? Are they pairwise independent?

7. Let the joint probability distribution function of X, Y, and Z be given by

$$F(x, y, z) = (1 - e^{-\lambda_1 x})(1 - e^{-\lambda_2 y})(1 - e^{-\lambda_3 z}), \quad x, y, z > 0,$$

where $\lambda_1, \lambda_2, \lambda_3 > 0$.
(a) Are X, Y, and Z independent?
(b) Find the joint probability density function of X, Y, and Z.
(c) Find $P(X < Y < Z)$.

8. (a) Show that the following is a joint probability density function.

$$f(x, y, z) = \begin{cases} -\dfrac{\ln x}{xy} & \text{if } 0 < z \le y \le x \le 1 \\ 0 & \text{otherwise.} \end{cases}$$

(b) Suppose that f is the joint probability density function of X, Y, and Z. Find $f_{X,Y}(x, y)$ and $f_Y(y)$.

9. Inside a circle of radius R, n points are selected at random and independently. Find the probability that the distance of the nearest point to the center is at least r.

10. A point is selected at random from the cube

$$\Omega = \{(x, y, z) : -a \le x \le a, \ -a \le y \le a, \ -a \le z \le a\}.$$

What is the probability that it is inside the sphere inscribed in the cube?

11. Is the following a joint probability density function?

$$f(x_1, x_2, \ldots, x_n) = \begin{cases} e^{-x_n} & \text{if } 0 < x_1 < x_2 < \cdots < x_n \\ 0 & \text{otherwise.} \end{cases}$$

12. Suppose that the lifetimes of radio transistors are independent exponential random variables with mean five years. Arnold buys a radio and decides to replace its transistor upon failure two times: once when the original transistor dies and once when the replacement dies. He stops using the radio when the second replacement of the transistor goes out of order. Assuming that Arnold repairs the radio if it fails for any other reason, find the probability that he uses the radio at least 15 years.

13. Let $X_1, X_2, \ldots, X_n$ be independent exponential random variables with means $1/\lambda_1, 1/\lambda_2, \ldots, 1/\lambda_n$, respectively. Find the probability distribution function of $X = \min(X_1, X_2, \ldots, X_n)$.

B

14. An item has n parts, each with an exponentially distributed lifetime with mean $1/\lambda$. If the failure of one part makes the item fail, what is the average lifetime of the item?
    HINT: Use the result of Exercise 13.

15. Let $X_1, X_2, \ldots, X_n$ be n independent random numbers from the interval $(0, 1)$. Find $E(\max_{1 \le i \le n} X_i)$ and $E(\min_{1 \le i \le n} X_i)$.

16. Let F be a probability distribution function. Prove that the functions F^n and $1 - (1 - F)^n$ are also probability distribution functions.
    HINT: Let $X_1, X_2, \ldots, X_n$ be independent random variables each with the probability distribution function F. Look at $\max(X_1, X_2, \ldots, X_n)$ and $\min(X_1, X_2, \ldots, X_n)$.

17. Let $X_1, X_2, \ldots, X_n$ be n independent random numbers from $(0, 1)$, and $Y_n = n \cdot \min(X_1, X_2, \ldots, X_n)$. Prove that

$$\lim_{n \to \infty} P(Y_n > x) = e^{-x}, \qquad x \ge 0.$$

18. Suppose that h is the probability density function of a continuous random variable. Let the joint probability density function of $X, Y,$ and Z be

$$f(x, y, z) = h(x)h(y)h(z), \qquad x, y, z \in \mathbf{R}.$$

Prove that $P(X < Y < Z) = 1/6$.

19. A point is selected at random from the pyramid

$$V = \{(x, y, z) : x, y, z \ge 0, \ x + y + z \le 1\}.$$

Letting (X, Y, Z) be its coordinates, determine if $X, Y,$ and Z are independent.

HINT: Recall that the volume of a pyramid is $Bh/3$, where h is the height and B is the area of the base.

8.5   Multinomial Distributions

Multinomial distribution is a generalization of binomial.   Suppose that whenever an experiment is performed, one of the disjoint outcomes A_1, $A_2, \ldots, A_r$ will occur. Let $P(A_i) = p_i$, $1 \leq i \leq r$. Then $p_1 + p_2 + \cdots + p_r = 1$. If in n independent performances of this experiment, X_i, $i = 1, 2, 3, \ldots, r$, denotes the number of times that A_i occurs, then $p(x_1, \ldots, x_r)$, the joint probability function of $X_1, X_2, \ldots, X_r$ is called *multinomial joint probability function* and its distribution is said to be a *multinomial distribution*. For any set of nonnegative integers $\{x_1, x_2, \ldots, x_r\}$ with $x_1 + x_2 + \cdots + x_r = n$,

$$p(x_1, x_2, \ldots, x_r) = P\{X_1 = x_1, X_2 = x_2, \ldots, X_r = x_r\}$$

$$= \frac{n!}{x_1! \, x_2! \cdots x_r!} \, p_1^{x_1} p_2^{x_2} \cdots p_r^{x_r}. \tag{8.15}$$

To prove this relation, recall that by Theorem 2.4, the number of distinguishable permutations of n objects of r different types where x_1 are alike, x_2 are alike, $\cdots$, x_r are alike ($n = x_1 + \cdots + x_r$) is $n!/(x_1! \, x_2! \cdots x_r!)$. Hence there are $n!/(x_1! \, x_2! \cdots x_r!)$ sequences of $A_1, A_2, \ldots, A_r$ in which the number of A_i's is x_i, $i = 1, 2, \ldots, r$. Relation (8.15) follows since the probability of the occurrence of any of these sequences is $p_1^{x_1} p_2^{x_2} \cdots p_r^{x_r}$. By Theorem 2.6, $p(x_1, x_2, \ldots, x_r)$'s given by (8.15) are the terms in the expansion of $(p_1 + p_2 + \cdots + p_r)^n$. For this reason the multinomial distribution sometimes is called the *polynomial distribution*. The following relation guarantees that $p(x_1, x_2, \ldots, x_r)$ is a joint probability function.

$$\sum_{x_1 + x_2 + \cdots + x_r = n} p(x_1, x_2, \ldots, x_r)$$

$$= \sum_{x_1 + x_2 + \cdots + x_r = n} \frac{n!}{x_1! \, x_2! \cdots x_r!} \, p_1^{x_1} p_2^{x_2} \cdots p_r^{x_r} = (p_1 + p_2 + \cdots + p_r)^n = 1.$$

Note that for $r = 2$, the multinomial distribution coincides with the binomial distribution. This is because for $r = 2$ the experiment has only two possible outcomes, A_1 and A_2. Hence it is a Bernoulli trial.

Example 8.25   In a certain town, at 8:00 P.M., 30% of the viewing audience watch the news, 25% watch a certain comedy, and the rest watch other programs.   What is the probability that in a statistical survey, of seven randomly selected viewers, exactly three watch the news and at least two watch the comedy?

SOLUTION: Let in the random sample X_1, X_2, and X_3 be the numbers of the viewers who watch the news, the comedy, and other programs, respectively. Then the joint distribution of X_1, X_2, and X_3 is multinomial with $p_1 = 0.30$, $p_2 = 0.25$, and $p_3 = 0.45$. Therefore, for $i + j + k = 7$,

$$P(X_1 = i, X_2 = j, X_3 = k) = \frac{7!}{i!\, j!\, k!} (0.30)^i (0.25)^j (0.45)^k.$$

The desired probability equals

$$P(X_1 = 3, X_2 \geq 2) = P(X_1 = 3, X_2 = 2, X_3 = 2)$$
$$+ P(X_1 = 3, X_2 = 3, X_3 = 1) + P(X_1 = 3, X_2 = 4, X_3 = 0)$$
$$= \frac{7!}{3!\, 2!\, 2!} (0.30)^3 (0.25)^2 (0.45)^2 + \frac{7!}{3!\, 3!\, 1!} (0.30)^3 (0.25)^3 (0.45)^1$$
$$+ \frac{7!}{3!\, 4!\, 0!} (0.30)^3 (0.25)^4 (0.45)^0 \approx 0.103. \quad \blacklozenge$$

Example 8.26   In a warehouse, there are 500 TV sets, of which 25 are defective, 300 are in working condition but used, and the rest are brand new. What is the probability that in a random sample of five TV sets from this warehouse, there are exactly one defective and exactly two brand new sets?

SOLUTION: In the random sample let X_1 be the number of defective TV sets, X_2 be the number of TV sets in working condition but used, and X_3 be the number of brand new sets. The desired probability is given by

$$P(X_1 = 1, X_2 = 2, X_3 = 2) = \frac{\binom{25}{1} \binom{300}{2} \binom{175}{2}}{\binom{500}{5}} \approx 0.067.$$

Note that since the number of sets is large and the sample size is small, the joint distribution of X_1, X_2, and X_3 is *approximately* multinomial with $p_1 = 25/500 = 1/20$, $p_2 = 300/500 = 3/5$, and $p_3 = 175/500 = 7/20$. Hence

$$P(X_1 = 1, X_2 = 2, X_3 = 2) \approx \frac{5!}{1!\, 2!\, 2!} \left(\frac{1}{20}\right)^1 \left(\frac{3}{5}\right)^2 \left(\frac{7}{20}\right)^2 \approx 0.066. \quad \blacklozenge$$

Example 8.27   (Marginals of Multinomials)   Let $X_1, X_2, \ldots, X_r$ ($r \geq$ 4) have the joint multinomial probability function $p(x_1, x_2, \ldots, x_r)$ with parameters n and $p_1, p_2, \ldots, p_r$. Find the marginal probability functions p_{X_1} and p_{X_1, X_2, X_3}.

SOLUTION: From the definition of marginal probability functions:

$$
\begin{aligned}
p_{X_1}(x_1) &= \sum_{x_2 + x_3 + \cdots + x_r = n - x_1} \frac{n!}{x_1! \, x_2! \cdots x_r!} \, p_1^{x_1} p_2^{x_2} \cdots p_r^{x_r} \\
&= \frac{n!}{x_1! \, (n - x_1)!} \, p_1^{x_1} \sum_{x_2 + x_3 + \cdots + x_r = n - x_1} \frac{(n - x_1)!}{x_2! \, x_3! \cdots x_r!} \, p_2^{x_2} p_3^{x_3} \cdots p_r^{x_r} \\
&= \frac{n!}{x_1! \, (n - x_1)!} \, p_1^{x_1} \, (p_2 + p_3 + \cdots + p_r)^{n - x_1} \\
&= \frac{n!}{x_1! \, (n - x_1)!} \, p_1^{x_1} \, (1 - p_1)^{n - x_1},
\end{aligned}
$$

where the next-to-last equality follows from multinomial expansion (Theorem 2.6) and the last inequality follows from $p_1 + p_2 + \cdots + p_r = 1$. This shows that the marginal probability density function of X_1 is binomial with parameters n and p_1. To find $p_{X_1, X_2, X_3}(x_1, x_2, x_3)$, let $k = n - x_1 - x_2 - x_3$; then

$$
\begin{aligned}
&p_{X_1, X_2, X_3}(x_1, x_2, x_3) \\
&= \sum_{x_4 + \cdots + x_r = k} \frac{n!}{x_1! \, x_2! \, x_3! \, x_4! \cdots x_r!} \, p_1^{x_1} p_2^{x_2} p_3^{x_3} p_4^{x_4} \cdots p_r^{x_r} \\
&= \frac{n!}{x_1! \, x_2! \, x_3! \, k!} \, p_1^{x_1} p_2^{x_2} p_3^{x_3} \sum_{x_4 + \cdots + x_r = k} \frac{k!}{x_4! \cdots x_r!} \, p_4^{x_4} \cdots p_r^{x_r} \\
&= \frac{n!}{x_1! \, x_2! \, x_3! \, k!} \, p_1^{x_1} p_2^{x_2} p_3^{x_3} \, (p_4 + \cdots + p_r)^k \\
&= \frac{n!}{x_1! \, x_2! \, x_3! \, (n - x_1 - x_2 - x_3)!} \, p_1^{x_1} p_2^{x_2} p_3^{x_3} \, (1 - p_1 - p_2 - p_3)^{n - x_1 - x_2 - x_3}.
\end{aligned}
$$

This shows that the marginal joint probability density function of X_1, X_2, X_3 is multinomial with parameters n and $p_1, p_2, p_3, 1 - p_1 - p_2 - p_3$. ◆

REMARK: Let the joint distribution of $X_1, X_2, \ldots, X_r$ be multinomial with parameters n and $p_1, p_2, \ldots, p_r$. The method of Example 8.27 can be extended to prove that the marginal joint probability density function of a subset X_{i_1}, X_{i_2}, ..., X_{i_k} of k $(k > 1)$ of these n random variables is multinomial with parameters n and p_{i_1}, p_{i_2}, ..., p_{i_k}, $1 - p_{i_1} - p_{i_2} - \cdots - p_{i_k}$. ◆

EXERCISES

A

1. Light bulbs manufactured by a certain factory last a random time between 400 and 1200 hours. What is the probability that of eight such bulbs, three burn out before 550 hours, two burn out after 800 hours, and three burn out after 550 but before 800 hours?

2. An urn contains 100 chips of which 20 are blue, 30 are red, and 50 are green. We draw 20 chips at random and with replacement. Let B, R, and G be the number of blue, red, and green chips, respectively. Calculate the joint probability function of B, R, and G.

3. Suppose that each day the price of a stock moves up 1/8 of a point with probability 1/4, remains the same with probability 1/3, and moves down 1/8 of a point with probability 5/12. If the price fluctuations from one day to another are independent, what is the probability that after 6 days the stock has its original price?

4. At a certain college, 16% of the calculus students get A's, 34% B's, 34% C's, 14% D's, and 2% F's. What is the probability that of 15 calculus students selected at random five get B's, five C's, two D's, and at least two A's?

5. Suppose that 50% of the watermelons grown on a farm are classified as large, 30% as medium, and 20% as small. Joanna buys five watermelons at random from this farm. What is the probability that (a) at least two of them are large; (b) two of them are large, two are medium, and one is small; (c) exactly two of them are medium if it is given that at least two of them are large?

6. Suppose that 30% of the teachers of a country are over 50, 20% are between 40 and 50, and 50% are below 40. In a random committee of 10 teachers from this country, two are above 50. What is the probability function of those who are below 40?

7. In diploid organisms, which we all are, each hereditary character of each individual is carried by a pair of genes. A gene is either recessive a or dominant A, so that the possible pairs of genes are AA, Aa (same as aA), and aa. For the first generation of a population, let $r = P(AA)$, $s = P(Aa)$, and $t = P(aa)$, $r + s + t = 1$. Then by the Hardy–Weinberg law in genetics (see Example 3.25), the proportion

of AA, Aa, and aa in all future generations remains the same as the second generation: $P(AA) = (r + s/2)^2$, $P(Aa) = 2(r + s/2)(t + s/2)$, $P(aa) = (t + s/2)^2$. Let $p = r + s/2$, then $1 - p = t + s/2$ and hence $P(AA) = p^2$, $P(Aa) = 2p(1 - p)$, and $P(aa) = (1 - p)^2$, $0 < p < 1$. A group of six members of the population is selected randomly. Determine the value of p that maximizes the probability of the event of obtaining two AA's, two Aa's, and two aa's.

B

8. Customers enter a department store at the rate of three per minute in accordance with a Poisson process. If 30% of them buy nothing, 20% pay cash, 40% use charge cards, and 10% write personal checks, what is the probability that in 5 operating minutes of the store, five customers use charge cards, two write personal checks, and three pay cash?

8.6   Transformations of Two Random Variables

In our discussions of random variables throughout the book, cases have arisen where we have calculated the distribution and the density functions of a function of a random variable X. Functions such as X^2, e^X, $\cos X$, $X^3 + 1$, and so on. In particular, in Section 6.2 we explained how, in general, density functions and distribution functions of such functions can be obtained. In this section we demonstrate a method for finding the joint density function of functions of two random variables. The key is the following, which is the analog of the change of variable theorem for functions of several variables.

THEOREM 8.9   *Let X and Y be continuous random variables with joint probability density function $f(x, y)$. Let g and h be real-valued functions of two variables, $U = g(X, Y)$ and $V = h(X, Y)$. If*

(a) *the system of two equations in two unknowns*

$$\begin{cases} g(x, y) = u \\ h(x, y) = v \end{cases} \tag{8.16}$$

has a unique solution for x and y in terms of u and v, and

(b) *the functions g and h have continuous partial derivatives, and the Jacobian of the transformation $u = g(x, y)$, $v = h(x, y)$ is nonzero at all points (x, y); that is, the following 2×2 determinant is nonzero:*

$$\frac{\partial(g, h)}{\partial(x, y)} = \begin{vmatrix} \dfrac{\partial g}{\partial x} & \dfrac{\partial g}{\partial y} \\[2mm] \dfrac{\partial h}{\partial x} & \dfrac{\partial h}{\partial y} \end{vmatrix} = \frac{\partial g}{\partial x}\frac{\partial h}{\partial y} - \frac{\partial g}{\partial y}\frac{\partial h}{\partial x} \neq 0,$$

then the random variables U and V are jointly continuous with the joint probability density function l(u, v) given by

$$l(u, v) = f(x, y) \left| \frac{\partial(g, h)}{\partial(x, y)} \right|^{-1}, \tag{8.17}$$

where in this formula, for given (u, v), we have that (x, y) is the unique solution of the system (8.16).

Theorem 8.9 is a result of the change of variable theorem in double integrals. To see this, let R be the region of interest in the uv-plane and Q denote the region in the xy-plane that is transformed to R by (8.16). Then $(U, V) \in R$ if and only if $(X, Y) \in Q$. Hence the events $(U, V) \in R$ and $(X, Y) \in Q$ are equiprobable and

$$P\{(U, V) \in R\} = P\{(X, Y) \in Q\} = \iint\limits_{Q} f(x, y)\, dx\, dy.$$

Using the change of variable formula for double integrals, we obtain

$$\iint\limits_{Q} f(x, y)\, dx\, dy = \iint\limits_{R} f(x, y) \left| \frac{\partial(g, h)}{\partial(x, y)} \right|^{-1} du\, dv.$$

Hence

$$P\{(U, V) \in R\} = \iint\limits_{R} f(x, y) \left| \frac{\partial(g, h)}{\partial(x, y)} \right|^{-1} du\, dv.$$

This shows that $l(u, v)$, the joint density of u and v, is given by (8.17).

Example 8.28   Let X and Y be positive independent random variables with the identical probability density function e^{-x} for $x > 0$. Find the joint probability density function of $U = X + Y$ and $V = X/Y$.

SOLUTION: Let $f(x, y)$ be the joint probability density function of X and Y. Then

$$f_X(x) = \begin{cases} e^{-x} & \text{if } x > 0 \\ 0 & \text{if } x \le 0, \end{cases}$$

$$f_Y(y) = \begin{cases} e^{-y} & \text{if } y > 0 \\ 0 & \text{if } y \le 0. \end{cases}$$

Therefore,

$$f(x, y) = f_X(x) f_Y(y) = \begin{cases} e^{-(x+y)} & \text{if } x > 0 \text{ and } y > 0 \\ 0 & \text{elsewhere.} \end{cases}$$

Let $g(x, y) = x + y$ and $h(x, y) = x/y$. Then the system of equations

$$\begin{cases} x + y = u \\ \dfrac{x}{y} = v \end{cases}$$

has the unique solution $x = (uv)/(v + 1)$, $y = u/(v + 1)$, and

$$\frac{\partial(g, h)}{\partial(x, y)} = \begin{vmatrix} 1 & 1 \\ \dfrac{1}{y} & -\dfrac{x}{y^2} \end{vmatrix} = -\frac{x + y}{y^2} \ne 0,$$

since $x > 0$ and $y > 0$. Hence by Theorem 8.9, $l(u, v)$, the joint probability density function of U and V, is given by

$$l(u, v) = \frac{y^2}{x + y} f(x, y) = \frac{\left(\dfrac{u}{v + 1}\right)^2}{\dfrac{uv}{v + 1} + \dfrac{u}{v + 1}} e^{-u} = \frac{u e^{-u}}{(v + 1)^2}.$$

Note that $x > 0$ and $y > 0$ imply that $u > 0$ and $v > 0$. ◆

The following example proves a well-known theorem called *Box–Muller's theorem*, which as we explain in Section 12.4, is used to simulate normal random variables (see Theorem 12.3).

Example 8.29   Let X and Y be two independent uniform random variables over $(0, 1)$; show that the random variables

$$U = \cos(2\pi X)\sqrt{-2 \ln Y}$$

and

$$V = \sin(2\pi X)\sqrt{-2 \ln Y}$$

are independent standard normal random variables.

SOLUTION: Let $g(x, y) = \cos(2\pi x)\sqrt{-2 \ln y}$ and $h(x, y) = \sin(2\pi x)$ $\sqrt{-2 \ln y}$. Then the system of equations

$$\begin{cases} \cos(2\pi x)\sqrt{-2 \ln y} = u \\ \sin(2\pi x)\sqrt{-2 \ln y} = v \end{cases}$$

can be solved uniquely in terms of x and y. To see this, square both sides of these equations and sum them up. We obtain $-2 \ln y = u^2 + v^2$, which gives $y = \exp[-(u^2 + v^2)/2]$. Putting $-2 \ln y = u^2 + v^2$ back into these equations, we get

$$\cos 2\pi x = \frac{u}{\sqrt{u^2 + v^2}} \quad \text{and} \quad \sin 2\pi x = \frac{v}{\sqrt{u^2 + v^2}},$$

which enable us to determine the unique value of x. For example, if $u > 0$ and $v > 0$, then $2\pi x$ is uniquely determined in the first quadrant from $2\pi x = \arccos u/\sqrt{u^2 + v^2}$. Hence the first condition of Theorem 8.9 is satisfied. To check the second condition, note that

$$\frac{\partial(g, h)}{\partial(x, y)} = \begin{vmatrix} -2\pi \sin(2\pi x)\sqrt{-2 \ln y} & \dfrac{-\cos(2\pi x)}{y\sqrt{-2 \ln y}} \\ 2\pi \cos(2\pi x)\sqrt{-2 \ln y} & \dfrac{-\sin(2\pi x)}{y\sqrt{-2 \ln y}} \end{vmatrix} = \frac{2\pi}{y} \neq 0,$$

since $y > 0$. Now Theorem 8.9 implies that

$$l(u, v) = \frac{y}{2\pi} f(x, y) = \begin{cases} \dfrac{y}{2\pi} & \text{if } 0 < x < 1, \ 0 < y < 1 \\ 0 & \text{elsewhere,} \end{cases}$$

where

$$f(x, y) = f_X(x) f_Y(y) = \begin{cases} 1 & \text{if } 0 < x < 1, \ 0 < y < 1 \\ 0 & \text{elsewhere} \end{cases}$$

and $y = \exp[-(u^2 + v^2)/2]$; therefore,

$$l(u, v) = \frac{1}{2\pi} \exp\left(-\frac{u^2 + v^2}{2}\right), \qquad -\infty < u < +\infty, \ -\infty < v < +\infty.$$

The probability density function of U is given by

$$l_U(u) = \int_{-\infty}^{\infty} \frac{1}{2\pi} \exp\left(-\frac{u^2 + v^2}{2}\right) dv$$

$$= \frac{1}{2\pi} \exp\left(\frac{-u^2}{2}\right) \int_{-\infty}^{\infty} \exp\left(\frac{-v^2}{2}\right) dv,$$

where $\int_{-\infty}^{\infty} 1/\sqrt{2\pi} \exp(-v^2/2)\, dv = 1$ implies that $\int_{-\infty}^{\infty} \exp(-v^2/2)\, dv = \sqrt{2\pi}$. Therefore,

$$l_U(u) = \frac{1}{\sqrt{2\pi}} \exp\left(\frac{-u^2}{2}\right),$$

which shows that U is standard normal. Similarly,

$$l_V(v) = \frac{1}{\sqrt{2\pi}} \exp\left(\frac{-v^2}{2}\right).$$

Since $l(u, v) = l_U(u)l_V(v)$, U and V are independent standard normal random variables. ◆

EXERCISES

A

1. Let X and Y be independent random variables with common probability density function

$$f(x) = \begin{cases} \dfrac{1}{x^2} & \text{if } x \geq 1 \\ 0 & \text{elsewhere.} \end{cases}$$

Calculate the joint probability density function of $U = X/Y$ and $V = XY$.

2. Let X and Y be independent random numbers from the interval $(0, 1)$. Find the joint probability density function of $U = -2 \ln X$ and $V = -2 \ln Y$.

3. Let X and Y be independent random variables with common probability density function

$$f(x) = \begin{cases} e^{-x} & \text{if } x > 0 \\ 0 & \text{elsewhere.} \end{cases}$$

Find the joint probability density function of $U = X + Y$ and $V = e^X$.

4. Let X and Y be two positive independent continuous random variables with the probability density functions $f_1(x)$ and $f_2(y)$, respectively. Find the probability density function of $U = X/Y$.
   HINT: Let $V = X$; find the joint probability density function of U and V. Then calculate the marginal probability density function of U.

5. Let $X \sim N(0, 1)$ and $Y \sim N(0, 1)$ be independent random variables. Find the joint probability density function of $R = \sqrt{X^2 + Y^2}$ and $\Theta = \arctan(Y/X)$. Show that R and Θ are independent. Note that (R, Θ) is the polar coordinate representation of (X, Y).

B

6. Prove that if X and Y are independent standard normal random variables, then $X + Y$ and $X - Y$ are independent random variables. This is a special case of the following important theorem.

   Let X and Y be independent random variables with a common distribution F. The random variables $X + Y$ and $X - Y$ are independent if and only if F is a normal distribution function.

7. Let X and Y be independent gamma random variables with parameters (r_1, λ) and (r_2, λ), respectively. Define $U = X + Y$ and $V = X/(X+Y)$.
   (a) Find the joint probability density function of U and V.
   (b) Prove that U and V are independent.
   (c) Show that U is gamma and V is beta.

8. Let X and Y be independent exponential random variables each with parameter λ. Are the random variables $X + Y$ and X/Y independent?

Review Problems

1. The joint probability function of X and Y is given by the following table.

		x	
y	1	2	3
2	0.05	0.25	0.15
4	0.14	0.10	0.17
6	0.10	0.02	0.02

Find $P(XY \le 6)$.

2. A fair die is tossed twice. The sum of the outcomes is denoted by X and the largest value by Y. Calculate the joint probability function of X and Y, and the marginal probability functions of X and Y.

3. Calculate the probability function of the number of spades in a random bridge hand that includes exactly four hearts.

4. Suppose that three cards are drawn at random from an ordinary deck of 52 cards. If X and Y are the numbers of diamonds and clubs, respectively, calculate the joint probability function of X and Y.

5. Calculate the probability function of the number of spades in a random bridge hand that includes exactly four hearts and exactly three clubs.

6. An urn contains 100 chips of which 20 are blue, 30 are red, and 50 are green. Suppose that 20 chips are drawn at random and without replacement. Let B, R, and G be the number of blue, red, and green chips, respectively. Calculate the joint probability function of B, R, and G.

7. Let the joint probability density function of X and Y be given by

$$f(x, y) = \begin{cases} \dfrac{c}{x} & \text{if } 0 < y < x, \ 0 < x < 2 \\ 0 & \text{elsewhere.} \end{cases}$$

(a) Determine the value of c.
(b) Find the marginal probability density functions of X and Y.

8. Let the joint probability density function of X and Y be given by

$$f(x, y) = \begin{cases} \dfrac{3}{4}x^2y + \dfrac{1}{4}y & \text{if } 0 < x < 1 \text{ and } 0 < y < 2 \\ 0 & \text{elsewhere.} \end{cases}$$

Determine if $E(XY) = E(X)E(Y)$.

9. Prove that the following cannot be the joint probability distribution function of two random variables X and Y.

$$F(x, y) = \begin{cases} 1 & \text{if } x + y \geq 1 \\ 0 & \text{if } x + y < 1. \end{cases}$$

10. Three concentric circles of radii r_1, r_2, and r_3, $r_1 > r_2 > r_3$, are the boundaries of the regions that form a circular target. If a person fires a shot at random at the target, what is the probability that it lands in the middle region?

11. A fair coin is flipped 20 times. If the total number of heads is 12, what is the expected number of heads in the first 10 flips?

12. Let the joint probability distribution function of the lifetimes of two brands of light bulbs be given by

$$F(x, y) = \begin{cases} (1 - e^{-x^2})(1 - e^{-y^2}) & \text{if } x > 0, \ y > 0 \\ 0 & \text{otherwise.} \end{cases}$$

Find the probability that one light bulb lasts more than twice as long as the other one lasts.

13. For $\Omega = \{(x, y) : 0 < x + y < 1, \ 0 < x < 1, 0 < y < 1\}$, a region in the plane, let

$$f(x, y) = \begin{cases} 3(x + y) & \text{if } (x, y) \in \Omega \\ 0 & \text{otherwise.} \end{cases}$$

be the joint probability density function of the random variables X and Y. Find the marginal probability density functions of X and Y, and $P(X + Y > 1/2)$.

14. Let X and Y be continuous random variables with the joint probability density function

$$f(x, y) = \begin{cases} e^{-y} & \text{if } y > 0, \ \ 0 < x < 1 \\ 0 & \text{elsewhere.} \end{cases}$$

Find $E(X^n \mid Y = y)$, $n \geq 1$.

15. From an ordinary deck of 52 cards, cards are drawn successively and with replacement. Let X and Y denote the number of spades in the first 10 cards and in the second 15 cards, respectively. Calculate the joint probability function of X and Y.

16. Let X be the smallest number obtained in rolling a balanced die n times. Calculate the probability distribution function and the probability function of X.

17. Let the joint probability density function of X and Y be given by

$$f(x, y) = \begin{cases} cx(1 - x) & \text{if } 0 \leq x \leq y \leq 1 \\ 0 & \text{otherwise.} \end{cases}$$

(a) Determine the value of c.
(b) Determine if X and Y are independent.

18. A point is selected at random from the bounded region between the curves $y = x^2 - 1$ and $y = 1 - x^2$. Let X be the the x-coordinate, and let Y be the y-coordinate of the point selected. Determine if X and Y are independent.

19. Suppose that n points are selected at random and independently inside the cube

$$\Omega = \{(x, y, z) : -a \leq x \leq a, \ -a \leq y \leq a, \ -a \leq z \leq a\}.$$

Find the probability that the distance of the nearest point to the center is at least r $(r < a)$.

20. Let X and Y be two independent uniformly distributed random variables over the intervals $(0, 1)$ and $(0, 2)$, respectively. Find the probability density function of X/Y.

21. The joint probability density function of random variables X, Y, and Z is given by

$$f(x, y, z) = \begin{cases} c(x + y + 2z) & \text{if } 0 \leq x, y, z \leq 1 \\ 0 & \text{otherwise.} \end{cases}$$

(a) Determine the value of c.
(b) Find $P(X < 1/3 \mid Y < 1/2, \ Z < 1/4)$.

22. If F is the probability distribution function of a random variable X, is $G(x, y) = F(x) + F(y)$ a joint probability distribution function?

23. A fair die is tossed 18 times. What is the probability that each face appears three times?

24. Alvie, a marksman, fires at a target seven independent shots. Suppose that the probabilities that he hits the bull's-eye, he hits the target but not the bull's-eye, and he misses the target are 0.4, 0.35, and 0.25, respectively. What is the probability that he hits the bull's-eye three times, the target but not the bull's-eyes two times, and misses the target two times?

25. A bar of length ℓ is broken into three pieces at two random spots. What is the probability that the length of at least one piece is less than $\ell/20$?

26. There are prizes in 10% of the boxes of a certain type of cereal. Let X be the number of boxes of such cereal that Kim should buy to find a prize. Let Y be the number of additional boxes of such cereal that she should purchase to find another prize. Calculate the joint probability function of X and Y.

27. Let the joint probability density function of random variables X and Y be given by

$$f(x, y) = \begin{cases} 1 & \text{if } |y| < x, \quad 0 < x < 1 \\ 0 & \text{otherwise.} \end{cases}$$

Show that $E(Y|X = x)$ is a linear function of x while $E(X|Y = y)$ is not a linear function of y.

Chapter 9

SUMS OF RANDOM VARIABLES

9.1 Moment-Generating Functions

For a random variable X, we have demonstrated how important its first moment $E(X)$ and its second moment $E(X^2)$ are. For other values of n also, $E(X^n)$ is a valuable measure both in theory and practice. For example, letting $\mu = E(X)$, $\mu_X^{(r)} = E[(X - \mu)^r]$, we have that the quantity $\mu_X^{(3)}/\sigma_X^3$ is a measure of symmetry of the distribution of X. It is called the measure of *skewness* and is zero if the distribution of X is symmetric, negative if it is skewed to the left, and positive if it is skewed to the right (for examples of distributions that are, say, skewed to the right, see Figure 7.12). As another example, $\mu_X^{(4)}/\sigma_X^4$, the measure of *kurtosis*, indicates relative flatness of the distribution function of X. For a standard normal distribution function this quantity is 3. Therefore, if $\mu_X^{(4)}/\sigma_X^4 > 3$, the distribution function of X is more peaked than that of standard normal, and if $\mu_X^{(3)}/\sigma_X^3 < 3$, it is less peaked (flatter) than that of the standard normal. The moments of a random variable X give information of other sorts, too. For example, it can be proven that $E(|X|^k) < \infty$ implies that $\lim_{n \to \infty} n^k P(|X| > n) \to 0$. This shows that if $E(|X|^k) < \infty$, then $P(|X| > n)$ approaches 0 faster than $1/n^k$ as $n \to \infty$.

In this section we study moment-generating functions of random variables. These are real-valued functions with two major properties: They enable us to calculate the moments of random variables easily, and upon existence, they are unique. That is, no two different random variables have the same moment-generating function. Because of this, for proving that a random variable has a certain distribution F, it is often shown that its moment-generating function coincides with that of F. This method enables us to prove many interesting facts, including the celebrated central limit theorem (see Section 10.3).

DEFINITION   *For a random variable X, let*

$$M_X(t) = E\left(e^{tX}\right).$$

If $M_X(t)$ is defined for all values of t in some interval $(-\delta, \delta)$, $\delta > 0$, then $M_X(t)$ is called the moment-generating function of X.

Hence if X is a discrete random variable with set of possible values A and probability function $p(x)$, then

$$M_X(t) = \sum_{x \in A} e^{tx} p(x),$$

and if X is a continuous random variable with probability density function $f(x)$, then

$$M_X(t) = \int_{-\infty}^{\infty} e^{tx} f(x)\, dx.$$

As we mentioned in the beginning, and as can be guessed from its name, the moment-generating function of a random variable X can be used to calculate the moments of X. To show how, let $M_X^{(n)}(t)$ be the nth derivative of $M_X(t)$. If X is continuous with probability density function f, then

$$M_X'(t) = \frac{d}{dt}\left(\int_{-\infty}^{\infty} e^{tx} f(x)\, dx\right) = \int_{-\infty}^{\infty} x e^{tx} f(x)\, dx,$$

$$M_X''(t) = \frac{d}{dt}\left(\int_{-\infty}^{\infty} x e^{tx} f(x)\, dx\right) = \int_{-\infty}^{\infty} x^2 e^{tx} f(x)\, dx,$$

$$\vdots$$

$$M_X^{(n)}(t) = \int_{-\infty}^{\infty} x^n e^{tx} f(x)\, dx, \tag{9.1}$$

where we have assumed that the derivatives of these integrals are equal to the integrals of the derivatives of their integrands. This property is valid for sufficiently smooth densities. Letting $t = 0$ in (9.1), we get

$$M_X^{(n)}(0) = \int_{-\infty}^{\infty} x^n f(x)\, dx = E(X^n).$$

Note that since in some interval $(-\delta, \delta)$, $\delta > 0$, $M_X(t)$ is finite, $M_X^{(n)}(0)$ exists for all $n \geq 1$.

In a similar way, the relation $M_X^{(n)}(0) = E(X^n)$ can be shown for discrete random variables as well. We leave it as an exercise. One important implication of this relation is that the MacLaurin's series of $M_X(t)$ is given by

$$M_X(t) = \sum_{n=0}^{\infty} \frac{M_X^{(n)}(0)}{n!} t^n = \sum_{n=0}^{\infty} \frac{E(X^n)}{n!} t^n. \tag{9.2}$$

Therefore, $E(X^n)$ is the coefficient of $t^n/n!$ in the MacLaurin's series representation of $M_X(t)$.

Example 9.1 Let X be a Bernoulli random variable with parameter p, that is,

$$P(X = x) = \begin{cases} 1 - p & \text{if } x = 0 \\ p & \text{if } x = 1 \\ 0 & \text{otherwise.} \end{cases}$$

Determine $M_X(t)$ and $E(X^n)$.

Solution: From the definition of moment-generating function,

$$M_X(t) = E\left(e^{tX}\right) = (1 - p)e^{t \cdot 0} + pe^{t \cdot 1} = (1 - p) + pe^t.$$

Since $M_X^{(n)}(t) = pe^t$ for all $n > 0$, we have that $E(X^n) = M_X^{(n)}(0) = p.$ ♦

Example 9.2 Let X be a binomial random variable with parameters (n, p). Find the moment-generating function of X and use it to calculate $E(X)$ and $\text{Var}(X)$.

Solution: The probability function of X, $p(x)$, is given by

$$p(x) = \binom{n}{x} p^x q^{n-x}, \qquad x = 0, 1, 2, \ldots, n, \qquad q = 1 - p.$$

Hence

$$M_X(t) = E\left(e^{tX}\right) = \sum_{x=0}^{n} e^{tx} \binom{n}{x} p^x q^{n-x}$$

$$= \sum_{x=0}^{n} \binom{n}{x} \left(pe^t\right)^x q^{n-x} = \left(pe^t + q\right)^n,$$

where the last equality follows from the binomial expansion (see Theorem 2.5). To find the mean and the variance of X, note that

$$M_X'(t) = npe^t \left(pe^t + q\right)^{n-1}$$

$$M_X''(t) = npe^t \left(pe^t + q\right)^{n-1} + n(n-1)\left(pe^t\right)^2 \left(pe^t + q\right)^{n-2}.$$

Thus

$$E(X) = M_X'(0) = np$$

$$E(X^2) = M_X''(0) = np + n(n-1)p^2.$$

Therefore,

$$\mathrm{Var}(X) = E(X^2) - [E(X)]^2 = np + n(n-1)p^2 - n^2 p^2 = npq. \quad \blacklozenge$$

Example 9.3   Let X be an exponential random variable with parameter λ. Using moment-generating functions, calculate the mean and the variance of X.

SOLUTION: The probability density function of X is given by

$$f(x) = \lambda e^{-\lambda x}, \qquad x \ge 0.$$

Thus

$$M_X(t) = E\left(e^{tX}\right) = \int_0^\infty e^{tx} \lambda e^{-\lambda x}\, dx = \lambda \int_0^\infty e^{(t-\lambda)x}\, dx.$$

Since the integral $\int_0^\infty e^{(t-\lambda)x}\, dx$ converges if $t < \lambda$, restricting the domain of $M_X(t)$ to $(-\infty, \lambda)$, we get $M_X(t) = \lambda/(\lambda - t)$. Thus $M_X'(t) = \lambda/(\lambda - t)^2$ and $M_X''(t) = (2\lambda)/(\lambda - t)^3$. We obtain $E(X) = M_X'(0) = 1/\lambda$ and $E(X^2) = M_X''(0) = 2/\lambda^2$. Therefore,

$$\mathrm{Var}(X) = E\left(X^2\right) - [E(X)]^2 = \frac{2}{\lambda^2} - \frac{1}{\lambda^2} = \frac{1}{\lambda^2}. \quad \blacklozenge$$

Example 9.4   Let X be an exponential random variable with parameter λ. Using moment-generating functions, find $E(X^n)$, where n is a positive integer.

SOLUTION: By Example 9.3,

$$M_X(t) = \frac{\lambda}{\lambda - t} = \frac{1}{1 - (t/\lambda)}, \qquad t < \lambda.$$

Now on the one hand, by the geometric series theorem,

$$M_X(t) = \frac{1}{1 - (t/\lambda)} = \sum_{n=0}^{\infty} \left(\frac{t}{\lambda}\right)^n = \sum_{n=0}^{\infty} \left(\frac{1}{\lambda}\right)^n t^n,$$

and on the other hand, by (9.2),

$$M_X(t) = \sum_{n=0}^{\infty} \frac{M_X^{(n)}(0)}{n!} t^n = \sum_{n=0}^{\infty} \frac{E(X^n)}{n!} t^n.$$

Comparing these two relations, we obtain

$$\frac{E(X^n)}{n!} = \left(\frac{1}{\lambda}\right)^n,$$

which gives $E(X^n) = n!/\lambda^n$.   ◆

Example 9.5   Let Z be a standard normal random variable.

(a) Calculate the moment-generating function of Z.

(b) Use part (a) to find the moment-generating function of X, where X is a normal random variable with mean μ and variance σ^2.

(c) Use part (b) to calculate the mean and the variance of X.

SOLUTION: (a) From the definition,

$$M_Z(t) = E\left(e^{tZ}\right) = \int_{-\infty}^{\infty} e^{tz} \frac{1}{\sqrt{2\pi}} e^{-z^2/2}\, dz = \frac{1}{\sqrt{2\pi}} \int_{-\infty}^{\infty} e^{tz - z^2/2}\, dz.$$

Since

$$tz - \frac{z^2}{2} = \frac{t^2}{2} - \frac{(z-t)^2}{2},$$

we obtain

$$M_Z(t) = \frac{1}{\sqrt{2\pi}} \int_{-\infty}^{\infty} \exp\left[\frac{t^2}{2} - \frac{(z-t)^2}{2}\right] dz$$

$$= \frac{1}{\sqrt{2\pi}} e^{t^2/2} \int_{-\infty}^{\infty} \exp\left[-\frac{(z-t)^2}{2}\right] dz.$$

Let $u = z - t$. Then $du = dz$ and

$$M_Z(t) = \frac{1}{\sqrt{2\pi}} e^{t^2/2} \int_{-\infty}^{\infty} e^{-u^2/2} \, du$$

$$= e^{t^2/2} \frac{1}{\sqrt{2\pi}} \int_{-\infty}^{\infty} e^{-u^2/2} \, du = e^{t^2/2},$$

where the last equality follows since the function $(1/\sqrt{2\pi})e^{-u^2/2}$ is the probability density function of a standard normal random variable, and hence its integral on $(-\infty, +\infty)$ is 1.

(b) Letting $Z = (X - \mu)/\sigma$, we have that Z is $N(0, 1)$ and $X = \sigma Z + \mu$. Thus

$$M_X(t) = E\left(e^{tX}\right) = E\left(e^{t\sigma Z + t\mu}\right) = E\left(e^{t\sigma Z} e^{t\mu}\right)$$

$$= e^{t\mu} E\left(e^{t\sigma Z}\right) = e^{t\mu} M_Z(t\sigma) = e^{t\mu} \exp\left(\frac{1}{2} t^2 \sigma^2\right)$$

$$= \exp\left(t\mu + \frac{1}{2}\sigma^2 t^2\right).$$

(c) Differentiating $M_X(t)$, we obtain

$$M_X'(t) = (\mu + \sigma^2 t) \exp\left(t\mu + \frac{1}{2}\sigma^2 t^2\right),$$

which upon differentiation gives

$$M_X''(t) = (\mu + \sigma^2 t)^2 \exp\left(t\mu + \frac{1}{2}\sigma^2 t^2\right) + \sigma^2 \exp\left(t\mu + \frac{1}{2}\sigma^2 t^2\right).$$

Therefore, $E(X) = M_X'(0) = \mu$ and $E(X^2) = M_X''(0) = \mu^2 + \sigma^2$. Thus

$$\text{Var}(X) = E(X^2) - [E(X)]^2 = \mu^2 + \sigma^2 - \mu^2 = \sigma^2. \quad \blacklozenge$$

Example 9.6   A positive random variable X is called *lognormal* with parameters μ and σ^2 if $\ln X \sim N(\mu, \sigma^2)$. Calculate the rth moment of a lognormal variable X with parameters μ and σ.

SOLUTION: Let $Y = \ln X$, then $X = e^Y$. Now $Y \sim N(\mu, \sigma^2)$ gives

$$E(X^r) = E\left(e^{rY}\right) = M_Y(r) = \exp\left(\mu r + \frac{1}{2}\sigma^2 r^2\right). \quad \blacklozenge$$

Another important property of a moment-generating function is its uniqueness, which we will now state without proof.

THEOREM 9.1   *Let X and Y be two random variables with moment-generating functions $M_X(t)$ and $M_Y(t)$. If for some $\delta > 0$, $M_X(t) = M_Y(t)$ for all values of t in $(-\delta, \delta)$, then X and Y have the same distribution.*

Note that the converse of Theorem 9.1 is trivially true.

Example 9.7   Let the moment-generating function of a random variable X be

$$M_X(t) = \frac{1}{7}e^t + \frac{3}{7}e^{3t} + \frac{2}{7}e^{5t} + \frac{1}{7}e^{7t}.$$

Since the moment-generating function of a discrete random variable with the probability function

i	1	3	5	7	Other values
$p(i)$	1/7	3/7	2/7	1/7	0

is $M_X(t)$ given above, by Theorem 9.1, the probability function of X is $p(i)$. $\blacklozenge$

Example 9.8   Let X be a random variable with moment-generating function $M_X(t) = e^{2t^2}$. Find $P(0 < X < 1)$.

SOLUTION: Comparing $M_X(t) = e^{2t^2}$ with $\exp\left[\mu t + (1/2)\sigma^2 t^2\right]$, the moment-generating function of $N(\mu, \sigma^2)$ (see Example 9.5), we have that $X \sim N(0, 4)$ by the uniqueness of the moment-generating function. Let $Z = (X - 0)/2$. Then $Z \sim N(0, 1)$, so

$$P(0 < X < 1) = P\left(0 < \frac{X}{2} < \frac{1}{2}\right) = P(0 < Z < 0.5)$$

$$= \Phi(0.5) - \Phi(0) \approx 0.6915 - 0.5 = 0.1915,$$

by Table 1 of the Appendix. ♦

One of the most important properties of moment-generating functions is that they enable us to find distribution functions of sums of independent random variables. We discuss this important property in Section 9.2 and use it in Section 10.3 to prove the celebrated central limit theorem.

Often it happens that the moment-generating function of a random variable is known but its distribution function is not. In such cases Table 2 of the Appendix might help us identify the distribution.

EXERCISES

A

1. Let X be a discrete random variable with probability function $p(i) = 1/5$, $i = 1, 2, \ldots, 5$, zero elsewhere. Find $M_X(t)$.

2. Let X be a random variable with probability density function

$$f(x) = \begin{cases} \dfrac{1}{4} & \text{if } x \in (-1, 3) \\ 0 & \text{otherwise.} \end{cases}$$

(a) Find $M_X(t)$, $E(X)$, and $\text{Var}(X)$.
(b) Using $M_X(t)$, calculate $E(X)$.
HINT: Note that by the definition of derivative,

$$M'_X(0) = \lim_{h \to 0} \frac{M_X(h) - M_X(0)}{h}.$$

3. Let X be a discrete random variable with the probability function

$$p(i) = 2\left(\frac{1}{3}\right)^i, \qquad i = 1, 2, 3, \ldots; \qquad \text{zero, elsewhere.}$$

Find $M_X(t)$ and $E(X)$.

4. Let X be a continuous random variable with probability density function $f(x) = 2x$, if $0 \leq x \leq 1$, zero elsewhere. Find the moment-generating function of X.

5. Let X be a continuous random variable with the probability density function $f(x) = 6x(1 - x)$, if $0 \leq x \leq 1$, zero elsewhere.
   (a) Find $M_X(t)$
   (b) Using $M_X(t)$, find $E(X)$.

6. Let X be a discrete random variable. Prove that $E(X^n) = M_X^{(n)}(0)$.

7. (a) Find $M_X(t)$, the moment-generating function of a Poisson random variable X with parameter λ.
   (b) Use $M_X(t)$ to find $E(X)$ and Var(X).

8. Let X be a uniform random variable over the interval (a, b). Find the moment-generating function of X.

9. Let X be a geometric random variable with parameter p. Show that the moment-generating function of X is given by

$$M_X(t) = \frac{pe^t}{1 - qe^t}, \qquad q = 1 - p, \qquad t < -\ln q.$$

Use $M_X(t)$ to find $E(X)$ and Var(X).

10. Let $M_X(t) = (1/21) \sum_{n=1}^{6} ne^{nt}$. Find the probability function of X.

11. Suppose that the moment-generating function of a random variable X is given by

$$M_X(t) = \frac{1}{3}e^t + \frac{4}{15}e^{3t} + \frac{2}{15}e^{4t} + \frac{4}{15}e^{5t}.$$

Find the probability function of X.

12. Let $M_X(t) = 1/(1 - t)$, $t < 1$ be the moment-generating function of a random variable X. Find the moment-generating function of the random variable $Y = 2X + 1$.

13. For a random variable X, $M_X(t) = [2/(2 - t)]^3$. Find $E(X)$ and Var(X).

14. Suppose that the moment-generating function of X is given by

$$M_X(t) = \frac{e^t + e^{-t}}{6} + \frac{2}{3}, \qquad -\infty < t < \infty.$$

Find $E(X^r)$, $r \geq 1$.

15. Prove that the function $t/(1 - t)$, $t < 1$, cannot be the moment-generating function of a random variable.

16. In each of the following cases $M_X(t)$, the moment-generating function of X, is given. Determine the distribution of X.

   (a) $M_X(t) = \left(\frac{1}{4}e^t + \frac{3}{4}\right)^7$.
   (b) $M_X(t) = e^t/(2 - e^t)$.
   (c) $M_X(t) = [2/(2 - t)]^r$.
   (d) $M_X(t) = \exp[3(e^t - 1)]$.
   HINT: Use Table 2 of the Appendix.

17. For a random variable X, $M_X(t) = (1/81)\left(e^t + 2\right)^4$. Find $P(X \le 2)$.

18. Suppose that for a random variable X, $E(X^n) = 2^n$, $n = 1, 2, 3, \ldots$. Calculate the moment-generating function and the probability function of X.
   HINT: Use (9.2).

19. Let X be a uniform random variable over $(0, 1)$. Using moment-generating functions, show that $Y = aX + b$ is uniformly distributed over $(a, a + b)$.

B

20. Let $Z \sim N(0, 1)$. Use $M_Z(t) = e^{t^2/2}$ to calculate $E(Z^n)$, where n is a positive integer.
   HINT: Use (9.2).

21. Let X be a gamma random variable with parameters r and λ. Derive a formula for $M_X(t)$ and use it to calculate $E(X)$ and $\text{Var}(X)$.

22. Let X be a continuous random variable whose probability density function, f, is even; that is, $f(-x) = f(x)$, $\forall x$. Prove that (a) the random variables X and $-X$ have the same probability distribution function; (b) the function $M_X(t)$ is an even function.

23. Let X be a discrete random variable with probability function

$$p(i) = \frac{6}{\pi^2 i^2}, \qquad i = 1, 2, 3, \ldots; \qquad \text{zero, elsewhere.}$$

   Show that the moment-generating function of X does not exist.
   HINT: Show that $M_X(t)$ is a divergent series on $(0, \infty)$. This implies that on no interval of the form $(-\delta, \delta)$, $\delta > 0$, $M_X(t)$ exists.

24. Suppose that $\forall n \ge 1$, the nth moment of a random variable X, is given by $E(X^n) = (n + 1)! \, 2^n$. Find the distribution of X.

9.2 Sums of Independent Random Variables

In this section, using moment-generating functions, we study the distribution functions of sums of random variables. To begin, we prove the following theorem, which together with the uniqueness property of moment-generating functions (see Theorem 9.1) are two of the main tools of this section.

THEOREM 9.2 *Let $X_1, X_2, \ldots, X_n$ be independent random variables with moment-generating functions $M_{X_1}(t)$, $M_{X_2}(t)$, $\ldots$, $M_{X_n}(t)$. The moment-generating function of $X_1 + X_2 + \cdots + X_n$ is given by*

$$M_{X_1+X_2+\cdots+X_n}(t) = M_{X_1}(t)M_{X_2}(t) \cdots M_{X_n}(t).$$

PROOF: Let $W = X_1 + X_2 + \cdots + X_n$; by definition,

$$
\begin{aligned}
M_W(t) &= E\left(e^{tW}\right) = E\left(e^{tX_1+tX_2+\cdots+tX_n}\right) \\
&= E\left(e^{tX_1}e^{tX_2}\cdots e^{tX_n}\right) \\
&= E\left(e^{tX_1}\right)E\left(e^{tX_2}\right)\cdots E\left(e^{tX_n}\right) \\
&= M_{X_1}(t)M_{X_2}(t)\cdots M_{X_n}(t),
\end{aligned}
$$

where the next-to-last equality follows from the independence of $X_1, X_2, X_3,$ $\ldots, X_n$. ♦

We now prove that sums of independent binomial random variables are binomial. Let X and Y be two independent binomial random variables with parameters (n, p) and (m, p), respectively. Then X and Y are the numbers of successes in n and m independent Bernoulli trials with parameter p, respectively. Thus $X + Y$ is the number of successes in $n + m$ independent Bernoulli trials with parameter p. Therefore, $X + Y$ is a binomial random variable with parameters $(n + m, p)$. An alternative for this probabilistic argument is the following analytic proof.

THEOREM 9.3 *Let $X_1, X_2, \ldots, X_r$ be independent binomial random variables with parameters (n_1, p), (n_2, p), $\ldots$, (n_r, p), respectively. Then $X_1 + X_2 + \cdots + X_r$ is a binomial random variable with parameters $n_1 + n_2 + \cdots + n_r$ and p.*

PROOF: Let, as usual, $q = 1 - p$. We know that

$$M_{X_i}(t) = \left(pe^t + q\right)^{n_i}, \qquad i = 1, 2, 3, \ldots, n$$

(see Example 9.2). Let $W = X_1 + X_2 + \cdots + X_r$; then, by Theorem 9.2,

$$M_W(t) = M_{X_1}(t)M_{X_2}(t) \cdots M_{X_r}(t)$$

$$= \left(pe^t + q\right)^{n_1} \left(pe^t + q\right)^{n_2} \cdots \left(pe^t + q\right)^{n_r}$$

$$= \left(pe^t + q\right)^{n_1 + n_2 + \cdots + n_r}.$$

Since $\left(pe^t + q\right)^{n_1 + n_2 + \cdots + n_r}$ is the moment-generating function of a binomial random variable with parameters $(n_1 + n_2 + \cdots + n_r, p)$, the uniqueness property of moment-generating functions implies that $W = X_1 + X_2 + \cdots + X_r$ is binomial with parameters $(n_1 + n_2 + \cdots + n_r, p)$. ◆

We just showed that if X and Y are independent binomial random variables with parameters (n, p) and (m, p), respectively, then $X + Y$ is a binomial random variable with parameters $(n + m, p)$. Since for large values of n and m, small p, and moderate $\lambda_1 = np$ and $\lambda_2 = mp$, Poisson probability functions with parameters λ_1, λ_2, and $\lambda_1 + \lambda_2$ approximate probability functions of X, Y, and $X + Y$, respectively, it is reasonable to expect that for independent Poisson random variables X and Y with parameters λ_1 and λ_2, $X + Y$ is a Poisson random variable with parameter $\lambda_1 + \lambda_2$. The proof of this interesting fact follows for n random variables.

THEOREM 9.4   *Let* $X_1, X_2, \ldots, X_n$ *be independent Poisson random variables with means* $\lambda_1, \lambda_2, \ldots, \lambda_n$, *respectively. Then* $X_1 + X_2 + \cdots + X_n$ *is a Poisson random variable with mean* $\lambda_1 + \lambda_2 + \cdots + \lambda_n$.

PROOF: Let Y be a Poisson random variable with mean λ. Then

$$M_Y(t) = E\left(e^{tY}\right) = \sum_{y=0}^{\infty} e^{ty} \frac{e^{-\lambda}\lambda^y}{y!} = e^{-\lambda} \sum_{y=0}^{\infty} \frac{\left(e^t\lambda\right)^y}{y!}$$

$$= e^{-\lambda}e^{\lambda e^t} = e^{\lambda(e^t - 1)}.$$

Let $W = X_1 + X_2 + \cdots + X_n$; then, by Theorem 9.2,

$$M_W(t) = M_{X_1}(t)M_{X_2}(t) \cdots M_{X_n}(t)$$

$$= e^{\lambda_1(e^t - 1)}e^{\lambda_2(e^t - 1)} \cdots e^{\lambda_n(e^t - 1)}$$

$$= e^{(\lambda_1 + \lambda_2 + \cdots + \lambda_n)(e^t - 1)}.$$

Now since $e^{(\lambda_1 + \lambda_2 + \cdots + \lambda_n)(e^t - 1)}$ is the moment-generating function of a Poisson random variable with mean $\lambda_1 + \lambda_2 + \cdots + \lambda_n$, the uniqueness property of moment-generating functions implies that $X_1 + X_2 + \cdots + X_n$ is Poisson with mean $\lambda_1 + \lambda_2 + \cdots + \lambda_n$. ◆

THEOREM 9.5 *Let* $X_1 \sim N(\mu_1, \sigma_1^2)$, $X_2 \sim N(\mu_2, \sigma_2^2)$, $\ldots$, $X_n \sim N(\mu_n, \sigma_n^2)$ *be independent random variables. Then*

$$X_1 + X_2 + \cdots + X_n \sim N(\mu_1 + \mu_2 + \cdots + \mu_n, \sigma_1^2 + \sigma_2^2 + \cdots + \sigma_n^2).$$

PROOF: In Example 9.5 we showed that if X is normal with parameters μ and σ^2, then $M_X(t) = \exp[\mu t + (1/2)\sigma^2 t^2]$. Let $W = X_1 + X_2 + \cdots + X_n$; then

$$M_W(t) = M_{X_1}(t) M_{X_2}(t) \cdots M_{X_n}(t)$$

$$= \exp\left(\mu_1 t + \frac{1}{2}\sigma_1^2 t^2\right) \exp\left(\mu_2 t + \frac{1}{2}\sigma_2^2 t^2\right) \cdots \exp\left(\mu_n t + \frac{1}{2}\sigma_n^2 t^2\right)$$

$$= \exp\left[(\mu_1 + \mu_2 + \cdots + \mu_n)t + \frac{1}{2}(\sigma_1^2 + \sigma_2^2 + \cdots + \sigma_n^2)t^2\right].$$

This implies that

$$X_1 + X_2 + \cdots + X_n \sim N(\mu_1 + \mu_2 + \cdots + \mu_n, \sigma_1^2 + \sigma_2^2 + \cdots + \sigma_n^2). \; \blacklozenge$$

The technique used to show Theorems 9.3, 9.4, and 9.5 can also be used to prove that sums of independent geometric random variables are negative binomial, sums of independent negative binomial random variables are negative binomial, sums of independent exponential random variables are gamma, sums of independent gamma random variables are gamma, and so on. These important facts are discussed in the exercises.

If X is a normal random variable with parameters (μ, σ^2), then for $\alpha \in \mathbf{R}$, we can easily prove that $M_{\alpha X}(t) = \exp[\alpha \mu t + (1/2)\alpha^2 \sigma^2 t^2]$, implying that $\alpha X \sim N(\alpha \mu, \alpha^2 \sigma^2)$ (see Exercise 1). This and Theorem 9.5 imply that linear combinations of sets of independent normal random variables are normal.

THEOREM 9.6 *Let* $\{X_1, X_2, \ldots, X_n\}$ *be a set of independent random variables and* $X_i \sim N(\mu_i, \sigma_i^2)$ *for* $i = 1, 2, \ldots, n$; *then for constants* $\alpha_1, \alpha_2, \ldots, \alpha_n$,

$$\sum_{i=1}^{n} \alpha_i X_i \sim N\left(\sum_{i=1}^{n} \alpha_i \mu_i, \; \sum_{i=1}^{n} \alpha_i^2 \sigma_i^2\right).$$

In particular, this theorem implies that if $X_1, X_2, \ldots, X_n$ are independent normal random variables all with the same mean μ, and the same variance σ^2, then $S_n = X_1 + X_2 + \cdots + X_n$ is $N(n\mu, n\sigma^2)$ and the sample mean, $\overline{X} = S_n/n$, is $N(\mu, \sigma^2/n)$. That is, $\overline{X} = (1/n) \sum_{i=1}^{n} X_i$, the sample mean of n independent $N(\mu, \sigma^2)$, is $N(\mu, \sigma^2/n)$.

Example 9.9   Suppose that the distribution of the grades of students in a probability test is normal with mean 72 and variance 25.

(a) What is the probability that the average of the grades of such a probability class with 25 students is 75 or more?

(b) If a professor teaches two different sections of this course, each of which has 25 students, what is the probability that the average of one class is at least 3 more than the average of the other class?

SOLUTION: (a) Let $X_1, X_2, \ldots, X_{25}$ denote the grades of the 25 students. Then $X_1, X_2, \ldots, X_{25}$ are independent random variables all being normal with $\mu = 72$ and $\sigma^2 = 25$. The average of the grades of the class, $\overline{X} = (1/25) \sum_{i=1}^{25} X_i$, is normal with mean $\mu = 72$ and variance $\sigma^2/n = 25/25 = 1$. Hence

$$P(\overline{X} \geq 75) = P\left(\frac{\overline{X} - 72}{1} \geq \frac{75 - 72}{1} \right)$$

$$= P(\overline{X} - 72 \geq 3) = 1 - \Phi(3) \approx 0.0013.$$

(b) Let $\overline{X}$ and $\overline{Y}$ denote the means of the grades of the two classes. Then, as seen in (a), $\overline{X}$ and $\overline{Y}$ are both $N(72, 1)$. Since a student does not take two sections of the same course, $\overline{X}$ and $\overline{Y}$ are independent random variables. By Theorem 9.6, $\overline{X} - \overline{Y}$ is $N(0, 2)$. Hence, by symmetry,

$$P(|\overline{X} - \overline{Y}| > 3) = P(\overline{X} - \overline{Y} > 3 \text{ or } \overline{Y} - \overline{X} > 3)$$

$$= 2P(\overline{X} - \overline{Y} > 3)$$

$$= 2P\left(\frac{\overline{X} - \overline{Y} - 0}{\sqrt{2}} > \frac{3 - 0}{\sqrt{2}} \right)$$

$$= 2P\left(\frac{\overline{X} - \overline{Y}}{\sqrt{2}} > 2.12 \right) = 2[1 - \Phi(2.12)] \approx 0.034. \quad \blacklozenge$$

As we mentioned above, using moment-generating functions, it can be easily shown that a sum of n independent exponential random variables, each with parameter λ, is gamma with parameters n and λ (see Exercise 3). We now present an alternative proof for this important fact: Let X be a gamma random variable with parameters (n, λ), where n is a positive integer.

Consider a Poisson process $\{N(t) : t \geq 0\}$ with rate λ. $\{X_1, X_2, X_3, \ldots\}$, the set of interarrival times of this process [X_i is the time between the $(i-1)$st event and the ith event, $i = 1, 2, 3, \ldots$], is an independent, identically distributed sequence of exponential random variables with $E(X_i) = 1/\lambda$ for $i = 1, 2, 3, \ldots$. Since the distribution of X is the same as the time of the occurrence of the nth event,

$$X = X_1 + X_2 + \cdots + X_n.$$

Hence a *gamma random variable with parameters* (n, λ) *is the sum of* n *independent exponential random variables each with mean* $1/\lambda$, *and vice versa.*

Using moment-generating functions, we can also easily prove that if $X_1, X_2, \ldots, X_n$ are n independent gamma random variables with parameters (r_1, λ), (r_2, λ), $\ldots$, (r_n, λ), respectively, then $X_1 + X_2 + \cdots + X_n$ is gamma with parameters $(r_1 + r_2 + \cdots + r_n, \lambda)$ (see Exercise 5). The following example is an application of this important fact. It gives an interesting class of gamma random variables with parameters $(r, 1/2)$, where r is not necessarily an integer.

Example 9.10 Let $X_1, X_2, \ldots, X_n$ be independent standard normal random variables. Then $X = X_1^2 + X_2^2 + \cdots + X_n^2$, referred to as *chi-squared random variable with n degrees of freedom*, is gamma with parameters $(n/2, 1/2)$. An example of such a gamma random variable is the error of hitting a target in n-dimensional Euclidean space when the error of each coordinate is individually normally distributed.

PROOF: Since the sum of n independent gamma random variables each with parameters $(1/2, 1/2)$ is gamma with parameters $(n/2, 1/2)$, it suffices to prove that for all i, $1 \leq i \leq n$, X_i^2 is gamma with parameters $(1/2, 1/2)$. To prove this, note that

$$P(X_i^2 \leq t) = P(-\sqrt{t} \leq X_i \leq \sqrt{t}) = \Phi(\sqrt{t}) - \Phi(-\sqrt{t}).$$

Let the probability density function of X_i^2 be f. Differentiating this equation yields

$$f(t) = \left(\frac{1}{2\sqrt{t}} \frac{1}{\sqrt{2\pi}} e^{-t/2} \right) - \left(-\frac{1}{2\sqrt{t}} \frac{1}{\sqrt{2\pi}} e^{-t/2} \right).$$

Therefore,

$$f(t) = \frac{1}{\sqrt{2\pi t}} e^{-t/2} = \frac{(1/2)e^{-t/2}(t/2)^{1/2-1}}{\sqrt{\pi}}.$$

Since $\sqrt{\pi} = \Gamma(1/2)$ (see Exercise 6 of Section 7.4), this relation shows that X_i^2 is a gamma random variable with parameters $(1/2, 1/2)$. ◆

As an application of Theorem 8.9, we now prove the following theorem which gives excellent tools for calculation of density and distribution functions of sums of continuous random variables.

THEOREM 9.7   (Convolution Theorem)   *Let X and Y be continuous independent random variables with probability density functions f_1 and f_2 and probability distribution functions F_1 and F_2, respectively. Then g and G, the probability density and distribution functions of $X + Y$, respectively, are given by*

$$g(t) = \int_{-\infty}^{\infty} f_1(x) f_2(t - x) \, dx;$$

$$G(t) = \int_{-\infty}^{\infty} f_1(x) F_2(t - x) \, dx.$$

PROOF: Let $f(x, y)$ be the joint probability density function of X and Y. Then $f(x, y) = f_1(x) f_2(y)$. Let $U = X + Y$, $V = X$, $h_1(x, y) = x + y$, and $h_2(x, y) = x$. Then the system of equations

$$\begin{cases} x + y = u \\ x = v \end{cases}$$

has the unique solution $x = v$, $y = u - v$, and

$$\frac{\partial(h_1, h_2)}{\partial(x, y)} = \begin{vmatrix} 1 & 1 \\ 1 & 0 \end{vmatrix} = -1 \neq 0.$$

Hence by Theorem 8.9, the joint probability density function of U and V, $l(u, v)$, is given by

$$l(u, v) = f_1(v) f_2(u - v).$$

Therefore, the marginal probability density function of $U = X + Y$ is

$$g(u) = \int_{-\infty}^{\infty} l(u, v) \, dv = \int_{-\infty}^{\infty} f_1(v) f_2(u - v) \, dv,$$

which is the same as

$$g(t) = \int_{-\infty}^{\infty} f_1(x) f_2(t - x) \, dx.$$

To find $G(t)$, the distribution function of $X + Y$, note that

$$G(t) = \int_{-\infty}^{t} g(u) \, du = \int_{-\infty}^{t} \left(\int_{-\infty}^{\infty} f_1(x) f_2(u - x) \, dx \right) du$$

$$= \int_{-\infty}^{\infty} \left(\int_{-\infty}^{t} f_2(u - x) \, du \right) f_1(x) \, dx$$

$$= \int_{-\infty}^{\infty} F_2(t - x) f_1(x) \, dx,$$

where letting $s = u - x$, the last equality follows from

$$\int_{-\infty}^{t} f_2(u - x) \, du = \int_{-\infty}^{t-x} f_2(s) \, ds = F_2(t - x). \quad \blacklozenge$$

Note that by symmetry we can also write

$$g(t) = \int_{-\infty}^{\infty} f_2(y) f_1(t - y) \, dy,$$

$$G(t) = \int_{-\infty}^{\infty} f_2(y) F_1(t - y) \, dy.$$

Let f_1 and f_2 be two probability density functions. Then the function $g(t)$ defined by $g(t) = \int_{-\infty}^{\infty} f_1(x) f_2(t - x) \, dx$ is called the *convolution* of f_1 and f_2. Theorem 9.7 shows that if X and Y are independent continuous random variables, the probability density function of $X + Y$ is the convolution of the probability density functions of X and Y. Theorem 9.7 is also valid for discrete random variables. Let p_X and p_Y be probability functions of two discrete random variables X and Y. Then the function

$$p(z) = \sum_{x} p_X(x) p_Y(z - x)$$

is called the *convolution* of p_X and p_Y. It is readily seen that if X and Y are independent, the probability function of $X + Y$ is the convolution of the probability functions of X and Y:

$$P(X + Y = z) = \sum_{x} P(X = x, Y = z - x)$$

$$= \sum_{x} P(X = x)P(Y = z - x)$$

$$= \sum_{x} p_X(x)p_Y(z - x).$$

Exercise 17, a famous example given by W. J. Hall, shows that the converse of Theorem 9.7 is not valid. That is, it may happen that the probability function of two dependent random variables X and Y is the convolution of the probability functions of X and Y.

EXERCISES

A

1. Show that if X is a normal random variable with parameters (μ, σ^2), then for $\alpha \in \mathbf{R}$, we have that $M_{\alpha X}(t) = \exp[\alpha\mu t + (1/2)\alpha^2\sigma^2 t^2]$.

2. Let $X_1, X_2, \ldots, X_n$ be independent geometric random variables each with parameter p. Using moment-generating functions, prove that $X_1 + X_2 + \cdots + X_n$ is negative binomial with parameters (n, p).

3. Let $X_1, X_2, \ldots, X_n$ be n independent exponential random variables with the identical mean $1/\lambda$. Use moment-generating functions to find the probability distribution function of $X_1 + X_2 + \cdots + X_n$.

4. Using moment-generating functions, show that the sum of n independent negative binomial random variables with parameters (r_1, p), (r_2, p), ..., (r_n, p) is negative binomial with parameters (r, p), $r = r_1 + r_2 + \cdots + r_n$.

5. Let $X_1, X_2, \ldots, X_n$ be n independent gamma random variables with parameters (r_1, λ), (r_2, λ), ..., (r_n, λ), respectively. Use moment-generating functions to find the probability distribution function of $X_1 + X_2 + \cdots + X_n$.

6. The probability that a bottle of a certain soda is underfilled is 0.15 independent of the amount of soda in other bottles. If machine 1 fills 100 bottles and machine 2 fills 80 bottles of this soda per hour, what is the probability that tomorrow between 10:00 A.M. and 11:00 A.M., both of these machines underfill altogether exactly 27 bottles?

7. Let X and Y be independent binomial random variables with parameters (n, p) and (m, p), respectively. Calculate $P(X = i \mid X + Y = j)$ and interpret the result.

8. Let X, Y, and Z be three independent Poisson random variables with parameters λ_1, λ_2, and λ_3, respectively. For $y = 0, 1, 2, \ldots, t$, calculate $P(Y = y \mid X + Y + Z = t)$.

9. Mr. Watkins is at a train station waiting to make a phone call. There is only one public telephone booth and it is occupied by a person. There is also another person ahead of Mr. Watkins waiting to call. If the duration of each telephone call is an exponential random variable with $\lambda = 1/8$, find the probability that Mr. Watkins should wait at least 12 minutes before being able to call.

10. Let $X \sim N(1, 2)$ and $Y \sim N(4, 7)$ be independent random variables. Find the probability of the following events: (a) $X + Y > 0$; (b) $X - Y < 2$; (c) $3X + 4Y > 20$.

11. The distribution of the IQ of a randomly selected student from a certain college is $N(110, 16)$. What is the probability that the average of the IQs of 10 randomly selected students from this college is at least 112?

12. Vicki has two department stores. Delinquent charge accounts at store 1 show a normal distribution with mean \$90 and standard deviation \$30, whereas at store 2 they show a normal distribution with mean \$100 and standard deviation \$50. If 10 delinquent accounts are selected randomly at store 1 and 15 at store 2, what is the probability that the average of the accounts selected at store 1 exceeds the average of those selected at store 2?

13. The capacity of an elevator is 2700 pounds. If the weight of a random athlete is normal with mean 225 pounds and standard deviation 25, what is the probability that the elevator can safely carry 12 random athletes?

14. The distributions of the grades of the students of probability and calculus at a certain university are $N(65, 418)$, and $N(72, 448)$, respectively. Dr. Olwell teaches a calculus section with 28 and a probability section with 22 students. What is the probability that the difference between the averages of the final grades of these two classes is at least 2?

15. Suppose that car mufflers last random times that are normally distributed with mean three years and standard deviation one year. If a certain family buys two new cars at the same time, what is the probability that: (a) they should change the muffler of one car at least $1\frac{1}{2}$ years before the muffler of the other car; (b) one car does not need a new muffler for a period during which the other car needs two new mufflers?

16. From the interval $(0, 1)$, 2 random numbers are selected independently. Show that the probability density function of their sum is given by

$$g(t) = \begin{cases} t & \text{if } 0 \le t < 1 \\ 2 - t & \text{if } 1 \le t < 2 \\ 0 & \text{otherwise.} \end{cases}$$

17. Let $-1/9 < c < 1/9$ be a constant. Let the joint probability function of the random variables X and Y [$p(x, y)$] be given by the following table:

		y	
x	-1	0	1
-1	$1/9$	$1/9 - c$	$1/9 + c$
0	$1/9 + c$	$1/9$	$1/9 - c$
1	$1/9 - c$	$1/9 + c$	$1/9$

(a) Show that the probability function of $X + Y$ is the convolution function of the probability functions of X and Y for all c.
(b) Show that X and Y are independent if and only if $c = 0$.

B

18. An elevator can carry up to 3500 pounds. The manufacturer has included a safety margin of 500 pounds and lists the capacity as 3000 pounds. The building management seeks to avoid accidents by limiting the number of passengers on the elevator. If the weight of the passengers using the elevator is $N(155, 625)$, what is the maximum number of passengers who can use the elevator if the odds of exceeding the rated capacity are to be less than 2 in 10,000?

19. Let the joint probability function of $X_1, X_2, \ldots, X_r$ be multinomial, that is,

$$p(x_1, x_2, \ldots, x_r) = \frac{n!}{x_1! \, x_2! \cdots x_r!} \, p_1^{x_1} p_2^{x_2} \cdots p_r^{x_r},$$

where $x_1 + x_2 + \cdots + x_r = n$, and $p_1 + p_2 + \cdots + p_r = 1$. Show that for $k < r$, $X_1 + X_2 + \cdots + X_k$ has a binomial distribution.

20. Kim is at a train station waiting to make a phone call. There are two public telephone booths next to each other occupied by two persons and there are 11 persons waiting in a single line ahead of Kim to call. If the duration of each telephone call is an exponential random variable with $\lambda = 1/3$, what are the distribution and the expectation of the time that Kim should wait until being able to call?

9.3   Expected Values of Sums of Random Variables

As we have seen, for a discrete random variable X with set of possible values A and probability function p, $E(X)$ is defined by $\sum_{x \in A} x p(x)$. For a continuous random variable X with probability density function f, the same quantity, $E(X)$, is defined by $\int_{-\infty}^{\infty} x f(x)\, dx$. Recall that for a random variable X, the expected value, $E(X)$, might not exist (see Examples 4.18 and 4.19 and Exercise 10 in Section 6.3). In the following discussion we always assume that the expected value of a random variable exists.

To begin, we prove the linearity property of expectation for continuous random variables.

THEOREM 9.8   *For random variables $X_1, X_2, \ldots, X_n$ defined on the same sample space,*

$$E(X_1 + X_2 + \cdots + X_n) = E(X_1) + E(X_2) + \cdots + E(X_n).$$

PROOF: For convenience, we prove this for continuous random variables. For discrete random variables the proof is similar. Let $f(x_1, x_2, \ldots, x_n)$ be the joint probability density function of $X_1, X_2, \ldots, X_n$; then, by Theorem 8.8,

$$E(X_1 + X_2 + \cdots + X_n)$$
$$= \int_{-\infty}^{\infty} \int_{-\infty}^{\infty} \cdots \int_{-\infty}^{\infty} (x_1 + x_2 + \cdots + x_n) f(x_1, x_2, \ldots, x_n)\, dx_1 dx_2 \cdots dx_n$$
$$= \int_{-\infty}^{\infty} \int_{-\infty}^{\infty} \cdots \int_{-\infty}^{\infty} x_1 f(x_1, x_2, \ldots, x_n)\, dx_1\, dx_2 \cdots dx_n$$
$$+ \int_{-\infty}^{\infty} \int_{-\infty}^{\infty} \cdots \int_{-\infty}^{\infty} x_2 f(x_1, x_2, \ldots, x_n)\, dx_1\, dx_2 \cdots dx_n + \cdots$$
$$+ \int_{-\infty}^{\infty} \int_{-\infty}^{\infty} \cdots \int_{-\infty}^{\infty} x_n f(x_1, x_2, \ldots, x_n)\, dx_1\, dx_2 \cdots dx_n$$
$$= E(X_1) + E(X_2) + \cdots + E(X_n). \quad \blacklozenge$$

In Sections 4.4 and 6.3 we have shown that for a random variable X and a constant α, $E(\alpha X) = \alpha E(X)$. This and Theorem 9.8 imply the following theorem.

THEOREM 9.9   *Let $X_1, X_2, \ldots, X_n$ be random variables on the same sample space and $\alpha_1, \alpha_2, \ldots, \alpha_n$ be constants. Then*

$$E(\alpha_1 X_1 + \alpha_2 X_2 + \cdots + \alpha_n X_n)$$
$$= \alpha_1 E(X_1) + \alpha_2 E(X_2) + \cdots + \alpha_n E(X_n).$$

Example 9.11   A die is rolled 15 times. What is the expected value of the sum of the outcomes?

SOLUTION: Let X be the sum of the outcomes, and for $i = 1, 2, \ldots, 15$, let X_i be the outcome of the ith roll. Then

$$X = X_1 + X_2 + \cdots + X_{15}.$$

Thus

$$E(X) = E(X_1) + E(X_2) + \cdots + E(X_{15}).$$

Now since for all values of i, X_i is 1, 2, 3, 4, 5, and 6 with probability of 1/6 for all six values,

$$E(X_i) = 1 \cdot \frac{1}{6} + 2 \cdot \frac{1}{6} + 3 \cdot \frac{1}{6} + 4 \cdot \frac{1}{6} + 5 \cdot \frac{1}{6} + 6 \cdot \frac{1}{6} = \frac{7}{2}.$$

Hence $E(X) = 15(7/2) = 52.5$.   ◆

Example 9.11 and the following examples show the power of Theorem 9.9. Note that in Example 9.11, while the formula

$$E(X_1 + X_2 + \cdots + X_n) = E(X_1) + E(X_2) + \cdots + E(X_n)$$

enables us to compute $E(X)$ very easily, computing $E(X)$ directly from probability function of X is not easy. This is because calculation of the probability function of X is time consuming and cumbersome.

Example 9.12   A well-shuffled ordinary deck of 52 cards is divided randomly into four piles of 13 each. Counting jack, queen, and king as 11, 12, and 13, respectively, we say that a match occurs in a pile if the jth card is j. What is the expected value of the total number of matches in all four piles?

SOLUTION: Let X_i, $i = 1, 2, 3, 4$ be the number of matches in the ith pile. $X = X_1 + X_2 + X_3 + X_4$ is the total number of matches and $E(X) = E(X_1) + E(X_2) + E(X_3) + E(X_4)$. To calculate $E(X_i)$, $1 \leq i \leq 4$, let A_{ij} be the event that the jth card in the ith pile is j ($1 \leq i \leq 4$, $1 \leq j \leq 13$). Then by defining

$$X_{ij} = \begin{cases} 1 & \text{if } A_{ij} \text{ occurs} \\ 0 & \text{otherwise,} \end{cases}$$

we have that

$$X_i = \sum_{j=1}^{13} X_{ij}.$$

Now $P(A_{ij}) = 4/52 = 1/13$ implies that

$$E(X_{ij}) = 1 \cdot P(A_{ij}) + 0 \cdot P(A_{ij}^c) = P(A_{ij}) = \frac{1}{13}.$$

Hence

$$E(X_i) = E\left(\sum_{j=1}^{13} X_{ij}\right) = \sum_{j=1}^{13} E(X_{ij}) = \sum_{j=1}^{13} \frac{1}{13} = 1.$$

Thus on average there is one match in every pile. From this we get

$$E(X) = E(X_1) + E(X_2) + E(X_3) + E(X_4) = 1 + 1 + 1 + 1 = 4,$$

showing that on average there are four matches altogether.   ◆

Example 9.13   In a small town, exactly n married couples are living. What is the expected number of intact couples after m deaths occur among the couples? Assume that the deaths occur at random, there are no divorces, and there are no new marriages.

SOLUTION: Let X be the number of intact couples after m deaths and for $i = 1, 2, \ldots, n$ define

$$X_i = \begin{cases} 1 & \text{if the } i\text{th couple is left intact} \\ 0 & \text{otherwise.} \end{cases}$$

Then

$$X = X_1 + X_2 + \cdots + X_n,$$

and hence

$$E(X) = E(X_1) + E(X_2) + \cdots + E(X_n),$$

where

$$E(X_i) = 1 \cdot P(X_i = 1) + 0 \cdot P(X_i = 0) = P(X_i = 1).$$

The event $\{X_i = 1\}$ occurs if the ith couple is left intact and thus all of the m deaths are among the remaining $n - 1$ couples. Knowing that the deaths occur at random among these people, we can write

$$P(X_i = 1) = \frac{\binom{2n-2}{m}}{\binom{2n}{m}} = \frac{\dfrac{(2n-2)!}{m!\,(2n-m-2)!}}{\dfrac{(2n)!}{(2n-m)!\,m!}} = \frac{(2n-m)(2n-m-1)}{2n(2n-1)}.$$

Thus the desired quantity equals

$$E(X) = E(X_1) + E(X_2) + \cdots + E(X_n) = n\frac{(2n-m)(2n-m-1)}{2n(2n-1)}$$

$$= \frac{(2n-m)(2n-m-1)}{2(2n-1)}.$$

To have a numerical feeling for this interesting example, let $n = 1000$. Then the following table shows the effect of the number of deaths on the expected value of the number of intact couples.

m	100	300	600	900	1200	1500	1800
$E(X)$	902.48	722.44	489.89	302.38	159.88	62.41	9.95

This example was posed by Daniel Bernoulli (1700–1782).   ♦

Example 9.14 Dr. Windler's secretary accidentally threw a patient's file into the wastebasket. A few minutes later, the janitor cleaned up the entire clinic, dumped the wastebasket containing the patient's file randomly into one of the 7 garbage cans outside the clinic, and left. Determine the expected number of cans that Dr. Windler should empty to find the file.

SOLUTION: Let X be the number of garbage cans that Dr. Windler should empty to find the patient's file. For $i = 1, 2, \ldots, 7$, let $X_i = 1$ if the patient's file is in the ith garbage can that Dr. Windler will empty, and $X_i = 0$, otherwise. Then

$$X = X_1 + X_2 + \cdots + X_7,$$

and therefore,

$$E(X) = E(X_1) + E(X_2) + \cdots + E(X_7)$$

$$= 1 \cdot \frac{1}{7} + 2 \cdot \frac{1}{7} + \cdots + 7 \cdot \frac{1}{7} = 4. \quad ♦$$

Example 9.15 A box contains nine light bulbs, of which two are defective. What is the expected value of the number of light bulbs that one will have to test (at random and without replacement) to find both defective bulbs?

SOLUTION: For $i = 1, 2, \cdots, 8$ and $j > i$, let $X_{ij} = j$ if the ith and jth light bulbs to be examined are defective, and $X_{ij} = 0$, otherwise. Then $\sum_{i=1}^{8} \sum_{j=i+1}^{9} X_{ij}$ is the number of light bulbs to be examined. Therefore,

$$E(X) = \sum_{i=1}^{8} \sum_{j=i+1}^{9} E(X_{ij}) = \sum_{i=1}^{8} \sum_{j=i+1}^{9} j \, \frac{1}{\binom{9}{2}}$$

$$= \frac{1}{36} \sum_{i=1}^{8} \sum_{j=i+1}^{9} j = \frac{1}{36} \sum_{i=1}^{8} \frac{90 - i^2 - i}{2} \approx 6.67,$$

where the next-to-last equality follows from

$$\sum_{j=i+1}^{9} j = \sum_{j=1}^{9} j - \sum_{j=1}^{i} j = \frac{9 \times 10}{2} - \frac{i(i+1)}{2} = \frac{90 - i^2 - i}{2}. \quad \blacklozenge$$

An elegant application of Theorem 9.8 is that it can be used to calculate the expected values of random variables, such as binomial, negative binomial, and hypergeometric. The following examples demonstrate such applications.

Example 9.16 Let X be a binomial random variable with parameters (n, p). Recall that X is the number of successes in n independent Bernoulli trials. Thus, for $i = 1, 2, \ldots, n$, letting

$$X_i = \begin{cases} 1 & \text{if the } i\text{th trial is a success} \\ 0 & \text{otherwise,} \end{cases}$$

we get

$$X = X_1 + X_2 + \cdots + X_n, \tag{9.3}$$

where X_i is a Bernoulli random variable for $i = 1, 2, \ldots, n$. Now since $\forall i, \ 1 \le i \le n$,

$$E(X_i) = 1 \cdot p + 0 \cdot (1 - p) = p,$$

(9.3) implies that

$$E(X) = E(X_1) + E(X_2) + \cdots + E(X_n) = np. \quad \blacklozenge$$

Example 9.17 Let X be a negative binomial random variable with parameters (r, p). Then in a sequence of independent Bernoulli trials each with success probability p, X is the number of trials until the rth success. Let X_1 be the number of trials until the first success, X_2 be the number of additional trials to get the second success, X_3 be the number of additional ones to obtain the third success, and so on. Then, clearly,

$$X = X_1 + X_2 + \cdots + X_r,$$

where for $i = 1, 2, \ldots, n$, the random variable X_i is geometric with parameter p. This is because $P(X_i = n) = (1 - p)^{n-1} p$ by the independence of the trials. Since $E(X_i) = 1/p$ $(i = 1, 2, \ldots, r)$,

$$E(X) = E(X_1) + E(X_2) + \cdots + E(X_r) = \frac{r}{p}.$$

This formula shows that, for example, in the experiment of throwing a fair die successively, on the average, it takes $5/(1/6) = 30$ trials to get five 6's. ◆

Example 9.18 Let X be a hypergeometric random variable with probability function

$$p(x) = P(X = x) = \frac{\dbinom{D}{x}\dbinom{N - D}{n - x}}{\dbinom{N}{n}},$$

$$\max(0, n + D - N) \leq x \leq \min(D, n).$$

Then X is the number of defective items among n items drawn at random and without replacement from an urn containing D defective and $N - D$ nondefective items. To calculate $E(X)$, let A_i be the event that the ith item drawn is defective. Also for $i = 1, 2, \ldots, n$, let

$$X_i = \begin{cases} 1 & \text{if } A_i \text{ occurs} \\ 0 & \text{otherwise,} \end{cases}$$

then

$$X = X_1 + X_2 + \cdots + X_n.$$

Hence

$$E(X) = E(X_1) + E(X_2) + \cdots + E(X_n),$$

where for $i = 1, 2, \ldots, n$:

$$E(X_i) = 1 \cdot P(X_i = 1) + 0 \cdot P(X_i = 0) = P(X_i = 1) = P(A_i) = \frac{D}{N}.$$

This follows since the ith item can be any of the N items with equal probabilities. Therefore,

$$E(X) = \frac{nD}{N}.$$

This shows that, for example, the expected number of spades in a random bridge hand is $(13 \times 13)/52 = 13/4 \approx 3.25$.   ◆

REMARK: For $n = \infty$, Theorem 9.8 is not necessarily true.   That is, $E(\sum_{i=1}^{\infty} X_i)$ might not be equal to $\sum_{i=1}^{\infty} E(X_i)$.   For example, for $i = 1, 2, 3, \ldots$, let

$$Y_i = \begin{cases} i & \text{with probability } \dfrac{1}{i} \\ 0 & \text{otherwise,} \end{cases}$$

and $X_i = Y_{i+1} - Y_i$. Then since $E(Y_i) = 1$ for $i = 1, 2, \ldots$, $E(X_i) = E(Y_{i+1}) - E(Y_i) = 0$. Hence $\sum_{i=1}^{\infty} E(X_i) = 0$. However, since $\sum_{i=1}^{\infty} X_i = -Y_1$, we have that $E(\sum_{i=1}^{\infty} X_i) = E(-Y_1) = -1$. Therefore,

$$E\left(\sum_{i=1}^{\infty} X_i\right) \neq \sum_{i=1}^{\infty} E(X_i).   ◆$$

It can be shown that, in general, if $\sum_{i=1}^{\infty} E(|X_i|) < \infty$ or if X_i is nonnegative for all i, that is, $P(X_i \geq 0) = 1$ for $i \geq 1$, then

$$E\left(\sum_{i=1}^{\infty} X_i\right) = \sum_{i=1}^{\infty} E(X_i). \tag{9.4}$$

As an application of (9.4), we prove the following important theorem.

THEOREM 9.10   *Let N be a discrete random variable with set of possible values $\{1, 2, 3, \ldots\}$. Then*

$$E(N) = \sum_{i=1}^{\infty} P(N \geq i).$$

PROOF: For $i \geq 1$, let

$$X_i = \begin{cases} 1 & \text{if } N \geq i \\ 0 & \text{otherwise,} \end{cases}$$

then

$$\sum_{i=1}^{\infty} X_i = \sum_{i=1}^{N} X_i + \sum_{i=N+1}^{\infty} X_i = \sum_{i=1}^{N} 1 + \sum_{i=N+1}^{\infty} 0 = N.$$

Since $P(X_i \geq 0) = 1$, for $i \geq 1$,

$$E(N) = E\left(\sum_{i=1}^{\infty} X_i\right) = \sum_{i=1}^{\infty} E(X_i) = \sum_{i=1}^{\infty} P(N \geq i). \quad \blacklozenge$$

Note that Theorem 9.10 is the analog of the fact that if X is a continuous nonnegative random variable, then

$$E(X) = \int_0^{\infty} P(X > t)\, dt.$$

This is shown in Remark 3 of Section 6.3.

We now prove an important inequality called the *Cauchy–Schwarz inequality*.

THEOREM 9.11   (Cauchy–Schwarz Inequality)   *For random variables X and Y,*

$$E(XY) \leq \sqrt{E(X^2)E(Y^2)}.$$

PROOF: For all real numbers λ, $(X - \lambda Y)^2 \geq 0$. Hence for all values of λ, $X^2 - 2XY\lambda + \lambda^2 Y^2 \geq 0$. Since nonnegative random variables have nonnegative expectations,

$$E\left(X^2 - 2XY\lambda + \lambda^2 Y^2\right) \geq 0,$$

which implies that

$$E(X^2) - 2E(XY)\lambda + \lambda^2 E(Y^2) \geq 0.$$

Rewriting this as a polynomial in λ of degree 2, we get

$$E(Y^2)\lambda^2 - 2E(XY)\lambda + E(X^2) \geq 0.$$

It is a well-known fact that if a polynomial of degree 2 is positive, its discriminant is negative. Therefore,

$$4[E(XY)]^2 - 4E(X^2)E(Y^2) \leq 0$$

or

$$[E(XY)]^2 \leq E(X^2)E(Y^2).$$

This gives

$$E(XY) \leq \sqrt{E(X^2)E(Y^2)}. \quad \blacklozenge$$

COROLLARY   *For a random variable X,*

$$[E(X)]^2 \leq E(X^2).$$

PROOF: In Cauchy–Schwarz's inequality, let $Y = 1$; then

$$E(X) = E(XY) \leq \sqrt{E(X^2)E(1)} = \sqrt{E(X^2)};$$

thus

$$[E(X)]^2 \leq E(X^2). \quad \blacklozenge$$

EXERCISES

A

1. Let the probability density function of a random variable X be given by

$$f(x) = \begin{cases} |x - 1| & \text{if } 0 \leq x \leq 2 \\ 0 & \text{otherwise.} \end{cases}$$

Find $E(X^2 + X)$.

2. A calculator is able to generate random numbers from the interval $(0, 1)$. We need five random numbers from $(0, 2/5)$. Using this calculator, on the average, how many independent random numbers should we generate to find the five numbers needed?

3. Let X, Y, and Z be three independent random variables such that $E(X) = E(Y) = E(Z) = 0$, and $\text{Var}(X) = \text{Var}(Y) = \text{Var}(Z) = 1$. Calculate $E\left[X^2(Y + 5Z)^2\right]$.

4. Let the joint probability density function of random variables X and Y be

$$f(x, y) = \begin{cases} 2e^{-(x+2y)} & \text{if } x \geq 0, \ y \geq 0 \\ 0 & \text{otherwise.} \end{cases}$$

Find $E(X)$, $E(Y)$, and $E(X^2 + Y^2)$.

5. A company puts five different types of prizes into their boxes of cereals, one in each box and in equal proportions. If a person decides to collect all five prizes, what is the expected number of the boxes of cereals that he or she should buy?

6. An absentminded professor wrote n letters and sealed them in envelopes without writing the addresses on the envelopes. Having forgotten which letter he had put in which envelope, he wrote the n addresses on the envelopes at random. What is the expected number of the letters addressed correctly?

   HINT: For $i = 1, 2, \ldots, n$, let

$$X_i = \begin{cases} 1 & \text{if the } i\text{th letter is addressed correctly} \\ 0 & \text{otherwise.} \end{cases}$$

   Calculate $E(X_1 + X_2 + \cdots + X_n)$.

7. A cultural society is arranging a party for its members. The cost of a band to play music, the amount that the caterer will charge, the rent of a hall to give the party, and other expenses (in dollars) are uniform random variables over the intervals $(1300, 1800)$, $(1800, 2000)$, $(800, 1200)$, and $(400, 700)$, respectively. If the number of people who will participate in the party is a random integer from $(150, 200]$, what is the least amount that the society should charge each participant to have no loss on average?

B

8. Let $\{X_1, X_2, \ldots, X_n\}$ be a sequence of independent random variables with $P(X_j = i) = p_i$ $(1 \leq j \leq n$ and $i \geq 1)$. Let $h_k = \sum_{i=k}^{\infty} p_i$. Using Theorem 9.10, prove that

$$E[\min(X_1, X_2, \ldots, X_n)] = \sum_{k=1}^{\infty} h_k^n.$$

9. A coin is tossed n times. What is the expected number of exactly three consecutive heads?

HINT: Let E_i be the event that the outcome $(i - 1)$ is tails, the outcomes i, $(i + 1)$, and $(i + 2)$ are heads, and the outcome $(i + 3)$ is tails. Let

$$X_i = \begin{cases} 1 & \text{if } E_i \text{ occurs} \\ 0 & \text{otherwise.} \end{cases}$$

Then calculate the expected value of an appropriate sum of X_i's.

10. Suppose that 80 balls are placed into 40 boxes at random and independently. What is the expected number of the empty boxes?

HINT: Let

$$X_i = \begin{cases} 1 & \text{if the } i\text{th box is empty} \\ 0 & \text{otherwise,} \end{cases}$$

then find $E(X_1 + X_2 + \cdots + X_{40})$.

11. There are 25 students in a probability class. What is the expected number of birthdays that belong only to one student? Assume that the birth rates are constant throughout the year and that each year has 365 days.

HINT: Let $X_i = 1$ if the birthday of the ith student is not the birthday of any other student, and $X_i = 0$, otherwise. Find $E(X_1 + X_2 + \cdots + X_{25})$.

12. There are 25 students in a probability class. What is the expected number of the days of the year that are birthdays of at least two students? Assume that the birth rates are constant throughout the year and that each year has 365 days.

HINT: Let $X_i = 1$, if the birthdays of at least two students are on the ith day of the year, and $X_i = 0$, otherwise. Calculate $E(\sum_{i=1}^{365} X_i)$.

13. Let X and Y be nonnegative random variables with an arbitrary joint probability distribution function. Let

$$I(x, y) = \begin{cases} 1 & \text{if } X > x, \quad Y > y \\ 0 & \text{otherwise.} \end{cases}$$

(a) Show that

$$\int_0^\infty \int_0^\infty I(x, y) \, dx \, dy = XY.$$

(b) By calculating expected values of both sides of part (a), prove that

$$E(XY) = \int_0^\infty \int_0^\infty P(X > x, \ Y > y) \, dx \, dy.$$

Note that this is a generalization of the result explained in Remark 3 of Section 6.3.

14. Let $\{X_1, X_2, \ldots, X_n\}$ be a sequence of continuous, independent, and identically distributed random variables. Let

$$N = \min\{n : X_1 \geq X_2 \geq X_3 \geq \cdots \geq X_{n-1}, X_{n-1} < X_n\}.$$

Find $E(N)$.

15. From an urn that contains a large number of red and blue chips, mixed in equal proportions, 10 chips are removed one by one and at random. The chips that are removed before the first red chip are returned to the urn. The first red chip, together with all those that follow, are placed in another urn that is initially empty. Calculate the expected number of the chips in the second urn.

16. Under what condition does Cauchy–Schwarz's inequality become equality?

Review Problems

1. A nurse in a hospital mixes the 10 pills of 10 patients accidentally. Suppose that she gives a pill at random to each patient from the mixed-up batch. If none of the 10 pills are of the same type, what is the expected number of patients who will get their own pills?

2. Let the probability function of a random variable X be given by

$$f(x) = \begin{cases} 2x - 2 & \text{if } 1 < x < 2 \\ 0 & \text{elsewhere.} \end{cases}$$

Find $E(X^3 + 2X - 7)$.

3. Yearly wages paid to the salespeople employed by a certain company are normally distributed with mean \$27,000 and standard deviation \$4900. What is the probability that the average wage of a random sample of 10 employees of this company is at least \$30,000?

4. Let the joint probability density function of random variables X and Y be

$$f(x, y) = \begin{cases} \dfrac{3x^3 + xy}{3} & \text{if } 0 \le x \le 1, \quad 0 \le y \le 2 \\ 0 & \text{elsewhere.} \end{cases}$$

Find $E(X^2 + 2XY)$.

5. The moment-generating function of a random variable X is given by

$$M_X(t) = \left(\frac{1}{3} + \frac{2}{3}e^t \right)^{10}.$$

Find Var(X) and $P(X \ge 8)$.

6. The moment-generating function of a random variable X is given by

$$M_X(t) = \frac{1}{6}e^t + \frac{1}{3}e^{2t} + \frac{1}{2}e^{3t}.$$

Find the distribution function of X.

7. In a town there are n taxis. A woman takes one of these taxis every day at random and with replacement. On average, how long does it take before she can claim that she has been in every taxi in the town? HINT: The final answer is in terms of $a_n = 1 + 1/2 + \cdots + 1/n$.

8. For a random variable X, suppose that $M_X(t) = \exp(2t^2 + t)$. Find $E(X)$ and Var(X).

9. Let the moment-generating function of a random variable X be given by

$$M_X(t) = \begin{cases} \dfrac{1}{t}(e^{t/2} - e^{-t/2}) & \text{if } t \ne 0 \\ 1 & \text{if } t = 0. \end{cases}$$

Find the distribution function of X.

10. The moment-generating function of X is given by

$$M_X(t) = \exp\left(\frac{e^t - 1}{2} \right).$$

Find $P(X > 0)$.

11. The moment-generating function of a random variable X is given by

$$M_X(t) = \frac{1}{(1 - t)^2}, \qquad t < 1.$$

Find the moments of X.

12. Suppose that in a community the distributions of heights of men and women (in centimeters) are $N(173, 40)$ and $N(160, 20)$, respectively. Calculate the probability that the average height of 10 randomly selected men is at least 5 centimeters larger than the average height of 6 randomly selected women.

13. Find the moment-generating function of a random variable X with *Laplace density function* defined by

$$f(x) = \frac{1}{2} e^{-|x|}, \qquad -\infty < x < \infty.$$

14. Let X and Y be independent Poisson random variables with parameters λ and μ, respectively.
    (a) Show that

$$P(X + Y = n) = \sum_{i=0}^{n} P(X = i) P(Y = n - i).$$

    (b) Use part (a) to prove that $X + Y$ is a Poisson random variable with parameter $\lambda + \mu$.

Chapter 10

LIMIT THEOREMS

10.1 Markov and Chebyshev Inequalities

So far, we have seen that to calculate probabilities, we need to know either probability distribution functions, or probability functions, or probability density functions. In studies of probabilistic phenomena, it frequently happens that for some random variables, we cannot determine any of these three functions, but we can calculate their expected values and/or variances. In such cases, although we cannot calculate exact probabilities, using Markov and Chebyshev inequalities, we are able to derive bounds on probabilities. These inequalities, which are very useful in applications, have significant theoretical values as well. Moreover, Chebyshev's inequality is a further indication of the importance of the concept of variance. Historically, Chebyshev's inequality was first discovered by the French mathematician Irénée Bienaymé (1796–1878). For this reason some authors call it the Chebyshev–Bienaymé inequality. In the middle of the nineteenth century, Chebyshev discovered the inequality independently in connection with the laws of large numbers (see Section 10.2). He used it to give an elegant and short proof for the law of large numbers discovered by James Bernoulli early in the eighteenth century. Since the usefulness and applicability of the inequality were demonstrated by Chebyshev, most authors call it Chebyshev's inequality.

THEOREM 10.1   (Markov's Inequality)   *Let X be a nonnegative random variable; then for any t > 0,*

$$P(X \geq t) \leq \frac{E(X)}{t}.$$

PROOF: We prove the theorem for a discrete random variable X with probability function $p(x)$. For continuous random variables the proof is similar. Let A be the set of possible values of X and $B = \{x \in A : x \geq t\}$. Then

$$E(X) = \sum_{x \in A} xp(x) \geq \sum_{x \in B} xp(x) \geq t \sum_{x \in B} p(x) = tP(X \geq t).$$

Thus

$$P(X \geq t) \leq \frac{E(X)}{t}. \quad \blacklozenge$$

Example 10.1   A post office, on average, handles 10,000 letters per day. What can be said about the probability that it handles (a) at least 15,000 letters tomorrow; (b) less than 15,000 letters tomorrow?

SOLUTION: Let X be the number of letters that this post office handles tomorrow. Then $E(X) = 10,000$.

(a) By Markov's equality,

$$P(X \geq 15,000) \leq \frac{E(X)}{15,000} = \frac{10,000}{15,000} = \frac{2}{3}.$$

(b) Using the inequality obtained in (a), we have

$$P(X < 15,000) = 1 - P(X \geq 15,000) \geq 1 - \frac{2}{3} = \frac{1}{3}. \quad \blacklozenge$$

THEOREM 10.2   (Chebyshev's Inequality)   *If X is a random variable with expected value μ and variance σ^2, then for any t > 0,*

$$P(|X - \mu| \geq t) \leq \frac{\sigma^2}{t^2}.$$

PROOF: Since $(X - \mu)^2 \geq 0$, by Markov's inequality

$$P\{(X - \mu)^2 \geq t^2\} \leq \frac{E[(X - \mu)^2]}{t^2} = \frac{\sigma^2}{t^2}.$$

Chebyshev's inequality follows since $(X - \mu)^2 \geq t^2$ is equivalent to $|X - \mu| \geq t$. ♦

Letting $t = k\sigma$ in Chebyshev's inequality, we get that $P(|X - \mu| \geq k\sigma) \leq 1/k^2$. That is, the probability that X deviates from its expected value at least k standard deviations is less than $1/k^2$. Thus, for example,

$$P(|X - \mu| \geq 2\sigma) \leq \frac{1}{4},$$

$$P(|X - \mu| \geq 4\sigma) \leq \frac{1}{16},$$

$$P(|X - \mu| \geq 10\sigma) \leq \frac{1}{100}.$$

On the other hand, $P(|X - \mu| \geq k\sigma) \leq 1/k^2$ implies that $P(|X - \mu| < k\sigma) \geq 1 - 1/k^2$. Therefore, the probability that X deviates from its mean less than k standard deviations is at least $1 - 1/k^2$. In particular, this implies that for any set of data, at least a fraction $1 - 1/k^2$ of the data are within k standard deviations on either side of the mean. Thus for any data, at least $1 - 1/2^2 = 3/4$, or 75% of the data, lie within two standard deviations on either side of the mean. This implication of Chebyshev's inequality is true for any set of real numbers. That is, let $\{x_1, x_2, \ldots, x_n\}$ be a set of real numbers, and define

$$\bar{x} = \frac{1}{n} \sum_{i=1}^{n} x_i, \qquad s^2 = \frac{1}{n-1} \sum_{i=1}^{n} (x_i - \bar{x})^2;$$

then at least a fraction $1 - 1/k^2$ of the x_i's are between $\bar{x} - ks$ and $\bar{x} + ks$ (see Exercise 13).

Example 10.2   Suppose that, on average, a post office handles 10,000 letters a day with a variance of 2000. What can be said about the probability that this post office will handle between 8000 and 12,000 letters tomorrow?

SOLUTION: Let X denote the number of letters that this post office will handle tomorrow. Then $\mu = E(X) = 10,000$, $\sigma^2 = \text{Var}(X) = 2000$. We want to calculate $P(8000 < X < 12,000)$. Since

$$P(8000 < X < 12,000) = P(-2000 < X - 10,000 < 2000)$$
$$= P(|X - 10,000| < 2000)$$
$$= 1 - P(|X - 10,000| \geq 2000),$$

by Chebyshev's inequality:

$$P(|X - 10,000| \geq 2000) \leq \frac{2000}{(2000)^2} = 0.0005.$$

Hence

$$P(8000 < X < 12,000) = P(|X - 10,000| < 2000) \geq 1 - 0.0005$$
$$= 0.9995.$$

Note that this answer is consistent with our intuitive understanding of the concepts of expectation and variance. ◆

Example 10.3   A blind fits Myra's bedroom's window if its width is between 41.5 and 42.5 inches. Myra buys a blind from a store that has 30 such blinds. What can be said about the probability that it fits her bedroom's window if the average of the widths of the blinds of the store is 42 inches with standard deviation 0.25?

SOLUTION: Let X be the width of the blind that Myra purchased. We know that

$$P(|X - \mu| < k\sigma) \geq 1 - \frac{1}{k^2}.$$

Therefore,

$$P(41.5 < X < 42.5) = P\{|X - 42| < 2(0.25)\} \geq 1 - \frac{1}{4} = 0.75. \quad ◆$$

It should be mentioned that the bounds that are obtained on probabilities by Markov's and Chebyshev's inequalities are not usually very close to the actual probabilities. The following example demonstrates this.

Example 10.4   Roll a die and let X be the outcome. Clearly,

$$E(X) = \frac{1}{6}(1 + 2 + 3 + 4 + 5 + 6) = \frac{21}{6},$$
$$E\left(X^2\right) = \frac{1}{6}(1^2 + 2^2 + 3^2 + 4^2 + 5^2 + 6^2) = \frac{91}{6}.$$

Thus $\text{Var}(X) = 91/6 - 441/36 = 35/12$. By Markov's inequality:

$$P(X \geq 6) \leq \frac{21/6}{6} \approx 0.583.$$

By Chebyshev's inequality,

$$P\left(\left|X - \frac{21}{6}\right| \geq \frac{3}{2}\right) \leq \frac{35/12}{9/4} \approx 1.296,$$

a trivial bound because we already know that $P(|X - 21/6| \geq 3/2) \leq 1$. However, the exact values of these probabilities are much smaller than these bounds: $P(X \geq 6) = 1/6 \approx 0.167$, and

$$P\left(\left|X - \frac{21}{6}\right| \geq \frac{3}{2}\right) = P(X \leq 2 \text{ or } X \geq 5) = \frac{4}{6} \approx 0.667. \quad \blacklozenge$$

The following example is an elegant proof of the fact shown previously that if $\text{Var}(X) = 0$, then X is constant. It is a good application of Chebyshev's inequality.

Example 10.5 Let X be a random variable with mean μ and variance 0. We show that

$$P(X = \mu) = 1.$$

To prove this, let $E_i = \{|X - \mu| < 1/i\}$; then

$$E_1 \supseteq E_2 \supseteq E_3 \supseteq \cdots \supseteq E_n \supseteq E_{n+1} \supseteq \cdots.$$

That is, $\{E_i, i = 1, 2, 3, \ldots\}$ is a decreasing sequence of events. Therefore,

$$\lim_{n \to \infty} E_n = \bigcap_{n=1}^{\infty} E_n = \bigcap_{n=1}^{\infty} \left\{|X - \mu| < \frac{1}{n}\right\} = \{X = \mu\}.$$

Hence by the continuity property of the probability function,

$$\lim_{n \to \infty} P(E_n) = P(\lim_{n \to \infty} E_n) = P(X = \mu).$$

Now by Chebyshev's inequality,

$$P\left(|X - \mu| \geq \frac{1}{n}\right) \leq \frac{\text{Var}(X)}{1/n^2} = 0,$$

implying that $\forall n \geq 1$,

$$P(E_n) = P\left(|X - \mu| < \frac{1}{n}\right) = 1.$$

Thus

$$P(X = \mu) = \lim_{n \to \infty} P(E_n) = 1. \quad \blacklozenge$$

EXERCISES

A

1. Let X be a random variable with $E(X) = 5$ and $E(X^2) = 42$. Find an upper bound for $P(X \geq 11)$ using (a) Markov's inequality, (b) Chebyshev's inequality.

2. The average and standard deviation of lifetimes of light bulbs manufactured by a certain factory are, respectively, 800 hours and 50 hours. What can be said about the probability that a random light bulb lasts at most 700 hours?

3. Suppose that the average number of accidents in an intersection is two per day.
   (a) Use Markov's inequality to find a bound for the probability that at least five accidents occur tomorrow.
   (b) Using Poisson random variables, calculate the probability that at least five accidents occur tomorrow. Compare this value with the bound obtained in part (a).
   (c) Let the variance of the number of accidents be two per day. Use Chebyshev's inequality to find a bound on the probability that tomorrow at least five accidents occur.

4. The average IQ score on a certain campus is 110. If the variance of these scores is 15, what can be said about the percentage of students with an IQ above 140?

5. The waiting period from the time a book is ordered until it is received is a random variable with mean seven days and standard deviation two days. If Helen wants to be 95% sure that she receives a book on a certain date, how early should she order the book?

6. Show that for a nonnegative random variable X with mean μ, $P(X \geq 2\mu) \leq 1/2$.

7. Suppose that X is a random variable with $E(X) = \text{Var}(X) = \mu$. What does Chebyshev's inequality say about $P(X > 2\mu)$?

8. Let X be a random variable with mean μ. Show that if $E[(X - \mu)^{2n}] < \infty$, then for $\alpha > 0$,

$$P(|X - \mu| \geq \alpha) \leq \frac{1}{\alpha^{2n}} E[(X - \mu)^{2n}].$$

9. Let X be a random variable and k be a constant. Prove that

$$P(X > t) \leq \frac{E\left(e^{kX}\right)}{e^{kt}}.$$

10. Prove that if the random variables X and Y satisfy $E[(X - Y)^2] = 0$, then with probability 1, $X = Y$.

11. Let X be a random variable; show that for $\alpha > 1$ and $t > 0$,

$$P\left(X \geq \frac{1}{t} \ln \alpha\right) \leq \frac{1}{\alpha} M_X(t)$$

B

12. Let the probability density function of a random variable X be

$$f(x) = \frac{x^n}{n!} e^{-x}, \qquad x \geq 0.$$

Show that

$$P(0 < X < 2n + 2) > \frac{n}{n + 1}.$$

HINT: Note that $\int_0^\infty x^n e^{-x}\, dx = \Gamma(n + 1) = n!$. Use this to calculate $E(X)$ and $\text{Var}(X)$. Then apply Chebyshev's inequality.

13. Let $\{x_1, x_2, \ldots, x_n\}$ be a set of real numbers and define

$$\bar{x} = \frac{1}{n} \sum_{i=1}^{n} x_i, \qquad s^2 = \frac{1}{n-1} \sum_{i=1}^{n} (x_i - \bar{x})^2.$$

Prove that at least a fraction $1 - 1/k^2$ of the x_i's are between $\bar{x} - ks$ and $\bar{x} + ks$.

SKETCH OF A PROOF: Let N be the number of $x_1, x_2, \ldots, x_n$ that fall in $A = [\bar{x} - ks, \bar{x} + ks]$. Then

$$s^2 = \frac{1}{n-1}\sum_{i=1}^{n}(x_i - \overline{x})^2 \geq \frac{1}{n-1}\sum_{x_i \notin A}(x_i - \overline{x})^2$$

$$\geq \frac{1}{n-1}\sum_{x_i \notin A}k^2 s^2 = \frac{n-N}{n-1}k^2 s^2.$$

This gives $(N-1)/(n-1) \geq 1 - (1/k^2)$. The result follows since $N/n \geq (N-1)/(n-1)$.

10.2   Laws of Large Numbers

Let A be an event of some experiment that can be repeated. In Chapter 1 we mentioned that the mathematicians of eighteenth and nineteenth centuries observed that in a series of sequential or simultaneous repetitions of the experiment, the proportion of times that A occurs approaches a constant. As a result, they were motivated to define the probability of A to be the number $p = \lim_{n\to\infty} n(A)/n$, where $n(A)$ is the number of times that A occurs in the first n repetitions. We also mentioned that this relative frequency interpretation of probability, which to some extent satisfies one's intuition, is mathematically problematic and cannot be the basis of a rigorous probability theory. We now show that despite these facts, for repeatable experiments, the relative frequency interpretation is valid. It is the special case of one of the most celebrated theorems of probability and statistics: the *strong law of large numbers*.

Recall that a sequence $X_1, X_2, \ldots$ of random variables is called *identically distributed* if all of them have the same distribution function. If $X_1, X_2, \ldots$ are identically distributed, then since they have the same probability distribution function, their set of possible values is the same. Also, in the discrete case their probability functions, and in the continuous case their probability density functions, are identical. Moreover, in both cases X_i's have the same expectation, the same standard deviation, and the same variance.

THEOREM 10.3   **(Strong Law of Large Numbers)**   *Let $X_1, X_2, \ldots$ be an independent and identically distributed sequence of random variables with $\mu = E(X_i)$, $i = 1, 2, 3, \ldots$. Then*

$$P\left(\lim_{n\to\infty} \frac{X_1 + X_2 + \cdots + X_n}{n} = \mu\right) = 1.$$

Note that in the standard statistical models, $(X_1 + X_2 + \cdots + X_n)/n$ is the sample mean.

To show how the strong law of large numbers relates to the relative frequency interpretation of probability, in a repeatable experiment with sample space S, let A be an event of S. Suppose that the experiment is repeated independently and let $X_i = 1$ if A occurs on the ith repetition, and $X_i = 0$ if A does not occur on the ith repetition. Then for $i = 1, 2, 3, \ldots$,

$$E(X_i) = 1 \cdot P(A) + 0 \cdot P(A^c) = P(A)$$

and

$$X_1 + X_2 + \cdots + X_n = n(A),$$

where $n(A)$, as before, is the number of times that A occurs in the first n repetitions of the experiment. Thus according to the strong law of large numbers,

$$\frac{X_1 + X_2 + \cdots + X_n}{n} = \frac{n(A)}{n}$$

converges to $P(A)$ with probability 1 as $n \to \infty$.

REMARK: Note that in this theorem convergence of $(X_1 + X_2 + \cdots + X_n)/n$ to μ is not the usual pointwise convergence studied in calculus. This type of convergence is called *almost sure convergence* and is defined as follows: Let X and the sequence $X_1, X_2, X_3, \ldots$ be random variables defined on a sample space S. Let V be the set of all points ω, in S, at which $X_n(\omega)$ converges to $X(\omega)$. That is,

$$V = \{\omega \in S : \lim_{n \to \infty} X_n(\omega) = X(\omega)\}.$$

If $P(V) = 1$, then we say that X_n converges to X *almost surely.*

To clarify this definition, we now give an example of a sequence that converges almost surely but is not pointwise convergent. Consider an experiment in which we choose a random point from the interval $[0, 1]$. The sample space of this experiment is $[0, 1]$; for all $\omega \in [0, 1]$, let $X_n(\omega) = \omega^n$, and $X(\omega) = 0$. Then clearly,

$$V = \{\omega \in [0, 1] : \lim_{n \to \infty} X_n(\omega) = X(\omega)\} = [0, 1).$$

Since $P(V) = P\{[0, 1)\} = 1$, X_n converges almost surely to X. However, this convergence is not pointwise because at 1, X_n converges to 1 and not to $X(1) = 0$. ◆

Example 10.6 Suppose that an immortal monkey is constantly typing on a word processor that is not breakable, lasts forever, and has infinite memory. Suppose that the keyboard of the word processor has $m - 1$ keys, a space bar for blank spaces, and separate keys for different symbols. If each time the monkey presses one of the m symbols (including the space bar) at random, and if at the end of each line and at the end of each page the word processor advances to a new line and a new page by itself, what is the probability that the monkey eventually will produce the complete works of Shakespeare in chronological order with no errors?

SOLUTION: The complete works of Shakespeare in chronological order with no errors are the result of typing N specific symbols (including the space bar) for some large N. For $i = 1, 2, 3, 4, \ldots$, let A_i be the event that the symbols number $(i - 1)N + 1$ to number iN typed by the monkey form the complete works of Shakespeare in chronological order with no errors. Also for $i = 1, 2, 3, \ldots$, let

$$X_i = \begin{cases} 1 & \text{if } A_i \text{ occurs} \\ 0 & \text{otherwise,} \end{cases}$$

then $E(X_i) = P(A_i) = (1/m)^N$. Now since $\{X_1, X_2, \ldots\}$ is a sequence of independent and identically distributed random variables, by the strong law of large numbers we have that

$$P\left\{\lim_{n \to \infty} \frac{X_1 + X_2 + \cdots + X_n}{n} = \left(\frac{1}{m}\right)^N\right\} = 1.$$

This relation shows that $\sum_{i=1}^{\infty} X_i = \infty$, the reason being that otherwise $\sum_{i=1}^{\infty} X_i < \infty$ implies that $\lim_{n \to \infty} (X_1 + X_2 + \cdots + X_n)/n$ is 0 and not $(1/m)^N > 0$. Now $\sum_{i=1}^{\infty} X_i = \infty$ implies that an infinite number of X_i's are equal to 1, meaning that infinitely many of A_i's will occur. Therefore, not only once but an infinite number of times the monkey will produce the complete works of Shakespeare in chronological order with no errors. However, if the monkey types 200 symbols a minute, then for $m = 88$ and $N = 10^7$, it is easily shown that, on average, it will take him T years to produce the complete works of Shakespeare for the first time, where T is a number with tens of millions of digits. ◆

Example 10.7   Suppose that $f: [0, 1] \to [0, 1]$ is a continuous function. We will now present a probabilistic method for evaluation of $\int_0^1 f(x)\, dx$ using the strong law of large numbers.   Let $\{X_1, Y_1, X_2, Y_2, \ldots\}$ be a sequence of independent random numbers from the interval $[0, 1]$. Then $\{(X_1, Y_1), (X_2, Y_2), \ldots\}$ is a sequence of independent random points from $[0, 1] \times [0, 1]$. For $i = 1, 2, 3, \ldots$, let

$$Z_i = \begin{cases} 1 & \text{if } Y_i < f(X_i) \\ 0 & \text{otherwise.} \end{cases}$$

Note that Z_i is 1 if (X_i, Y_i) falls under the curve $y = f(x)$, and is 0 if it falls above $y = f(x)$. Therefore, from the first n points selected, $Z_1 + Z_2 + \cdots + Z_n$ of them fall below the curve $y = f(x)$, and thus $\lim_{n \to \infty} (Z_1 + Z_2 + \cdots + Z_n)/n$ is the limit of the relative frequency of the points that fall below the curve $y = f(x)$. We will show that this limit converges to $\int_0^1 f(x)\, dx$. To do so, note that Z_i's are identically distributed and independent. Moreover, since g, the joint probability density function of (X_i, Y_i), $i \geq 0$, is given by

$$g(x, y) = \begin{cases} 1 & \text{if } 0 \leq x \leq 1, \quad 0 \leq y \leq 1 \\ 0 & \text{otherwise,} \end{cases}$$

we have that

$$E(Z_i) = P\{Y_i < f(X_i)\} = \iint\limits_{\{(x,y):y<f(x)\}} g(x, y)\, dx\, dy$$

$$= \int_0^1 \left(\int_0^{f(x)} dy \right) dx = \int_0^1 f(x)\, dx < \infty.$$

Hence by the strong law of large numbers, with probability 1,

$$\lim_{n \to \infty} \frac{Z_1 + Z_2 + \cdots + Z_n}{n} = \int_0^1 f(x)\, dx.$$

This gives a new method for calculation of $\int_0^1 f(x)\, dx$. Namely, to find an approximate value for $\int_0^1 f(x)\, dx$, it suffices that for a large n, we choose n random points from $[0, 1] \times [0, 1]$ and calculate m, the number of those falling below the curve $y = f(x)$. The quotient m/n is then approximately $\int_0^1 f(x)\, dx$. This is elaborated further while discussing Monte Carlo procedure and simulation in Section 12.5. ◆

Example 10.8 Suppose that F, the probability distribution function of the elements of a class of random variables (say, a population), is unknown and we want to estimate it at a point x. For this purpose we take a random sample from the population; that is, we find independent random variables $X_1, X_2, \ldots, X_n$ each with distribution function F. Then we let $n(x) =$ the number of X_i's $\leq x$ and $\hat{F}_n(x) = n(x)/n$. Clearly, $\hat{F}_n(x)$ is the relative frequency of the number of data $\leq x$ [in statistics, $\hat{F}_n(x)$ is called the *empirical distribution function of the sample*]. We will show that $\lim_{n \to \infty} \hat{F}_n(x) = F(x)$. To do so, let

$$Y_i = \begin{cases} 1 & \text{if } X_i \leq x \\ 0 & \text{otherwise.} \end{cases}$$

Then Y_i's are identically distributed, independent, and

$$E(Y_i) = P(X_i \leq x) = F(x) \leq 1 < \infty.$$

Thus by the strong law of large numbers,

$$\lim_{n \to \infty} \hat{F}_n(x) = \lim_{n \to \infty} \frac{Y_1 + Y_2 + \cdots + Y_n}{n} = E(Y_i) = F(x).$$

Therefore, for large n, $\hat{F}_n(x)$ is an approximation of $F(x)$. ◆

Example 10.9 Suppose that there exist N families on the earth, and the maximum number of children a family has is c. Let α_j, $j = 0, 1, 2, \ldots,$ c, $\sum_{j=0}^{c} \alpha_j = 1$, be the fraction of families with j children. Determine the fraction of the children in the world who are from a family with k ($k = 1, 2, \ldots, c$) children.

SOLUTION: Let X_i denote the number of children of family i ($i = 1, 2, \ldots, N$) and

$$Y_i = \begin{cases} 1 & \text{if family } i \text{ has } k \text{ children} \\ 0 & \text{otherwise.} \end{cases}$$

Then the fraction of the children in the world who are from a family with k children is

$$\frac{k \sum_{i=1}^{N} Y_i}{\sum_{i=1}^{N} X_i} = \frac{k \dfrac{1}{N} \sum_{i=1}^{N} Y_i}{\dfrac{1}{N} \sum_{i=1}^{N} X_i}.$$

Now $(1/N) \sum_{i=1}^{N} Y_i$ is the fraction of the families with k children; therefore, it is equal to α_k. The quantity $(1/N) \sum_{i=1}^{N} X_i$ is the average number of children in a family; hence it is the same as $\sum_{j=0}^{c} j\alpha_j$. So the desired fraction is $k\alpha_k / \sum_{j=0}^{c} j\alpha_j$.

For large N, the strong law of large numbers can be used to show the same thing. Note that it is reasonable to assume that $\{X_1, X_2, \ldots\}$ and hence $\{Y_1, Y_2, \ldots\}$ are sequences of independent random variables. So we can write that

$$\lim_{N \to \infty} \frac{k \dfrac{1}{N} \displaystyle\sum_{i=1}^{N} Y_i}{\dfrac{1}{N} \displaystyle\sum_{i=1}^{N} X_i} = \frac{kE(Y_i)}{E(X_i)} = \frac{k\alpha_k}{\displaystyle\sum_{j=0}^{c} j\alpha_j}. \quad \blacklozenge$$

So far, we have explained the strong law of large numbers and some of its applications. It is important to know that historically, before the strong law of large numbers was proved, a weaker version, now called the *weak law of large numbers*, was shown in an ingenious way by Jacob Bernoulli (1654–1705) for Bernoulli random variables. Later, the weak law of large numbers was generalized by Chebyshev for a certain class of sequences of identically distributed and independent random variables. Chebyshev discovered his inequality in this connection. We now prove Chebyshev's version of the weak law of large numbers. This is weaker than the theorem that the Russian mathematician Khintchine proved; that is, its conclusions are the same as Khintchine's result, but its assumptions are more restrictive. In Khintchine's proof, it is not necessary to assume that $\text{Var}(X_i) < \infty$.

THEOREM 10.4   (Weak Law of Large Numbers)   *Let $X_1, X_2, X_3, \ldots$ be a sequence of independent and identically distributed random variables with $\mu = E(X_i)$ and $\sigma^2 = \text{Var}(X_i) < \infty$, $i = 1, 2, \ldots$. Then $\forall \varepsilon > 0$,*

$$\lim_{n \to \infty} P\left(\left| \frac{X_1 + X_2 + \cdots + X_n}{n} - \mu \right| > \varepsilon \right) = 0.$$

Let $\{X_1, X_2, \ldots, X_n\}$ be a random sample of a population. The weak law of large numbers implies that if the sample size is large enough ($n \to \infty$), then μ, the mean of the population, is very close to $\overline{X} = (X_1 + X_2 + \cdots + X_n)/n$. That is, for sufficiently large n, it is very likely that $|\overline{X} - \mu|$ is very small.

Note that the type of convergence in the weak law of large numbers is neither the usual pointwise convergence discussed in calculus, nor it is almost sure convergence. This type of convergence is called *convergence in probability* and is defined as follows.

DEFINITION   *Let $X_1, X_2, \ldots$ be a sequence of random variables defined on a sample space S. We say that X_n converges to a random variable X in probability if for each $\varepsilon > 0$,*

$$\lim_{n \to \infty} P(|X_n - X| > \varepsilon) = 0.$$

So if X_n converges to X in probability, then for sufficiently large n, it is very likely that $|X_n - X|$ is very small. By this definition, the weak law of large numbers states that if X_1, X_2, X_3, ... is a sequence of independent and identically distributed random variables with $\mu = E(X_i)$ and $\sigma^2 = \text{Var}(X_i) < \infty$, $i = 1, 2, \ldots$, then $(X_1 + X_2 + \cdots + X_n)/n$ converges to μ in probability. It is not hard to show that if X_n converges to X almost surely, it converges to X in probability. However, the converse of this is not true (see Exercise 6). To prove Theorem 10.4, we use the fact that the variance of a sum of independent random variables is the sum of the variances of the random variables. We show this in Chapter 11 [see (11.6)].

PROOF OF THEOREM 10.4: Since

$$E\left(\frac{X_1 + X_2 + \cdots + X_n}{n}\right) = \frac{1}{n}[E(X_1) + E(X_2) + \cdots + E(X_n)]$$

$$= \frac{1}{n}n\mu = \mu$$

and

$$\text{Var}\left(\frac{X_1 + X_2 + \cdots + X_n}{n}\right) = \frac{1}{n^2}[\text{Var}(X_1) + \text{Var}(X_2) + \cdots + \text{Var}(X_n)]$$

$$= \frac{1}{n^2}n\sigma^2 = \frac{\sigma^2}{n},$$

by Chebyshev's inequality we get

$$P\left(\left|\frac{X_1 + X_2 + \cdots + X_n}{n} - \mu\right| > \varepsilon\right) \le \frac{\sigma^2/n}{\varepsilon^2} = \frac{\sigma^2}{\varepsilon^2 n}.$$

This shows that

$$\lim_{n\to\infty} P\left(\left|\frac{X_1 + X_2 + \cdots + X_n}{n} - \mu\right| > \varepsilon\right) = 0. \quad \blacklozenge$$

EXERCISES

A

1. Let $\{X_1, X_2, \ldots\}$ be a sequence of nonnegative independent random variables, and for all i, suppose that the probability density function of X_i is

$$f(x) = \begin{cases} 4x(1-x) & \text{if } 0 \leq x \leq 1 \\ 0 & \text{otherwise.} \end{cases}$$

Find

$$\lim_{n\to\infty} \frac{X_1 + X_2 + \cdots + X_n}{n}.$$

2. Let $\{X_1, X_2, \ldots\}$ be a sequence of independent, identically distributed random variables with positive expected value. Show that for all $M > 0$,

$$\lim_{n\to\infty} P(X_1 + X_2 + \cdots + X_n > M) = 1.$$

3. Let X be a nonnegative continuous random variable with probability density function $f(x)$. Define

$$Y_n = \begin{cases} 1 & \text{if } X > n \\ 0 & \text{otherwise.} \end{cases}$$

Prove that Y_n converges to 0 in probability.

4. A random number is selected from the interval $[0, 1)$ and its decimal expansion is denoted by $0.X_1X_2X_3X_4\cdots$. Since the probability of selecting a terminating decimal is zero, this expansion is unique with probability 1. It is clear that X_1, X_2, X_3, ... are random variables. For fixed m, $0 \leq m \leq 9$, let

$$Y_i = \begin{cases} 1 & \text{if } X_i = m \\ 0 & \text{otherwise.} \end{cases}$$

(a) Prove that $\{Y_1, Y_2, \ldots\}$ is a sequence of independent random variables.

(b) Show that with probability 1,

$$\lim_{n\to\infty} \frac{Y_1 + Y_2 + \cdots + Y_n}{n} = \frac{1}{10}.$$

That is, for all integers m, $0 \leq m \leq 9$, the proportion of m's in the decimal expansion of the selected random number is $1/10$.

5. Let $\{X_1, X_2, \ldots\}$ be a sequence of independent, identically distributed random variables.   In other words, for all n, let $X_1, X_2, \ldots, X_n$ be a random sample from a distribution with mean $\mu < \infty$. Let $S_n = X_1 + X_2 + \cdots + X_n$, $\overline{X}_n = S_n/n$. Show that S_n grows at rate n. That is,

$$\lim_{n \to \infty} P\{n(\mu - \varepsilon) \le S_n \le n(\mu + \varepsilon)\} = 1.$$

B

6. For a positive integer n, let $\tau(n) = (2^k, i)$, where i is the remainder when we divide n by 2^k, the largest possible power of 2. For example, $\tau(10) = (2^3, 2)$, $\tau(12) = (2^3, 4)$, $\tau(19) = (2^4, 3)$, and $\tau(69) = (2^6, 5)$. In an experiment a point is selected at random from $[0, 1]$. For $n \ge 1$, $\tau(n) = (2^k, i)$, let

$$X_n = \begin{cases} 1 & \text{if the outcome is in } \left[\dfrac{i}{2^k}, \dfrac{i+1}{2^k} \right] \\ 0 & \text{otherwise.} \end{cases}$$

Show that X_n converges to 0 in probability while it does not converge at any point, let alone almost sure convergence.

10.3   Central Limit Theorem

Let X be a binomial random variable with parameters n and p. In Section 7.2 we discussed the De Moivre–Laplace theorem, which states that for real numbers a and b, $a < b$,

$$\lim_{n \to \infty} P\left(a < \frac{X - np}{\sqrt{np(1 - p)}} < b \right) = \frac{1}{\sqrt{2\pi}} \int_a^b e^{-t^2/2} \, dt.$$

We mentioned that for values of n and p for which $np(1 - p) \ge 10$, this theorem can be used to approximate binomial by standard normal. Since this approximation is very useful, it is highly desirable to obtain similar ones for distributions other than binomial. However, the form of the De Moivre–Laplace theorem above does not readily suggest any generalizations. Therefore, we restate it in an alternative way that can be generalized easily. To do so, note that if X is a binomial random variable with parameters n and p, then X is the sum of n independent Bernoulli random variables each with parameter p. That is, $X = X_1 + X_2 + \cdots + X_n$, where

$$X_i = \begin{cases} 1 & \text{if the } i\text{th trial is a success} \\ 0 & \text{if the } i\text{th trial is a failure.} \end{cases}$$

With this representation, the De Moivre–Laplace theorem is restated as follows:

Let X_1, X_2, X_3, ... be a sequence of independent Bernoulli random variables each with parameter p. Then $\forall i$, $E(X_i) = p$, $\text{Var}(X_i) = p(1 - p)$, and

$$\lim_{n \to \infty} P\left(\frac{X_1 + X_2 + \cdots + X_n - np}{\sqrt{np(1 - p)}} \leq x\right) = \frac{1}{\sqrt{2\pi}} \int_{-\infty}^{x} e^{-y^2/2}\, dy.$$

It is this version of the De Moivre–Laplace theorem that motivates elegant generalizations. The first of such generalizations was given in 1887 by Chebyshev. But since its proof was not rigorous enough, his student Markov worked on the proof and made it both rigorous and simpler. In 1901, Lyapunov, another student of Chebyshev, weakened the conditions of Chebyshev and proved the following generalization of the De Moivre–Laplace theorem, which is now called *the central limit theorem*. It can be argued that this is the most celebrated theorem in both probability and statistics.

THEOREM 10.5 (Central Limit Theorem) *Let X_1, X_2, X_3, ... be a sequence of independent and identically distributed random variables, each with expectation μ and variance σ^2. Then the distribution of*

$$Z_n = \frac{X_1 + X_2 + \cdots + X_n - n\mu}{\sigma\sqrt{n}}$$

converges to the distribution of a standard normal random variable. That is,

$$\lim_{n \to \infty} P(Z_n \leq x) = \frac{1}{\sqrt{2\pi}} \int_{-\infty}^{x} e^{-y^2/2}\, dy.$$

Note that Z_n is simply $X_1 + X_2 + \cdots + X_n$ standardized.

The central limit theorem shows that many natural phenomena obey approximately a standard normal distribution. In natural processes, the changes that take place are often the result of a sum of many insignificant random factors. Although the effect of each of these factors, separately, may possibly be ignored, the effect of their sum in general may not. Therefore, the study of distributions of sums of a large number of independent random variables is essential. In this connection, the central limit theorem shows that in an enormous number of random phenomena, the distribution functions of such sums are approximately normal. Some examples are the weight of a man, the height of a woman, the error made in a measurement, the position and velocity of a molecule of a gas, the quantity of diffused material, the growth distribution of plants, animals or their organs, and so on. The weight of a man, for example, is the result of a large number of environmental and genetic factors which are more or less unrelated but each of which contributes only a small amount to the weight.

The proof to the central limit theorem given here is a standard one based on the following result of Lévy, which we state without proof.

THEOREM 10.6   (Lévy Continuity Theorem)   *Let X_1, X_2, X_3, ... be a sequence of random variables with distribution functions F_1, F_2, F_3, ... and moment-generating functions $M_{X_1}(t)$, $M_{X_2}(t)$, $M_{X_3}(t)$, ... Let X be a random variable with distribution function F and moment-generating function $M_X(t)$. If for all values of t, $M_{X_n}(t)$ converges to $M_X(t)$, then at the points of continuity of F, F_n converges to F.*

PROOF OF THEOREM 10.5 (CENTRAL LIMIT THEOREM): Let $Y_n = X_n - \mu$; then $E(Y_n) = 0$ and $\text{Var}(Y_n) = \sigma^2$. We want to prove that the distribution function of $Z_n = (Y_1 + Y_2 + \cdots + Y_n)/(\sigma \sqrt{n})$ converges to the distribution of Z. Since $Y_1, Y_2, Y_3, \ldots$ are identically distributed, they have the same moment-generating function M. By independence of $Y_1, Y_2, \ldots, Y_n$:

$$
\begin{aligned}
M_{Z_n}(t) &= E\left[\exp\left(\frac{Y_1 + Y_2 + \cdots + Y_n}{\sigma \sqrt{n}} t\right)\right] = M_{Y_1 + Y_2 + \cdots + Y_n}\left(\frac{t}{\sigma \sqrt{n}}\right) \\
&= M_{Y_1}\left(\frac{t}{\sigma \sqrt{n}}\right) M_{Y_2}\left(\frac{t}{\sigma \sqrt{n}}\right) \cdots M_{Y_n}\left(\frac{t}{\sigma \sqrt{n}}\right) \\
&= \left[M\left(\frac{t}{\sigma \sqrt{n}}\right)\right]^n .
\end{aligned}
\tag{10.1}
$$

By the Lévy continuity theorem, it suffices to prove that $M_{Z_n}(t)$ converges to $\exp(t^2/2)$, the moment-generating function of Z. Equivalently, it is enough to show that

$$
\lim_{n \to \infty} \ln M_{Z_n}(t) = \frac{t^2}{2}.
\tag{10.2}
$$

Let $h = t/(\sigma \sqrt{n})$; then $n = t^2/(\sigma^2 h^2)$. Thus (10.1) gives

$$
\ln M_{Z_n}(t) = n \ln M(h) = \frac{t^2}{\sigma^2 h^2} \ln M(h) = \frac{t^2}{\sigma^2}\left[\frac{\ln M(h)}{h^2}\right].
$$

Therefore,

$$
\lim_{n \to \infty} \ln M_{Z_n}(t) = \frac{t^2}{\sigma^2} \lim_{h \to 0}\left[\frac{\ln M(h)}{h^2}\right].
\tag{10.3}
$$

Since $M(0) = 1$, $\lim_{h\to 0}[\ln M(h)/h^2]$ is indeterminate. To determine its value, we apply L'Hôpital's rule twice. We get

$$\lim_{h\to 0}\left[\frac{\ln M(h)}{h^2}\right] = \lim_{h\to 0}\frac{M'(h)/M(h)}{2h} = \lim_{h\to 0}\frac{M'(h)}{2hM(h)}$$

$$= \lim_{h\to 0}\frac{M''(h)}{2M(h) + 2hM'(h)} = \frac{M''(0)}{2M(0)} = \frac{\sigma^2}{2},$$

where $M''(0) = \sigma^2$ since $E(Y_i) = 0$. Thus, by (10.3),

$$\lim_{n\to\infty}\ln M_{Z_n}(t) = \frac{t^2}{\sigma^2}\frac{\sigma^2}{2} = \frac{t^2}{2}.$$

This establishes (10.2) and proves the theorem.   ◆

We now show some of the many useful applications of the central limit theorem.

Example 10.10   If 20 random numbers are selected independently from the interval $(0, 1)$, what is the approximate probability that the sum of these numbers is at least 8?

SOLUTION: Let X_i be the ith number selected, $i = 1, 2, \ldots, 20$. To calculate $P(\sum_{i=1}^{20} X_i \geq 8)$, we use the central limit theorem. Since $\forall i$, $E(X_i) = (0+1)/2 = 1/2$ and $\text{Var}(X) = (1-0)^2/12 = 1/12$,

$$P\left(\sum_{i=1}^{20} X_i \geq 8\right) = P\left\{\frac{\sum_{i=1}^{20} X_i - 20(1/2)}{\sqrt{1/12}\,\sqrt{20}} \geq \frac{8 - 20(1/2)}{\sqrt{1/12}\,\sqrt{20}}\right\}$$

$$= P\left(\frac{\sum_{i=1}^{20} X_i - 10}{\sqrt{5/3}} \geq -1.55\right)$$

$$\approx 1 - \Phi(-1.55) = \Phi(1.55) \approx 0.9394.   ◆$$

Example 10.11   Customers arrive at a mall in accordance with a Poisson process with rate 4000 persons per day. Find an approximate value for the probability that tomorrow at least 3850 customers will enter the mall.

SOLUTION: Let X be the number of customers who will enter the mall tomorrow. Using the probability function of Poisson random variable with mean 4000, we get

$$P(X \geq 3850) = e^{-4000}\sum_{i=3850}^{\infty}\frac{(4000)^i}{i!},$$

which cannot easily be calculated. However, since we know that the sum of 4000 independent Poisson random variables each with mean 1 is a Poisson random variable with mean 4000 (see Theorem 9.4), we can use the central limit theorem to calculate $P(X \geq 3850)$ approximately. For $i = 1, 2, \ldots, 4000$, let X_i be a Poisson random variable with mean 1; then $X = \sum_{i=1}^{4000} X_i$, $E(X_i) = 1$, $n = 4000$, $\text{Var}(X_i) = 1$, and $\sigma_{X_i} = 1$. Thus by the central limit theorem,

$$P(X \geq 3850) = P\left(\frac{X - 4000}{\sqrt{4000}} \geq \frac{3850 - 4000}{\sqrt{4000}}\right)$$

$$= P\left(\frac{X - 4000}{\sqrt{4000}} \geq -2.37\right)$$

$$\approx 1 - \Phi(-2.37) = \Phi(2.37) \approx 0.9911. \quad \blacklozenge$$

Example 10.12   A biologist wants to estimate l, the life expectancy of a certain type of insect. To do so, he takes a sample of size n and measures the lengths from birth to death of each of them. Then he finds the average of these numbers. If he believes that the lifetimes of these insects are independent random variables with variance 1.5 days, how big should he choose the sample to be 98% sure that his average is accurate within ± 0.2 (± 4.8 hours)?

SOLUTION: For $i = 1, 2, \ldots, n$, let X_i be the lifetime of the ith insect of the sample. We want to determine n, so that

$$P\left(-0.2 < \frac{X_1 + X_2 + \cdots + X_n}{n} - l < 0.2\right) \approx 0.98.$$

Since $E(X_i) = l$ and $\text{Var}(X_i) = 1.5$, by the central limit theorem we have that

$$P\left(-0.2 < \frac{\sum_{i=1}^{n} X_i}{n} - l < 0.2\right) = P\left\{(-0.2)n < \sum_{i=1}^{n} X_i - nl < (0.2)n\right\}$$

$$= P\left\{\frac{-(0.2)n}{\sqrt{1.5n}} < \frac{\sum_{i=1}^{n} X_i - nl}{\sqrt{1.5n}} < \frac{(0.2)n}{\sqrt{1.5n}}\right\}$$

$$\approx \Phi\left[\frac{(0.2)n}{\sqrt{1.5n}}\right] - \Phi\left[\frac{(-0.2)n}{\sqrt{1.5n}}\right]$$

$$= 2\Phi\left(\frac{0.2\sqrt{n}}{\sqrt{1.5}}\right) - 1.$$

Thus the quantity $2\Phi(0.2\sqrt{n}/\sqrt{1.5}) - 1$ should approximately equal 0.98, that is, $\Phi(0.2\sqrt{n}/\sqrt{1.5}) \approx 0.99$. From Table 1 of the Appendix, we find that $0.2\sqrt{n}/\sqrt{1.5} = 2.33$. This gives $n = 203.58$. Therefore, the biologist should choose a sample of size 204.   ♦

REMARK: In this solution we have assumed implicitly that for a sample of size 204, normal distribution is a good approximation. Although usually such assumptions are reasonable, the size of n depends on the distribution function of X_i's, which is not known in most cases. If the biologist wants to be absolutely sure about the results of an analysis, he or she should use Chebyshev's inequality instead. The results obtained by applying Chebyshev's inequality are safe but in most cases much larger than needed. For this example it can be shown that $\text{Var}(\sum_{i=1}^{n} X_i/n) = 1.5/n$. Since $E(\sum_{i=1}^{n} X_i/n) = l$, by Chebyshev's inequality,

$$P\left(\left|\sum_{i=1}^{n} \frac{X_i}{n} - l\right| > 0.2\right) \leq \frac{1.5/n}{(0.2)^2} = \frac{37.5}{n}.$$

Letting $37.5/n = 0.02$ gives $n = 1875$. So the answer obtained by applying Chebyshev's inequality is 1875, a number much larger than the 204 obtained using the central limit theorem.   ♦

Example 10.13   Suppose that customers enter a post office at a Poisson rate of 1/2 per minute. Find the approximate probability that in the 8 working hours next Monday, the 250th customer enters within the last hour.

SOLUTION: Let $N(t)$ denote the number of customers entering the post office up to and prior to t. We are given that $\{N(t), t \geq 0\}$ is a Poisson process with $\lambda = E[N(1)] = 1/2$. The arrival time of the 250th customer, X, is a gamma random variable with parameters $(250, 1/2)$. The probability that this arrival occurs within the last hour of the 8 working hours is

$$P(420 < X < 480) = \int_{420}^{480} \frac{\left(\frac{1}{2}e^{-x/2}\right)(x/2)^{249}}{\Gamma(250)} \, dx.$$

Clearly, it is not practical to calculate this integral. However, noting that X is the sum of 250 independent exponential random variables, each with mean $1/\lambda = 2$, we can use the central limit theorem to approximate this probability. For $i = 1, 2, 3, \ldots, 250$, let X_i be an exponential random variable with mean 2; then $X = \sum_{i=1}^{250} X_i$, $E(X_i) = 2$, $\text{Var}(X_i) = 4$, and $\sigma_{X_i} = 2$. The central limit theorem yields

$$P(420 < X < 480) = P\left\{ \frac{420 - 250(2)}{2\sqrt{250}} < \frac{X - 250(2)}{2\sqrt{250}} < \frac{480 - 250(2)}{2\sqrt{250}} \right\}$$

$$= P\left(-2.53 < \frac{X - 500}{2\sqrt{250}} < -0.63 \right)$$

$$\approx \Phi(-0.63) - \Phi(-2.53)$$

$$= \Phi(2.53) - \Phi(0.63) \approx 0.9943 - 0.7357 = 0.2586. \quad \blacklozenge$$

EXERCISES

A

1. What is the probability that the average of 150 random points from the interval $(0, 1)$ is within 0.02 of the midpoint of the interval?

2. Let $X_1, X_2, \ldots, X_n$ be independent and identically distributed random variables, and let $S_n = X_1 + X_2 + \cdots + X_n$. For large n, what is the approximate probability that S_n is between $E(S_n) - \sigma_{S_n}$ and $E(S_n) + \sigma_{S_n}$?

3. Each time that Jim charges an item to his credit card, he will round the amount to the nearest dollar in his records. If he has used his credit card 300 times in the last 12 months, what is the probability that his record differs from the total expenditure by at most 10 dollars?

4. A physical quantity is measured 50 times and the average of these measurements is taken as the result. If each measurement has a random error uniformly distributed over $(-1, 1)$, what is the probability that our result differs from the actual value by less than 0.25?

5. Passengers make reservations on a particular flight 24 hours a day at a Poisson rate of three per hour. If for the flight 400 seats are available, what is the probability that at the end of the fifth day all the seats are reserved?
   HINT: It is reasonable to assume that X, the number of requests for reservations at the end of the fifth day, is Poisson with rate 360.

6. Suppose that the number of rainy days in a year in a certain area has a Poisson distribution with mean $\lambda = 110$. What is the probability that it will rain at least 100 days next year?

7. Suppose that whenever invited to a party, the probability that a person attends with his or her guest is 1/3, attends alone is 1/3, and does not attend is 1/3. A company has invited all 300 of its employees and their guests to a Christmas party. What is the probability that at least 320 will attend?

8. The number of phone calls to a certain exchange is a Poisson process with rate 23 per hour. Using the central limit theorem, calculate the probability that the time until the 91st call is at least 4 hours.

   HINT: Note that the time until the nth call is a gamma random variable and hence is the sum of n independent exponential random variables.

B

9. An investor buys 1000 shares of the XYZ Corporation at $50.00 per share. Subsequently the stock price varies by $0.125 (1/8) every day, but unfortunately it is just as likely to move down as up. What is the most likely value of his holdings after 60 days?

   HINT: First calculate the distribution of the *change* in the stock price after 60 days.

10. A coin is tossed successively. What is an approximate probability of at least 25 heads before 50 tails?

11. Let $\{X_1, X_2, \ldots\}$ be a sequence of independent Poisson random variables, each with parameter 1. By applying the central limit theorem to this sequence, prove that

$$\lim_{n \to \infty} \frac{1}{e^n} \sum_{k=0}^{n} \frac{n^k}{k!} = \frac{1}{2}.$$

Review Problems

1. A psychologist wants to estimate μ, the mean IQ of the students of a university. To do so, she takes a sample of size n of the students and measures their IQs. Then she finds the average of these numbers. If she believes that the IQs of these students are independent random variables with variance 170, how big should she choose the sample to be 98% sure that her average is accurate within ± 0.2?

2. Each time that Ed charges an expense to his credit card, he chops off the cents and records only the dollar value. If this month he has charged his credit card 20 times, using Chebyshev's inequality, find an upper bound on the probability that his record shows at least $15 less than the actual amount charged.

3. In a multiple-choice test with false answers receiving negative scores, the mean of the grades of the students is 0 and its standard deviation

is 15. Find an upper bound for the probability that a student's grade is at least 45.

4. A randomly selected book from Vernon's library is X centimeters thick, where $X \sim N(3, 1)$. Vernon has an empty shelf 87 centimeters long. What is the probability that he can fit 31 randomly selected books in it?

5. A fair die is rolled 20 times. What is the approximate probability that the sum of the outcomes is between 65 and 75?

6. The number of phone calls to a certain exchange is a Poisson process having a rate of four per minute. What is the approximate probability that at least 250 calls arrive at this exchange within the next hour?

7. Show that for a nonnegative random variable X with mean μ, we have that $\forall n, nP(X \geq n\mu) \leq 1$.

Chapter 11

MORE EXPECTATIONS
AND VARIANCES

11.1   Covariance

In Sections 4.5 and 6.3 we studied the notion of the variance of a random variable X. We showed that $E[(X - E(X))^2]$, the variance of X, measures the average magnitude of the fluctuations of the random variable X from its expectation, $E(X)$. We mentioned that this quantity measures the dispersion or spread of the distribution of X about its expectation. Now suppose that X and Y are two jointly distributed random variables. Then $\text{Var}(X)$ and $\text{Var}(Y)$ determine the dispersions of X and Y independently rather than jointly. In fact, $\text{Var}(X)$ measures the spread or dispersion along the x-direction, and $\text{Var}(Y)$ measures the spread or dispersion along the y-direction in the plane. We now calculate $\text{Var}(aX + bY)$, the joint spread or dispersion of X and Y along the $(ax + by)$-direction for arbitrary real numbers a and b:

$$
\begin{aligned}
\text{Var}(aX + bY) &= E\left[(aX + bY) - E(aX + bY)\right]^2 \\
&= E\left[(aX + bY) - aE(X) - bE(Y)\right]^2 \\
&= E\left[a(X - E(X)) + b(Y - E(Y))\right]^2 \\
&= E\left[a^2(X - E(X))^2 + b^2(Y - E(Y))^2 \right. \\
&\quad \left. + 2ab(X - E(X))(Y - E(Y))\right] \\
&= a^2\text{Var}(X) + b^2\text{Var}(Y) \\
&\quad + 2abE\left[(X - E(X))(Y - E(Y))\right]. \qquad (11.1)
\end{aligned}
$$

This formula shows that the joint spread or dispersion of X and Y can be measured in *any* direction $(ax + by)$ if the quantities Var(X), Var(Y), and $E[(X - E(X))(Y - E(Y))]$ are known. On the other hand, the joint spread or dispersion of X and Y depends on these three quantities. However, Var(X) and Var(Y) determine the dispersions of X and Y independently, therefore $E[(X - E(X))(Y - E(Y))]$ is the quantity that gives information about the joint spread or dispersion of X and Y. It is called the *covariance* of X and Y, is denoted by Cov(X, Y), and determines how X and Y covary jointly. For example, by relation (11.1), if for random variables X, Y, and Z, Var(Y)=Var(Z) and $ab > 0$, then the joint dispersion of X and Y along the $(ax + by)$-direction is greater than the joint dispersion of X and Z along the $(ax + bz)$-direction if and only if Cov$(X, Y) >$ Cov(X, Z).

DEFINITION   *Let X and Y be jointly distributed random variables; then the covariance of X and Y is defined by*

$$Cov(X, Y) = E\left[(X - E(X))(Y - E(Y))\right].$$

Note that

$$\mathrm{Cov}(X, X) = \sigma_X^2 = \mathrm{Var}(X).$$

Also, by the Cauchy–Schwarz inequality (Theorem 9.11),

$$\begin{aligned}
\mathrm{Cov}(X, Y) &= E\left[(X - E(X))(Y - E(Y))\right] \\
&\leq \sqrt{E[X - E(X)]^2 E[Y - E(Y)]^2} \\
&= \sqrt{\sigma_X^2 \sigma_Y^2} = \sigma_X \sigma_Y,
\end{aligned}$$

which shows that if $\sigma_X < \infty$ and $\sigma_Y < \infty$, then Cov$(X, Y) < \infty$.

Rewriting relation (11.1) in terms of Cov(X, Y), we obtain the following important theorem:

THEOREM 11.1   *For random variables X and Y,*

$$\mathrm{Var}(aX + bY) = a^2 \mathrm{Var}(X) + b^2 \mathrm{Var}(Y) + 2ab\, \mathrm{Cov}(X, Y).$$

In particular, if $a = 1$, and $b = 1$, this gives

$$\mathrm{Var}(X + Y) = \mathrm{Var}(X) + \mathrm{Var}(Y) + \mathrm{Cov}(X, Y). \qquad (11.2)$$

Letting $\mu_X = E(X)$, and $\mu_Y = E(Y)$, an alternative formula for

$$\mathrm{Cov}(X, Y) = E[(X - E(X))(Y - E(Y))]$$

is calculated by the expansion of $E[(X - E(X))(Y - E(Y))]$:

$$\begin{aligned}
\text{Cov}(X, Y) &= E\left[(X - \mu_X)(Y - \mu_Y)\right] \\
&= E(XY - \mu_X Y - \mu_Y X + \mu_X \mu_Y) \\
&= E(XY) - \mu_X E(Y) - \mu_Y E(X) + \mu_X \mu_Y \\
&= E(XY) - \mu_X \mu_Y - \mu_Y \mu_X + \mu_X \mu_Y \\
&= E(XY) - \mu_X \mu_Y = E(XY) - E(X)E(Y).
\end{aligned}$$

Therefore,

$$\text{Cov}(X, Y) = E(XY) - E(X)E(Y). \tag{11.3}$$

Using this relation, we get

$$\begin{aligned}
&\text{Cov}(aX + b, cY + d) \\
&= E[(aX + b)(cY + d)] - E(aX + b)E(cY + d) \\
&= E(acXY + bcY + adX + bd) - [aE(X) + b][cE(Y) + d] \\
&= ac[E(XY) - E(X)E(Y)] = ac\,\text{Cov}(X, Y). \tag{11.4}
\end{aligned}$$

For random variables X and Y, $\text{Cov}(X, Y)$ might be positive, negative, or zero. It is positive if the expected value of $[X - E(X)][Y - E(Y)]$ is positive, that is, if X and Y decrease together or increase together. It is negative if X increases while Y decreases, or vice versa. If $\text{Cov}(X, Y) > 0$, we say that X and Y are *positively correlated*. If $\text{Cov}(X, Y) < 0$, we say that they are *negatively correlated*. If $\text{Cov}(X, Y) = 0$, we say that X and Y are *uncorrelated*. For example, the blood cholesterol level of a person is positively correlated with the amount of saturated fat consumed by the person, whereas the amount of alcohol in the blood is negatively correlated with motor coordination. The more saturated fat a person ingests, generally, the higher his or her blood cholesterol level will be. The more a person drinks alcohol, the poorer his or her level of motor coordination becomes. As another example, let X be the weight of a person before starting a health fitness program and Y be his or her weight afterward. Then X and Y are negatively correlated because the effect of fitness programs is that usually heavier persons lose weight, whereas lighter persons gain weight. The best examples for uncorrelated random variables are independent ones. If X and Y are independent, then

$$\text{Cov}(X, Y) = E(XY) - E(X)E(Y) = 0.$$

However, as the following example shows, the converse of this is not true, that is, two dependent random variables might be uncorrelated.

Example 11.1 Let X be uniformly distributed over $(-1, 1)$ and $Y = X^2$. Then

$$\text{Cov}(X, Y) = E(X^3) - E(X)E(X^2) = 0,$$

since $E(X) = 0$ and $E(X^3) = 0$. Thus the perfectly related random variables X and Y are uncorrelated.   ◆

Example 11.2 Let X be the lifetime of an electronic system and Y be the lifetime of one of its components. Suppose that the electronic system fails if the component does (but not necessarily vice versa). Furthermore, suppose that the joint probability density function of X and Y (in years) is given by

$$f(x, y) = \begin{cases} \dfrac{1}{49}e^{-y/7} & \text{if } 0 \le x \le y < \infty \\ 0 & \text{elsewhere.} \end{cases}$$

(a) Determine the expected value of the remaining lifetime of the component when the system dies.

(b) Find the covariance of X and Y.

SOLUTION: (a) The remaining lifetime of the component when the system dies is $Y - X$. So the desired quantity is

$$E(Y - X) = \int_0^\infty \int_0^y (y - x)\frac{1}{49}e^{-y/7} \, dx \, dy$$

$$= \frac{1}{49} \int_0^\infty e^{-y/7} \left(y^2 - \frac{y^2}{2} \right) dy$$

$$= \frac{1}{98} \int_0^\infty y^2 e^{-y/7} \, dy = 7,$$

where the last integral is calculated using integration by parts twice.

(b) To find $\text{Cov}(X, Y) = E(XY) - E(X)E(Y)$, note that

$$E(XY) = \int_0^\infty \int_0^y (xy)\frac{1}{49}e^{-y/7} \, dx \, dy$$

$$= \frac{1}{49} \int_0^\infty ye^{-y/7} \left(\int_0^y x \, dx \right) dy$$

$$= \frac{1}{98} \int_0^\infty y^3 e^{-y/7} \, dy = \frac{14,406}{98} = 147,$$

where the last integral is calculated using integration by parts three times. We also have

$$E(X) = \int_0^\infty \int_0^y x \frac{1}{49} e^{-y/7} \, dx \, dy = 7,$$

$$E(Y) = \int_0^\infty \int_0^y y \frac{1}{49} e^{-y/7} \, dx \, dy = 14.$$

Therefore, $\text{Cov}(X, Y) = 147 - 7(14) = 49$. Note that $\text{Cov}(X, Y) > 0$ is expected because X and Y are positively correlated. ♦

As relation (11.2) shows, one important application of the covariance of two random variables X and Y is that it enables us to find $\text{Var}(X + Y)$. This relation is generalized as follows:

$$\text{Var}\left(\sum_{i=1}^n X_i\right) = \sum_{i=1}^n \text{Var}(X_i) + 2 \sum_{j=2}^n \sum_{i=1}^{j-1} \text{Cov}(X_i, X_j)$$

$$= \sum_{i=1}^n \text{Var}(X_i) + 2 \sum_{i<j} \sum \text{Cov}(X_i, X_j). \quad (11.5)$$

By (11.5), if $X_1, X_2, \ldots, X_n$ are *pairwise independent*, or more generally, *pairwise uncorrelated*, then

$$\text{Var}\left(\sum_{i=1}^n X_i\right) = \sum_{i=1}^n \text{Var}(X_i), \quad (11.6)$$

the reason being that $\text{Cov}(X_i, X_j) = 0$, for all $i, j, i \neq j$.

Example 11.3   Let X be the number of 6's in n rolls of a fair die. Find $\text{Var}(X)$.

SOLUTION: Let $X_i = 1$ if in the ith roll the die lands 6, and $X_i = 0$, otherwise. Then $X = X_1 + X_2 + \cdots + X_n$. Since $X_1, X_2, \ldots, X_n$ are independent,

$$\text{Var}(X) = \text{Var}(X_1) + \text{Var}(X_2) + \cdots + \text{Var}(X_n).$$

But for $i = 1, 2, \ldots, n$,

$$E(X_i) = 1 \cdot \frac{1}{6} + 0 \cdot \frac{5}{6} = \frac{1}{6},$$

$$E(X_i^2) = (1)^2 \cdot \frac{1}{6} + 0^2 \cdot \frac{5}{6} = \frac{1}{6},$$

and hence

$$\text{Var}(X_i) = \frac{1}{6} - \frac{1}{36} = \frac{5}{36}.$$

Therefore, $\text{Var}(X) = n(5/36) = (5n)/36$. ◆

Example 11.4   Using relation (11.6), calculate the variance of a binomial random variable X with parameters (n, p).

SOLUTION: Recall that X is the number of successes in n independent Bernoulli trials. Thus for $i = 1, 2, \ldots, n$, letting

$$X_i = \begin{cases} 1 & \text{if the } i\text{th trial is a success} \\ 0 & \text{otherwise,} \end{cases}$$

we obtain

$$X = X_1 + X_2 + \cdots + X_n, \tag{11.7}$$

where X_i is a Bernoulli random variable for $i = 1, 2, \ldots, n$. Note that $E(X_i) = p$ and $\text{Var}(X_i) = p(1 - p)$. Since $\{X_1, X_2, \ldots, X_n\}$ is an independent set of random variables, by (11.6),

$$\text{Var}(X) = \text{Var}(X_1) + \text{Var}(X_2) + \cdots + \text{Var}(X_n) = np(1 - p). ◆$$

Example 11.5   Using relation (11.6), calculate the variance of a negative binomial random variable X, with parameter (r, p).

SOLUTION: Recall that in a sequence of independent Bernoulli trials, X is the number of trials until the rth success. Let X_1 be the number of trials until the first success, X_2 be the number of additional trials to get the second success, X_3 be the number of additional ones to obtain the third success, and so on. Then $X = X_1 + X_2 + \cdots + X_r$, where for $i = 1, 2, \ldots, r$, the random variable X_i is geometric with parameter p. Since $X_1, X_2, \ldots, X_r$ are independent,

$$\text{Var}(X) = \text{Var}(X_1) + \text{Var}(X_2) + \cdots + \text{Var}(X_r).$$

But $\text{Var}(X_i) = (1 - p)/p^2$, for $i = 1, 2, \ldots, r$ (see Section 5.3). Thus $\text{Var}(X) = r(1 - p)/p^2$. ◆

EXERCISES

A

1. For random variables X, Y, and Z prove that
   (a) $\text{Cov}(X + Y, Z) = \text{Cov}(X, Z) + \text{Cov}(Y, Z)$.
   (b) $\text{Cov}(X, Y + Z) = \text{Cov}(X, Y) + \text{Cov}(X, Z)$.

2. For random variables X and Y show that

$$\text{Cov}(X + Y, X - Y) = \text{Var}(X) - \text{Var}(Y).$$

3. Prove that

$$\text{Var}(X - Y) = \text{Var}(X) + \text{Var}(Y) - 2\,\text{Cov}(X, Y).$$

4. Let X and Y be two independent random variables.
   (a) Show that $X - Y$ and $X + Y$ are uncorrelated if and only if $\text{Var}(X) = \text{Var}(Y)$.
   (b) Show that $\text{Cov}(X, XY) = E(Y)\text{Var}(X)$.

5. In n independent Bernoulli trials, each with probability of success p, let X be the number of successes and Y be the number of failures. Calculate $E(XY)$ and $\text{Cov}(X, Y)$.

6. Prove that if Θ is a random number from the interval $[0, 2\pi]$, the dependent random variables $X = \sin\Theta$ and $Y = \cos\Theta$ are uncorrelated.

7. Let X and Y be the coordinates of a random point selected uniformly from the unit disk $\{(x, y) : x^2 + y^2 \le 1\}$. Are X and Y independent? Are they uncorrelated? Why or why not?

8. Mr. Jones has two jobs. Next year, he will get a salary raise of X thousand dollars from his first job and a salary raise of Y thousand dollars from his second job. Suppose that X and Y are independent random variables with probability density functions f and g, respectively, where

$$f(x) = \begin{cases} \dfrac{8x}{15} & \text{if } \dfrac{1}{2} < x < 2 \\ 0 & \text{elsewhere,} \end{cases}$$

$$g(y) = \begin{cases} \dfrac{6\sqrt{y}}{13} & \text{if } \dfrac{1}{4} < y < \dfrac{9}{4} \\ 0 & \text{elsewhere.} \end{cases}$$

What are the expected value and variance of the total raise that Mr. Jones will get next year?

9. Let X and Y be independent random variables with expected values μ_1 and μ_2, and variances σ_1^2 and σ_2^2, respectively. Show that

$$\text{Var}(XY) = \sigma_1^2 \sigma_2^2 + \mu_1^2 \sigma_2^2 + \mu_2^2 \sigma_1^2.$$

10. Let X and Y have the following joint probability density function

$$f(x, y) = \begin{cases} 8xy & \text{if } 0 < x \le y < 1 \\ 0 & \text{otherwise.} \end{cases}$$

(a) Calculate $\text{Var}(X + Y)$.
(b) Show that X and Y are not independent. Explain why this does not contradict Exercise 24 of Section 8.2.

11. Find the variance of a sum of n randomly and independently selected points from the interval $(0, 1)$.

12. Let X and Y be jointly distributed with joint probability density function

$$f(x, y) = \begin{cases} \dfrac{1}{3} x^3 e^{-xy-x} & \text{if } x > 0, \ y > 0 \\ 0 & \text{otherwise.} \end{cases}$$

Determine if X and Y are positively correlated, negatively correlated, or uncorrelated.
HINT: Note that for all $a > 0$, $\int_0^\infty x^n e^{-ax} \, dx = n!/a^{n+1}$.

13. Let X be a random variable. Prove that $\text{Var}(X) = \min_t E(X - t)^2$.
HINT: Let $\mu = E(X)$ and look at the expansion of $E(X - t)^2 = E(X - \mu + \mu - t)^2$.

B

14. Let S be the sample space of an experiment. Let A and B be two events of S. Let I_A and I_B be the indicator variables for A and B. That is,

$$I_A(\omega) = \begin{cases} 1 & \text{if } \omega \in A \\ 0 & \text{if } \omega \notin A, \end{cases}$$

$$I_B(\omega) = \begin{cases} 1 & \text{if } \omega \in B \\ 0 & \text{if } \omega \notin B. \end{cases}$$

Show that I_A and I_B are positively correlated if and only if $P(A \mid B) > P(A)$ and if an only if $P(B \mid A) > P(B)$.

15. Show that for random variables X, Y, Z, and W and constants a, b, c, and d,

$$\text{Cov}(aX + bY, cZ + dW)$$
$$= ac\text{Cov}(X, Z) + bc\text{Cov}(Y, Z) + ad\text{Cov}(X, W) + bd\text{Cov}(Y, W).$$

HINT: For a simpler proof, use the results of Exercise 1.

16. Prove the following generalization of Exercise 15:

$$\text{Cov}\left(\sum_{i=1}^{n} a_i X_i, \sum_{j=1}^{m} b_j Y_j\right) = \sum_{i=1}^{n} \sum_{j=1}^{m} a_i b_j \text{Cov}(X_i, Y_j).$$

17. A fair die is thrown n times. What is the covariance of the number of ones and the number of 6's obtained?
HINT: Use the result of Exercise 16.

18. Show that if $X_1, X_2, \ldots, X_n$ are random variables and $a_1, a_2, \ldots, a_n$ are constants, then

$$\text{Var}\left(\sum_{i=1}^{n} a_i X_i\right) = \sum_{i=1}^{n} a_i^2 \text{Var}(X_i) + 2 \sum \sum_{i<j} a_i a_j \text{Cov}(X_i, X_j).$$

In particular, if $X_1, X_2, \ldots, X_n$ are independent, then

$$\text{Var}\left(\sum_{i=1}^{n} a_i X_i\right) = \sum_{i=1}^{n} a_i^2 \text{Var}(X_i).$$

19. Let X be a hypergeometric random variable with probability function

$$p(x) = P(X = x) = \frac{\binom{D}{x}\binom{N - D}{n - x}}{\binom{N}{n}},$$

where $\max(0, n + D - N) \le x \le \min(D, n)$. Recall that X is the number of defective items among n items drawn randomly from a box containing D defective and $N - D$ nondefective items. Find $\text{Var}(X)$.
HINT: Let A_i be the event that the ith item drawn is defective. Also for $i = 1, 2, \ldots, n$, let

$$X_i = \begin{cases} 1 & \text{if } A_i \text{ occurs} \\ 0 & \text{otherwise.} \end{cases}$$

Then $X = X_1 + X_2 + \cdots + X_n$.

20. In a small town exactly n married couples are living. What is the variance of the intact couples after m deaths occur among the couples? Assume that the deaths occur at random, there are no divorces, and there are no new marriages.

NOTE: This involves the Daniel Bernoulli problem discussed in Example 9.13.

11.2   Correlation

While for random variables X and Y, $\text{Cov}(X, Y)$ provides information about how X and Y vary jointly, it has a major shortcoming: It is not independent of the units in which X and Y are measured. To see this, suppose that for random variables X and Y, when measured in (say) centimeters, $\text{Cov}(X, Y) = 0.15$. For the same random variables if we change the measurements to millimeters, then $X_1 = 10X$ and $Y_1 = 10Y$ will be the new observed values, and by relation (11.4) we get

$$\text{Cov}(X_1, Y_1) = \text{Cov}(10X, 10Y) = 100\text{Cov}(X, Y) = 15,$$

showing that $\text{Cov}(X, Y)$ is sensitive to the units of measurement. From Section 4.6 we know that for a random variable X, the standardized X, $X^* = [X - E(X)]/\sigma_X$, is independent of the units in which X is measured. Thus, to define a measure of association between X and Y, independent of the scales of measurements, it is appropriate to consider $\text{Cov}(X^*, Y^*)$ rather than $\text{Cov}(X, Y)$. Using relation (11.4), we obtain

$$\begin{aligned}
\text{Cov}(X^*, Y^*) &= \text{Cov}\left(\frac{X - E(X)}{\sigma_X}, \frac{Y - E(Y)}{\sigma_Y}\right) \\
&= \text{Cov}\left(\frac{1}{\sigma_X}X - \frac{E(X)}{\sigma_X}, \frac{1}{\sigma_Y}Y - \frac{E(Y)}{\sigma_Y}\right) \\
&= \frac{1}{\sigma_X}\frac{1}{\sigma_Y}\text{Cov}(X, Y) = \frac{\text{Cov}(X, Y)}{\sigma_X\sigma_Y}.
\end{aligned}$$

DEFINITION   *Let X and Y be two random variables with $0 < \sigma_X^2 < \infty$ and $0 < \sigma_Y^2 < \infty$. The covariance between the standardized X and the standardized Y is called the correlation coefficient between X and Y and* *is denoted by $\rho = \rho(X, Y)$. Therefore,*

$$\rho = Cov\left(\frac{X - E(X)}{\sigma_X}, \frac{Y - E(Y)}{\sigma_Y}\right) = \frac{Cov(X, Y)}{\sigma_X \sigma_Y}.$$

The quantity $Cov(X, Y)/(\sigma_X\sigma_Y)$ gives all the important information that $Cov(X, Y)$ provides about how X and Y covary, and at the same time, it is not sensitive to the scales of measurement. Clearly, $\rho(X, Y) > 0$ if and only if X and Y are positively correlated; $\rho(X, Y) < 0$ if and only if X and Y are negatively correlated; and $\rho(X, Y) = 0$ if and only if X and Y are uncorrelated. Moreover, $\rho(X, Y)$ roughly measures the amount and the sign of linear relationship between X and Y. It is -1 if $Y = aX + b$, $a < 0$, and $+1$ if $Y = aX + b$, $a > 0$. Thus $\rho(X, Y) = \pm 1$ in the case of perfect linear relationship and 0 in the case of independence of X and Y. The key to the proof of "$\rho(X, Y) = \pm 1$ if and only if $Y = aX + b$" is the following lemma, which is interesting by itself.

LEMMA 11.1   *For random variables X and Y with correlation coefficient $\rho(X, Y)$,*

$$Var\left(\frac{X}{\sigma_X} + \frac{Y}{\sigma_Y}\right) = 2 + 2\rho(X, Y),$$

$$Var\left(\frac{X}{\sigma_X} - \frac{Y}{\sigma_Y}\right) = 2 - 2\rho(X, Y).$$

PROOF: We prove the first relation; the second can be shown similarly. By (11.2),

$$Var\left(\frac{X}{\sigma_X} + \frac{Y}{\sigma_Y}\right) = Var\left(\frac{X}{\sigma_X}\right) + Var\left(\frac{Y}{\sigma_Y}\right) + 2\,Cov\left(\frac{X}{\sigma_X}, \frac{Y}{\sigma_Y}\right)$$

$$= \frac{1}{\sigma_X^2}Var(X) + \frac{1}{\sigma_Y^2}Var(Y) + 2\frac{Cov(X, Y)}{\sigma_X\sigma_Y}$$

$$= 2 + 2\rho(X, Y). \quad \blacklozenge$$

THEOREM 11.2   *For random variables X and Y with correlation coefficient $\rho(X, Y)$:*

(a) $-1 \le \rho(X, Y) \le 1$.

(b) *With probability 1, $\rho(X, Y) = 1$ if and only if $Y = aX + b$ for some constants a, b, $a > 0$.*

(c) *With probability 1, $\rho(X, Y) = -1$ if and only if $Y = aX + b$ for some constants a, b, $a < 0$.*

Proof: (a) Since the variance of a random variable is nonnegative,

$$\text{Var}\left(\frac{X}{\sigma_X} + \frac{Y}{\sigma_Y}\right) \geq 0 \quad \text{and} \quad \text{Var}\left(\frac{X}{\sigma_X} - \frac{Y}{\sigma_Y}\right) \geq 0.$$

Therefore, by Lemma 11.1, $2 + 2\rho(X, Y) \geq 0$ and $2 - 2\rho(X, Y) \geq 0$. That is, $\rho(X, Y) \geq -1$ and $\rho(X, Y) \leq 1$.

(b) First, suppose that $\rho(X, Y) = 1$. In this case, by Lemma 11.1,

$$\text{Var}\left(\frac{X}{\sigma_X} - \frac{Y}{\sigma_Y}\right) = 2[1 - \rho(X, Y)] = 0.$$

Therefore, with probability 1,

$$\frac{X}{\sigma_X} - \frac{Y}{\sigma_Y} = c,$$

for some constant c. Hence, with probability 1,

$$Y = \frac{\sigma_Y}{\sigma_X} X - c\sigma_Y \equiv aX + b,$$

with $a = \sigma_Y/\sigma_X > 0$ and $b = -c\sigma_Y$. Next, assume that $Y = aX + b, a > 0$. We have that

$$\rho(X, Y) = \rho(X, aX + b) = \frac{\text{Cov}(X, aX + b)}{\sigma_X \sigma_{aX+b}}$$

$$= \frac{a\,\text{Cov}(X, X)}{\sigma_X a\sigma_X} = \frac{a\,\text{Var}(X)}{a\,\text{Var}(X)} = 1.$$

(c) The proof of this statement is similar to that of (b).   ◆

Example 11.6   Show that if X and Y are continuous random variables with the joint probability density function

$$f(x, y) = \begin{cases} x + y & \text{if } 0 < x < 1, \ 0 < y < 1 \\ 0 & \text{otherwise,} \end{cases}$$

then X and Y are not linearly related.

Solution: Since X and Y are linearly related if and only if $\rho(X, Y) = \pm 1$ with probability 1, it suffices to prove that $\rho(X, Y) \neq \pm 1$. To do so, note that

$$E(X) = \int_0^1 \int_0^1 x(x+y)\, dx\, dy = \frac{7}{12},$$

$$E(XY) = \int_0^1 \int_0^1 xy(x+y)\, dx\, dy = \frac{1}{3}.$$

Also by symmetry, $E(Y) = 7/12$; therefore,

$$\text{Cov}(X, Y) = E(XY) - E(X)E(Y) = \frac{1}{3} - \frac{7}{12}\frac{7}{12} = -\frac{1}{144}.$$

Similarly,

$$E(X^2) = \int_0^1 \int_0^1 x^2(x+y)\, dx\, dy = \frac{5}{12},$$

$$\sigma_X = \sqrt{E(X^2) - [E(X)]^2} = \sqrt{\frac{5}{12} - \left(\frac{7}{12}\right)^2} = \frac{\sqrt{11}}{12}.$$

Again by symmetry, $\sigma_Y = \sqrt{11}/12$. Thus

$$\rho(X, Y) = \frac{\text{Cov}(X, Y)}{\sigma_X \sigma_Y} = \frac{-1/144}{\sqrt{11}/12 \cdot \sqrt{11}/12} = -\frac{1}{11} \neq \pm 1. \quad \blacklozenge$$

The following example shows that even if X and Y are dependent through a nonlinear relationship such as $Y = X^2$, still, statistically, there might be a strong linear association between X and Y. That is, $\rho(X, Y)$ might be very close to 1 or -1, indicating that the points (x, y) are tightly clustered around a line.

Example 11.7   Let X be a random number from the interval $(0, 1)$, and $Y = X^2$. The probability density function of X is

$$f(x) = \begin{cases} 1 & \text{if } 0 < x < 1 \\ 0 & \text{elsewhere,} \end{cases}$$

and for $n \geq 1$,

$$E(X^n) = \int_0^1 x^n\, dx = \frac{1}{n+1}.$$

Thus $E(X) = 1/2$, $E(Y) = E(X^2) = 1/3$,

$$\sigma_X^2 = \text{Var}(X) = E(X^2) - [E(X)]^2 = \frac{1}{3} - \frac{1}{4} = \frac{1}{12},$$

$$\sigma_Y^2 = \text{Var}(Y) = E(X^4) - [E(X^2)]^2 = \frac{1}{5} - \left(\frac{1}{3}\right)^2 = \frac{4}{45},$$

and finally,

$$\text{Cov}(X, Y) = E(X^3) - E(X)E(X^2) = \frac{1}{4} - \frac{1}{2}\frac{1}{3} = \frac{1}{12}.$$

Therefore,

$$\rho(X, Y) = \frac{\text{Cov}(X, Y)}{\sigma_X \sigma_Y} = \frac{1/12}{1/2\sqrt{3} \cdot 2/3\sqrt{5}} = \frac{\sqrt{15}}{4} = 0.968. \quad \blacklozenge$$

EXERCISES

A

1. Let X and Y be jointly distributed with $\rho(X, Y) = 1/2$, $\sigma_X = 2$, $\sigma_Y = 3$. Find $\text{Var}(2X - 4Y + 3)$.

2. Let the joint probability density function of X and Y be given by

$$f(x, y) = \begin{cases} \sin x \sin y & \text{if } 0 \le x \le \pi/2, \quad 0 \le y \le \pi/2, \\ 0 & \text{otherwise.} \end{cases}$$

Calculate the correlation coefficient of X and Y.

3. A stick of length 1 is broken into two pieces at a random point. Find the correlation coefficient and the covariance of these pieces.

4. For real numbers α and β, let

$$\text{sgn}(\alpha\beta) = \begin{cases} 1 & \text{if } \alpha\beta > 0 \\ 0 & \text{if } \alpha\beta = 0 \\ -1 & \text{if } \alpha\beta < 0. \end{cases}$$

Prove that for random variables X and Y,

$$\rho(\alpha_1 X + \alpha_2, \beta_1 Y + \beta_2) = \rho(X, Y) \, \text{sgn}(\alpha_1 \beta_1).$$

5. Is it possible that for some random variables X and Y, $\rho(X, Y) = 3$, $\sigma_X = 2$, and $\sigma_Y = 3$?

6. Prove that if $\text{Cov}(X, Y) = 0$, then

$$\rho(X + Y, X - Y) = \frac{\text{Var}(X) - \text{Var}(Y)}{\text{Var}(X) + \text{Var}(Y)}.$$

B

7. Show that if the joint probability density function of X and Y is

$$f(x, y) = \begin{cases} \dfrac{1}{2} \sin(x + y) & \text{if } 0 \le x \le \dfrac{\pi}{2}, \quad 0 \le y \le \dfrac{\pi}{2}, \\ 0 & \text{elsewhere}, \end{cases}$$

then there exists no linear relation between X and Y.

8. Let the joint probability function of $X_1, X_2, \ldots, X_r$ be multinomial with parameters n and $p_1, p_2, \ldots, p_r$ $(p_1 + p_2 + \cdots + p_r = 1)$. Find $\rho(X_i, X_j)$, $1 \le i \ne j \le r$.

HINT: Note that by the remark following Example 8.27, X_i and X_j are binomial random variables and the joint marginal probability function of X_i and X_j is multinomial. To find $E(X_i X_j)$, calculate

$$M(t_1, t_2) = E\left(e^{t_1 X_i + t_2 X_j}\right)$$

for all values of t_1 and t_2 and note that

$$E(X_i X_j) = \frac{\partial^2 M(0, 0)}{\partial t_1 \, \partial t_2}.$$

The function $M(t_1, t_2)$ is called the *moment-generating function of the joint distribution of* X_i *and* X_j.

9. Let X and Y be two randomly selected numbers from the set of positive integers $\{1, 2, 3, \ldots, n\}$. Prove that $\rho(X, Y) = 1$ if and only if $Y = X$ with probability 1.

HINT: First prove that $E[(X - Y)^2] = 0$.

11.3 Bivariate Normal Distribution

Let $f(x, y)$ be the joint probability density function of continuous random variables X and Y; f is called a *bivariate normal probability density function* if

(a) The distribution function of X is normal. That is, the marginal probability density function of X is

$$f_X(x) = \frac{1}{\sigma_X \sqrt{2\pi}} \exp\left[-\frac{(x - \mu_X)^2}{2\sigma_X^2}\right], \qquad -\infty < x < \infty. \quad (11.8)$$

(b) The conditional distribution of Y, given that $X = x$, is normal for each $x \in (-\infty, \infty)$. That is, $f_{Y|X}(y|x)$ has a normal density for each $x \in \mathbf{R}$.

(c) The conditional expectation of Y, given that $X = x$, $E(Y \mid X = x)$, is a linear function of x. That is, $E(Y \mid X = x) = a + bx$ for some $a, b \in \mathbf{R}$.

(d) The conditional variance of Y, given that $X = x$, is constant. That is, $\sigma_{Y|X=x}^2$ is independent of the value of x.

As an example, let X be the height of a man and let Y be the height of his daughter. It is reasonable to assume, and is statistically verified, that the joint probability density function of X and Y satisfies (a) to (d) and hence is bivariate normal. As another example, let X and Y be grade-point averages of a student in his or her freshman and senior years, respectively. Then the joint probability density function of X and Y is bivariate normal.

We now prove that if $f(x, y)$ satisfies (a) to (d), it must be of the following form:

$$f(x, y) = \frac{1}{2\pi \sigma_X \sigma_Y \sqrt{1 - \rho^2}} \exp\left[-\frac{1}{2(1 - \rho^2)} Q(x, y)\right], \quad (11.9)$$

where ρ is the correlation coefficient of X and Y and

$$Q(x, y) = \left(\frac{x - \mu_X}{\sigma_X}\right)^2 - 2\rho \frac{x - \mu_X}{\sigma_X} \frac{y - \mu_Y}{\sigma_Y} + \left(\frac{y - \mu_Y}{\sigma_Y}\right)^2.$$

Figure 11.1 demonstrates the graph of $f(x, y)$ in the case where $\rho = 0$, $\mu_X = \mu_Y = 0$, and $\sigma_X = \sigma_Y = 1$. To prove (11.9), note that from Lemmas 11.2 and 11.3 below, we have that for each $x \in \mathbf{R}$ the expected value and the variance of the normal density function $f_{Y|X}(y|x)$ are $\mu_Y + \rho(\sigma_Y/\sigma_X)(x - \mu_X)$ and $(1 - \rho^2)\sigma_Y^2$, respectively. Therefore, for every real x,

$$f_{Y|X}(y|x)$$

$$= \frac{1}{\sigma_Y \sqrt{2\pi} \sqrt{1 - \rho^2}} \exp\left\{-\frac{[y - \mu_Y - \rho(\sigma_Y/\sigma_X)(x - \mu_X)]^2}{2\sigma_Y^2(1 - \rho^2)}\right\}, \quad (11.10)$$

$-\infty < y < \infty$. Now $f(x, y) = f_{Y|X}(y|x) f_X(x)$, and $f_X(x)$ has the normal density given by (11.8). Multiplying (11.8) by (11.10), we obtain (11.9). So

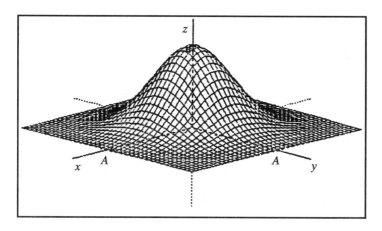

Figure 11.1   Bivariate normal probability density function.

what remains to be shown is the proofs of the following general lemmas, which are valid even if the joint density function of X and Y is not bivariate normal.

LEMMA 11.2   *Let X and Y be two random variables with probability density function $f(x, y)$. If $E(Y \mid X = x)$ is a linear function of x, that is, if $E(Y \mid X = x) = a + bx$ for some $a, b \in \mathbf{R}$, then*

$$E(Y \mid X = x) = \mu_Y + \rho \frac{\sigma_Y}{\sigma_X}(x - \mu_X).$$

PROOF: By definition,

$$E(Y \mid X = x) = \int_{-\infty}^{\infty} y \frac{f(x, y)}{f_X(x)} \, dy = a + bx.$$

Therefore,

$$\int_{-\infty}^{\infty} y f(x, y) \, dy = a f_X(x) + b x f_X(x). \tag{11.11}$$

This gives

$$\int_{-\infty}^{\infty} \int_{-\infty}^{\infty} y f(x, y) \, dy \, dx = a \int_{-\infty}^{\infty} f_X(x) \, dx + b \int_{-\infty}^{\infty} x f_X(x) \, dx, \tag{11.12}$$

which is equivalent to

$$\mu_Y = a + b\mu_X. \tag{11.13}$$

Now multiplying (11.11) by x and integrating both sides, we obtain

$$\int_{-\infty}^{\infty} \int_{-\infty}^{\infty} xy f(x, y)\, dy\, dx = a \int_{-\infty}^{\infty} x f_X(x)\, dx + b \int_{-\infty}^{\infty} x^2 f_X(x)\, dx,$$

which is equivalent to

$$E(XY) = a\mu_X + bE(X^2). \tag{11.14}$$

Solving (11.13) and (11.14) for a and b, we get

$$b = \frac{E(XY) - \mu_X \mu_Y}{E(X^2) - \mu_X^2} = \frac{\mathrm{Cov}(X, Y)}{\sigma_X^2} = \frac{\rho \sigma_X \sigma_Y}{\sigma_X^2} = \frac{\rho \sigma_Y}{\sigma_X}$$

and

$$a = \mu_Y - \frac{\rho \sigma_Y}{\sigma_X} \mu_X.$$

Therefore,

$$E(Y \mid X = x) = a + bX = \mu_Y + \rho \frac{\sigma_Y}{\sigma_X}(x - \mu_X). \quad \blacklozenge$$

LEMMA 11.3   *Let $f(x, y)$ be the joint probability density function of continuous random variables X and Y. If $E(Y \mid X = x)$ is a linear function of x and $\sigma_{Y|X=x}^2$ is constant, then*

$$\sigma_{Y|X=x}^2 = (1 - \rho^2)\sigma_Y^2.$$

PROOF: We have that

$$\sigma_{Y|X=x}^2 = \int_{-\infty}^{\infty} [y - E(Y \mid X = x)]^2 \, f_{Y|X}(y|x)\, dy.$$

Multiplying both sides of this equation by $f_X(x)$ and integrating on x gives

$$\sigma_{Y|X=x}^2 \int_{-\infty}^{\infty} f_X(x)\, dx$$

$$= \int_{-\infty}^{\infty} \int_{-\infty}^{\infty} [y - E(Y \mid X = x)]^2 \, f_{Y|X}(y|x) f_X(x)\, dy\, dx.$$

Now since $f(x, y) = f_{Y|X}(y|x) f_X(x)$, and

$$E(Y \mid X = x) = \mu_Y + \rho \frac{\sigma_Y}{\sigma_X}(x - \mu_X),$$

we get

$$\sigma^2_{Y|X=x} = \int_{-\infty}^{\infty} \int_{-\infty}^{\infty} \left[y - \mu_Y - \rho \frac{\sigma_Y}{\sigma_X}(x - \mu_X) \right]^2 f(x, y)\, dx\, dy,$$

or equivalently,

$$\sigma^2_{Y|X=x} = E \left[(Y - \mu_Y) - \rho \frac{\sigma_Y}{\sigma_X}(X - \mu_X) \right]^2.$$

Therefore,

$$\sigma^2_{Y|X=x} = E \left[(Y - \mu_Y)^2 \right] - 2\rho \frac{\sigma_Y}{\sigma_X} E \left[(Y - \mu_Y)(X - \mu_X) \right]$$

$$+ \rho^2 \frac{\sigma_Y^2}{\sigma_X^2} E \left[(X - \mu_X)^2 \right] = \sigma_Y^2 - 2\rho \frac{\sigma_Y}{\sigma_X} \rho \sigma_X \sigma_Y + \rho^2 \frac{\sigma_Y^2}{\sigma_X^2} \sigma_X^2$$

$$= (1 - \rho^2)\sigma_Y^2. \quad \blacklozenge$$

We know that if X and Y are independent random variables, their correlation coefficient is 0. We also know that, in general, the converse of this fact is not true. However, if X and Y have a bivariate normal distribution, the converse is also true. That is, $\rho = 0$ implies that X and Y are independent. This can be deduced from (11.9). For $\rho = 0$,

$$f(x, y) = \frac{1}{2\pi \sigma_X \sigma_Y} \exp \left[-\frac{(x - \mu_X)^2}{2\sigma_X^2} - \frac{(y - \mu_Y)^2}{2\sigma_Y^2} \right]$$

$$= \frac{1}{\sigma_X \sqrt{2\pi}} \exp \left[-\frac{(x - \mu_X)^2}{2\sigma_X^2} \right] \frac{1}{\sigma_Y \sqrt{2\pi}} \exp \left[-\frac{(y - \mu_Y)^2}{2\sigma_Y^2} \right]$$

$$= f_X(x) f_Y(y),$$

showing that X and Y are independent.

Example 11.8   At a certain university, the joint probability density function of X and Y, the grade-point averages of a student in his or her freshman and senior years, respectively, is bivariate normal. From the grades of past years it is known that $\mu_X = 3$, $\mu_Y = 2.5$, $\sigma_X = 0.5$, $\sigma_Y = 0.4$, and $\rho = 0.4$. Find the probability that a student with grade-point average 3.5 in his or her freshman year will earn a grade-point average of at least 3.2 in his or her senior year.

SOLUTION: The conditional probability density function of Y, given that $X = 3.5$, is normal with mean $2.5 + (0.4)(0.4/0.5)(3.5 - 3) = 2.66$, and standard deviation $(0.4)\sqrt{1 - (0.4)^2} = 0.37$. Therefore, the desired probability is calculated as follows:

$$P(Y \geq 3.2 \mid X = 3.5) = P\left(\frac{Y - 2.66}{0.37} \geq \frac{3.2 - 2.66}{0.37}\middle| X = 3.5\right)$$

$$= P(Z \geq 1.46 \mid X = 3.5) = 1 - \Phi(1.46)$$

$$\approx 1 - 0.9279 = 0.0721,$$

where Z is a standard normal random variable.   ◆

EXERCISES

1. Let X be the height of a man and let Y be the height of his daughter (both in inches). Suppose that the joint probability density function of X and Y is bivariate normal with the following parameters: $\mu_X = 71$, $\mu_Y = 60$, $\sigma_X = 3$, $\sigma_Y = 2.7$, and $\rho = 0.45$. Find the probability that the height of the daughter of a man who is 70 inches tall is at least 59 inches.

2. The joint probability density function of X and Y is bivariate normal with $\sigma_X = \sigma_Y = 9$, $\mu_X = \mu_Y = 0$, and $\rho = 0$. Find (a) $P(X \leq 6, Y \leq 12)$ and (b) $P(X^2 + Y^2 \leq 36)$.

3. Let the joint probability density function of X and Y be bivariate normal. For what values of α is the variance of $\alpha X + Y$ minimum?

4. Let $f(x, y)$ be a joint bivariate normal probability density function. Determine the point at which the maximum value of f is obtained.

5. Let the joint probability density function of two random variables X and Y be given by

$$f(x, y) = \begin{cases} 2 & \text{if } 0 < y < x, 0 < x < 1 \\ 0 & \text{elsewhere.} \end{cases}$$

Find $E(X|Y = y)$, $E(Y|X = x)$, and $\rho(X, Y)$.
HINT: To find ρ, use Lemma 11.2.

6. Let Z and W be independent standard normal random variables. Let X and Y be defined by

$$X = \sigma_1 Z + \mu_1,$$
$$Y = \sigma_2[\rho Z + \sqrt{1 - \rho^2} W] + \mu_2,$$

where $\sigma_1, \sigma_2 > 0$, $-\infty < \mu_1, \mu_2 < \infty$, and $-1 < \rho < 1$. Show that the joint probability density function of X and Y is bivariate normal and $\sigma_X = \sigma_1$, $\sigma_Y = \sigma_2$, $\mu_X = \mu_1$, $\mu_Y = \mu_2$, and $\rho(X, Y) = \rho$.
NOTE: By this exercise, if the joint probability density function of X and Y is bivariate normal, X and Y can be written as sums of independent standard normal random variables.

7. Let the joint probability density function of X and Y be bivariate normal. Prove that any linear combination of X and Y, $\alpha X + \beta Y$, is a normal random variable.
HINT: Use Theorem 9.6 and the result of Exercise 6.

8. For Exercise 1, using the result of Exercise 7, find the probability that a man is at least 8 inches taller than his daughter.

9. Let the joint probability density function of random variables X and Y be bivariate normal. Show that if $\sigma_X = \sigma_Y$, then $X + Y$ and $X - Y$ are independent random variables.
HINT: Show that the joint probability density function of $X + Y$ and $X - Y$ is bivariate normal with correlation coefficient 0.

11.4 Conditioning on Random Variables

An important application of conditional expectations is that ordinary expectations and probabilities can be calculated by conditioning on appropriate random variables. To explain this procedure, first we state a definition.

DEFINITION *Let X and Y be two random variables. By $E(X|Y)$ we mean a function of Y that is defined to be $E(X|Y = y)$ when $Y = y$.*

Recall that a function of a random variable Y, say $h(Y)$, is defined to be $h(a)$ at all sample points at which $Y = a$. For example, if $Z = \log Y$, then at a sample point ω where $Y(\omega) = a$, we have that $Z(\omega) = \log a$. In this definition $E(X|Y)$ is a function of Y, which at a sample point ω is defined to be $E(X \mid Y = y)$, where $y = Y(\omega)$. Since $E(X|Y)$ is defined only when $p_Y(y) > 0$, it is defined at all sample points ω where $p_Y(y) > 0$,

$y = Y(\omega)$. $E(X|Y)$, being a function of Y, is a random variable. Its expectation, whenever finite, is equal to the expectation of X. This extremely important property sometimes enables us to calculate expectations which are otherwise, if possible at all, very difficult to find.

THEOREM 11.3   *Let X and Y be two random variables. Then*

$$E[E(X \mid Y)] = E(X).$$

PROOF: We will prove the theorem in the case where X and Y are continuous random variables. In the discrete case the proof is similar.

$$
\begin{aligned}
E[E(X|Y)] &= \int_{-\infty}^{\infty} E(X|Y = y) f_Y(y)\, dy \\
&= \int_{-\infty}^{\infty} \left(\int_{-\infty}^{\infty} x f_{X|Y}(x|y)\, dx \right) f_Y(y)\, dy \\
&= \int_{-\infty}^{\infty} x \left(\int_{-\infty}^{\infty} f_{X|Y}(x|y) f_Y(y)\, dy \right) dx \\
&= \int_{-\infty}^{\infty} x \left(\int_{-\infty}^{\infty} \frac{f(x, y)}{f_Y(y)} f_Y(y)\, dy \right) dx \\
&= \int_{-\infty}^{\infty} x \left(\int_{-\infty}^{\infty} f(x, y)\, dy \right) dx \\
&= \int_{-\infty}^{\infty} x f_X(x)\, dx = E(X). \quad \blacklozenge
\end{aligned}
$$

Example 11.9   Suppose that $N(t)$, the number of people who pass by a museum at or prior to t, is a Poisson process having rate λ. If a person passing by enters the museum with probability p, what is the expected number of people who enter the museum at or prior to t?

SOLUTION: Let $M(t)$ denote the number of people who enter the museum at or prior to t; then

$$E[M(t)] = E[E(M(t)|N(t))] = \sum_{n=0}^{\infty} E[M(t) \mid N(t) = n] P\{N(t) = n\}.$$

Given that $N(t) = n$, the number of people who enter the museum at or prior to t, is a binomial random variable with parameters n and p. Thus $E[M(t)|N(t) = n] = np$. Therefore,

$$E[M(t)] = \sum_{n=1}^{\infty} np \frac{e^{-\lambda t}(\lambda t)^n}{n!} = pe^{-\lambda t}\lambda t \sum_{n=1}^{\infty} \frac{(\lambda t)^{n-1}}{(n-1)!}$$

$$= pe^{-\lambda t}\lambda t e^{\lambda t} = p\lambda t. \quad \blacklozenge$$

Example 11.10   Let X and Y be continuous random variables with joint probability density function

$$f(x, y) = \begin{cases} \dfrac{3}{2}(x^2 + y^2) & \text{if } 0 < x < 1, \quad 0 < y < 1 \\ 0 & \text{otherwise.} \end{cases}$$

Find $E(X|Y)$.

SOLUTION: $E(X|Y)$ is a random variable that is defined to be $E(X|Y = y)$ when $Y = y$. First we calculate $E(X \mid Y = y)$:

$$E(X \mid Y = y) = \int_0^1 x f_{X|Y}(x|y) \, dx = \int_0^1 x \frac{f(x, y)}{f_Y(y)} \, dx,$$

where

$$f_Y(y) = \int_0^1 \frac{3}{2}(x^2 + y^2) \, dx = \frac{3}{2}y^2 + \frac{1}{2}.$$

Hence

$$E(X \mid Y = y) = \int_0^1 x \frac{(3/2)(x^2 + y^2)}{(3/2)y^2 + 1/2} \, dx = \int_0^1 x \frac{3(x^2 + y^2)}{3y^2 + 1} \, dx$$

$$= \frac{3}{3y^2 + 1} \int_0^1 (x^3 + xy^2) \, dx = \frac{3(2y^2 + 1)}{4(3y^2 + 1)}.$$

Thus, if $Y = y$, then $E(X \mid Y = y) = [3(2y^2 + 1)]/[4(3y^2 + 1)]$. Now since the random variable $E(X|Y)$ coincides with $E(X|Y = y)$ if $Y = y$, we have

$$E(X|Y) = \frac{3(2Y^2 + 1)}{4(3Y^2 + 1)}. \quad \blacklozenge$$

Example 11.11   What is the expected number of random digits that should be generated to obtain three consecutive zeros?

SOLUTION: Let X be the number of random digits to be generated until three consecutive zeros are obtained. Let Y be the number of random digits to be generated until the first nonzero digit is obtained. Then

$$E(X) = E[E(X|Y)] = \sum_{i=1}^{\infty} E(X \mid Y = i)P(Y = i)$$

$$= \sum_{i=1}^{3} E(X \mid Y = i)P(Y = i) + \sum_{i=4}^{\infty} E(X \mid Y = i)P(Y = i)$$

$$= \sum_{i=1}^{3} [i + E(X)] \left(\frac{1}{10}\right)^{i-1} \left(\frac{9}{10}\right) + \sum_{i=4}^{\infty} 3 \left(\frac{1}{10}\right)^{i-1} \left(\frac{9}{10}\right),$$

which gives

$$E(X) = 1.107 + 0.999E(X) + 0.003.$$

Solving this for $E(X)$, we find that $E(X) = 1110$. ◆

We now explain a procedure for calculation of probabilities by conditioning on random variables. Let B be an event associated with an experiment and X be a discrete random variable with possible set of values A. Let

$$Y = \begin{cases} 1 & \text{if } B \text{ occurs} \\ 0 & \text{if } B \text{ does not occur.} \end{cases}$$

Then

$$E(Y) = E[E(Y|X)]. \tag{11.15}$$

But

$$E(Y) = 1 \cdot P(B) + 0 \cdot P(B^c) = P(B) \tag{11.16}$$

and

$$E[E(Y|X)] = \sum_{x \in A} E(Y \mid X = x)P(X = x) = \sum_{x \in A} P(B \mid X = x)P(X = x), \tag{11.17}$$

where the last equality follows since

$$E(Y \mid X = x) = 1 \cdot P(Y = 1 \mid X = x) + 0 \cdot P(Y = 0 \mid X = x)$$

$$= \frac{P(Y = 1, X = x)}{P(X = x)} = \frac{P(B \text{ and } X = x)}{P(X = x)} = P(B \mid X = x).$$

Hence by comparing (11.15), (11.16), and (11.17), the following theorem is obtained.

THEOREM 11.4   *Let B be an arbitrary event and X be a discrete random variable with possible set of values A; then*

$$P(B) = \sum_{x \in A} P(B \mid X = x) P(X = x). \tag{11.18}$$

If X is a continuous random variable, the relation analogous to (11.18) is

$$P(E) = \int_{-\infty}^{\infty} P(E \mid X = x) f(x) \, dx,$$

where f is the probability density function of X.

Example 11.12   Let $N(t)$ be the number of earthquakes that occur at or prior to time t worldwide. Suppose that $\{N(t) : t \geq 0\}$ is a Poisson process and the probability that the magnitude of an earthquake on the Richter scale is 5 or more is p. Find the probability of k earthquakes of such magnitudes at or prior to t worldwide.

SOLUTION: Let $X(t)$ be the number of earthquakes of magnitude 5 or more on the Richter scale at or prior to t worldwide. By Theorem 11.4,

$$P\{X(t) = k\} = \sum_{n=0}^{\infty} P\{X(t) = k \mid N(t) = n\} P\{N(t) = n\}.$$

Now clearly, $P\{X(t) = k \mid N(t) = n\} = 0$ if $n < k$. If $n > k$, the conditional probability function of $X(t)$ given that $N(t) = n$ is binomial with parameters n and p. Thus

$$
\begin{aligned}
P\{X(t) = k\} &= \sum_{n=k}^{\infty} \binom{n}{k} p^k (1-p)^{n-k} \frac{e^{-\lambda t}(\lambda t)^n}{n!} \\
&= \sum_{n=k}^{\infty} \frac{n!}{k!\,(n-k)!} p^k (1-p)^{n-k} \frac{e^{-\lambda t}(\lambda t)^k (\lambda t)^{n-k}}{n!} \\
&= \frac{e^{-\lambda t}(\lambda t p)^k}{k!} \sum_{n=k}^{\infty} \frac{1}{(n-k)!} [\lambda t (1-p)]^{n-k} \\
&= \frac{e^{-\lambda t}(\lambda t p)^k}{k!} \sum_{j=0}^{\infty} \frac{1}{(j)!} [\lambda t (1-p)]^j
\end{aligned}
$$

$$= \frac{e^{-\lambda t}(\lambda tp)^k}{k!} e^{\lambda t(1-p)}$$

$$= \frac{e^{-\lambda tp}(\lambda tp)^k}{k!}.$$

Therefore, $\{X(t) : t \geq 0\}$ is itself a Poisson process with mean λp. ◆

Example 11.13   The time between consecutive earthquakes in San Francisco and the time between consecutive earthquakes in Los Angeles are independent and exponentially distributed with means $1/\lambda_1$ and $1/\lambda_2$, respectively. What is the probability that the next earthquake occurs in Los Angeles?

SOLUTION: Let X and Y denote the times between now and the next earthquakes in San Francisco and Los Angeles, respectively. Because of the memoryless property of exponential distribution, X and Y are exponentially distributed with means $1/\lambda_1$ and $1/\lambda_2$, respectively. To calculate $P(X > Y)$, the desired probability, we will condition on Y:

$$P(X > Y) = \int_0^\infty P(X > Y \mid Y = y)\lambda_2 e^{-\lambda_2 y} \, dy$$

$$= \int_0^\infty P(X > y)\lambda_2 e^{-\lambda_2 y} \, dy = \int_0^\infty e^{-\lambda_1 y}\lambda_2 e^{-\lambda_2 y} \, dy$$

$$= \lambda_2 \int_0^\infty e^{-(\lambda_1 + \lambda_2)y} \, dy = \frac{\lambda_2}{\lambda_1 + \lambda_2},$$

where $P(X > y)$ is calculated from

$$P(X > y) = 1 - P(X \leq y) = 1 - \left(1 - e^{-\lambda_1 y}\right) = e^{-\lambda_1 y}. \quad ◆$$

EXERCISES

1. A fair coin is tossed until two tails occur successively. Find the expected number of the tosses required.
   HINT: Let

$$X = \begin{cases} 1 & \text{if the first toss results in tails} \\ 0 & \text{if the first toss results in heads,} \end{cases}$$

   and condition on X.

2. The orders received for grain by a farmer add up to X tons, where X is a continuous random variable uniformly distributed over $(4, 7)$. Every

ton of grain sold brings a profit of a, and every ton that is not sold is destroyed at a loss of $a/3$. How many tons of grain should the farmer produce to maximize his expected profit?

HINT: Let $Y(t)$ be the profit if the farmer produces t tons of grain. Then

$$E[Y(t)] = E\left[aX - \frac{a}{3}(t - X)\right]P(X < t) + E(at)P(X \geq t).$$

3. Lynn has b batteries in a box of which d are dead. She tests them randomly and one by one. Every time that a good battery is drawn, she will return it to the box; every time that a dead battery is drawn, she will replace it by a good one.

   (a) Determine the expected value of the number of good batteries in the box after n of them are checked.

   (b) Determine the probability that on the nth draw Lynn draws a good battery.

   HINT: Let X_n be the number of good batteries in the box after n of them are checked. Show that

   $$E(X_n \mid X_{n-1}) = 1 + \left(1 - \frac{1}{b}\right)X_{n-1}.$$

   Then by computing the expected value of this random variable, find a recursive relation between $E(X_n)$ and $E(X_{n-1})$. Use this relation and induction to prove that

   $$E(X_n) = b - d\left(1 - \frac{1}{b}\right)^n.$$

   Note that n should approach ∞ to get $E(X_n) = b$. For (b), let E_n be the event that on the nth draw she gets a good battery. By conditioning on X_n prove that $P(E_n) = E(X_{n-1})/b$.

4. A typist, on average, makes three typing errors in every two pages. If pages with more than two errors should be retyped, on the average how many pages must she type to prepare a report of 200 pages? Assume that the number of errors in a page is a Poisson random variable. Note that some of the retyped pages should be retyped, and so on.

   HINT: Find p, the probability that a page should be retyped. Let X_n be the number of pages that should be typed at least n times. Show that $E(X_1) = 200p$, $E(X_2) = 200p^2$, ..., $E(X_n) = 200p^n$. The desired quantity is $E(\sum_{i=1}^{\infty} X_i)$, which can be calculated using relation (9.4).

5. From an ordinary deck of 52 cards, cards are drawn at random, one by one, and without replacement until a heart is drawn. What is the expected value of the number of cards drawn?

   HINT: Consider a deck of cards with 13 hearts and $39 - n$ nonheart cards. Let X_n be the number of cards to be drawn *before* the first heart is drawn. Let

   $$Y = \begin{cases} 1 & \text{if the first card drawn is a heart} \\ 0 & \text{otherwise.} \end{cases}$$

   By conditioning on Y, find a recursive relation between $E(X_n)$ and $E(X_{n+1})$. Use $E(X_{39}) = 0$ to show that $E(X_i) = (39 - i)/14$. The answer is $1 + E(X_0)$.

6. Suppose that X and Y are independent random variables with probability density functions f and g, respectively. Use conditioning technique to calculate $P(X < Y)$.

7. Prove that for a Poisson random variable N, if the parameter λ is not fixed and is itself an exponential random variable with parameter 1, then

   $$P(N = i) = \left(\frac{1}{2}\right)^{i+1}.$$

Review Problems

1. Determine the expected number of tosses of a die required to obtain four consecutive 6's.

2. Let the joint probability density function of X, Y, and Z be given by

   $$f(x, y, z) = \begin{cases} 8xyz & \text{if } 0 < x < 1, \quad 0 < y < 1, \quad 0 < z < 1 \\ 0 & \text{otherwise.} \end{cases}$$

   Find $\rho(X, Y)$, $\rho(X, Z)$, and $\rho(Y, Z)$.

3. Let X and Y be jointly distributed with $\rho(X, Y) = 2/3$, $\sigma_X = 1$, $\text{Var}(Y) = 9$. Find $\text{Var}(3X - 5Y + 7)$.

4. Two green and two blue dice are rolled. If X and Y are the numbers of sixes on the green and on the blue dice, respectively, calculate the correlation coefficient of $|X - Y|$ and $X + Y$.

5. A random point (X, Y) is selected from the rectangle $[0, \pi/2] \times [0, 1]$. What is the probability that it lies below the curve $y = \sin x$?

6. Let the joint probability density function of X and Y be given by

$$f(x, y) = \begin{cases} e^{-x} & \text{if } 0 < y < x < \infty \\ 0 & \text{elsewhere.} \end{cases}$$

   (a) Find the marginal probability density functions of X and Y.
   (b) Determine the correlation coefficient of X and Y.

7. In terms of the means, variances, and the covariance of the random variables X and Y, find α and β for which $E(Y - \alpha - \beta X)^2$ is minimum. This is the *method of least squares*; it fits the "best" line $y = \alpha + \beta x$ to the distribution of Y.

8. Let the joint probability density function of X and Y be given by

$$f(x, y) = \begin{cases} ye^{-y(1+x)} & \text{if } x > 0, \quad y > 0 \\ 0 & \text{otherwise.} \end{cases}$$

   (a) Show that $E(X)$ does not exist.
   (b) Find $E(X|Y)$.

9. An ordinary deck of 52 cards is divided randomly into 26 pairs of two each. Using Chebyshev's inequality, find an upper bound for the probability that at most 10 pairs consist of a black and a red card. HINT: For $i = 1, 2, \ldots, 26$, let $X_i = 1$, if the ith red card is paired with a black card, and $X_i = 0$, otherwise. Find an upper bound for

$$P\left(\sum_{i=1}^{26} X_i \le 10\right).$$

10. Bus A arrives at a station at a random time between 10:00 A.M. and 10:30 A.M. tomorrow. Bus B arrives at the same station at a random time between 10:00 A.M. and the arrival time of bus A. Find the expected value of the arrival time of bus B.

11. Let $\{X_1, X_2, X_3, \ldots\}$ be a sequence of independent and identically distributed exponential random variables with parameter λ. Let N be a geometric random variable with parameter p independent of $\{X_1, X_2, X_3, \ldots\}$. Find the distribution function of $\sum_{i=1}^{N} X_i$.

Chapter 12

SIMULATION

12.1 Introduction

To solve a scientific or an industrial problem, usually a mathematical analysis and/or a simulation is used. To perform a simulation, we repeat an experiment a large number of times and measure the quantities in which we are interested. For example, to estimate the probability of at least one 6 in rolls of four dice, we may do a large number of experiments rolling four dice and calculate the fraction of times that at least one 6 is obtained. Similarly, to estimate the fraction of time that in a certain bank all the tellers are busy, we may measure the lengths of such time intervals over a long period X, add them, and then divide by X. Clearly, in simulations, the key to reliable answers is *to perform the experiment a large number of times or over a long period of time, whichever is applicable.* Since manually this is almost impossible, simulations are carried out by computers. Only computers can handle millions of operations in short periods of time.

To simulate a problem that involves random phenomena, generating random numbers from the interval $(0, 1)$ is essential. In almost every simulation of a probabilistic model, the selection of random points from $(0, 1)$ comes up. For example, to simulate the experiment of tossing a fair coin, we draw a random number from $(0, 1)$. If it is in $(0, 1/2)$, we say that the outcome is heads, and if it is in $[1/2, 1)$, we say that it is tails. Similarly, in the simulation of die tossing, the outcomes 1, 2, 3,

428

4, 5, and 6, respectively, correspond to the events that the random point from $(0, 1)$ is in $(0, 1/6)$, $[1/6, 1/3)$, $[1/3, 1/2)$, $[1/2, 2/3)$, $[2/3, 5/6)$, and $[5/6, 1)$. Most of the computer languages are equipped with subroutines that generate random numbers from $(0, 1)$. In Turbo Pascal, from Borland International (Turbo Pascal, Borland International, Scotts Valley, California), the command *random* selects a random number from $(0, 1)$. Therefore, for example, $a + (b - a) * random$ chooses a random number from the interval (a, b) and $m + trunc((n + 1) * random)$ picks up an integer from $\{m, m + 1, m + 2, \ldots, m + n\}$ randomly, where in Pascal the command $trunc(x)$ finds the greatest integer less than or equal to x. However, there are computer languages that are not equipped with a subroutine that generates random numbers and there are a few that are equipped with inefficient algorithms. In such cases one can easily write his or her own efficient subroutine to generate random numbers, say, based on the algorithm given by Stephen Park and Keith Miller in the paper "Random Number Generators: Good Ones Are Hard to Find", *Communications of ACM*, October 1988, Volume 31, Number 10.

At this point, let us emphasize that the main goal of scientists and engineers is always to solve a problem mathematically. It is a mathematical solution that is accurate, exact, and 100% reliable. Simulations cannot take the place of a rigorous mathematical solution. They are widely used, (a) to find good estimations for solutions of problems that either cannot be modeled mathematically or whose mathematical models are too difficult to solve, (b) to get a better understanding of the behavior of a complicated phenomenon, and (c) to obtain a mathematical solution by acquiring insight into the nature of the problem, its functions, and the magnitude and characteristics of its solution. Intuitively, it is clear why the results that are obtained by simulations are good. Theoretically, however, most of them are justified by the strong law of large numbers, discussed in Section 10.2.

For a simulation of each of the following examples we have presented an algorithm in English that can be translated into any programming language. Therefore, readers may use these algorithms to write and execute their own programs in their favorite computer languages.

Example 12.1 Two numbers are selected at random and without replacement from the set $\{1, 2, 3, \ldots, l\}$. Write an algorithm for a computer simulation of approximating the probability that the difference of these numbers is at least k.

SOLUTION: For a large number of times, say n, each time choose two distinct random numbers a and b from $\{1, 2, 3, \ldots, l\}$ and check to see if $|a - b| \geq k$. In all of these n experiments, let m be the number of those in which $|a - b| \geq k$. Then m/n is the desired approximation. An algorithm follows.

STEP 1:   *Set i = 1;*
STEP 2:   *Set m = 0;*
STEP 3:   *While i ≤ n, do steps 4 to 7.*
STEP 4:   *Generate a random number a from* {1, 2, 3, . . . , l}.
STEP 5:   *Generate a random number b from* {1, 2, 3, . . . , l}.
STEP 6:   *If b = a, Goto step 5.*
STEP 7:   *If* |a − b| ≥ k, *then*
                Set m = m + 1;
                Set i = i + 1;
                Goto step 3.
           else
                Set i = i + 1;
                Goto step 3.
STEP 8:   *Set p = m/n;*
STEP 9:   *Output(p).*
   STOP

Since the exact answer to this problem obtained by analytical methods is $\dfrac{(l-k)(l-k+1)}{2l(l-1)}$, any simulation result should be close to this number.   ◆

Example 12.2   An urn contains 25 white and 35 red balls. Balls are drawn from the urn successively and without replacement. Write an algorithm to approximate by simulation the probability that at some instant the number of red and white balls drawn are equal (a tie occurs).

SOLUTION: For a large number of times n, we will repeat the following experiment and count m, the number of times that at some instant a tie occurs. The desired quantity is m/n. To begin, let $m = 0$, $w = 25$, and $r = 35$. Generate a random number from $(0, 1)$. If it belongs to $\left(0, \dfrac{w}{w+r}\right]$, a white ball is selected. Thus change w to $w - 1$. Otherwise, a red ball is drawn, thus change r to $r - 1$. Repeat this procedure and check each time to see if $25 - w = 35 - r$. If at some instance these two quantities are equal, change m to $m + 1$ and start a new experiment. Otherwise, continue until either $r = 0$ or $w = 0$. An algorithm follows.

STEP 1:   *Set m = 0;*
STEP 2:   *Set i = 1;*
STEP 3:   *While i ≤ n, do steps 4 to 10.*
STEP 4:   *Set w = 25;*

STEP 5:   *Set r = 35;*

STEP 6:   *While r ≥ 0 and w ≥ 0 do steps 7 to 10.*

STEP 7:   *Generate a random number x from (0, 1).*

STEP 8:   *If x ≤ w/(w + r), then*
    Set w = w − 1;
else
    Set r = r − 1;

STEP 9:   *If 25 − w = 35 − r, then*
    Set m = m + 1;
    Set i = i + 1;
    Goto step 3.

STEP 10: *If r = 0 or w = 0, then*
    Set i = i + 1;
    Goto step 3.

STEP 11: *Set p = m/n;*

STEP 12: *Output(p);*

   STOP ◆

EXERCISES

1. A die is rolled successively until for the first time a number appears three consecutive times.   By simulation, determine the approximate probability that it takes at least 50 rolls before we accomplish this.

2. In an election, the Democratic candidate obtained 3586 and the Republican candidate obtained 2958 votes. Use simulations to find the approximate probability that the Democratic candidate was ahead during the entire process of counting the votes.
ANSWER: approximately 0.096.

3. The probability that a bank refuses to finance an applicant is 0.35. Using simulation, find the approximate probability that of 300 applicants more than 100 are refused.

4. There are five urns each containing 10 white and 15 red balls. A ball is drawn at random from the first urn and put into the second one. Then a ball is drawn at random from the second urn and put into the third one and the process is continued. By simulation, determine the approximate probability that the last ball is red.
ANSWER: 0.6.

5. In a small town of 1000 inhabitants, a person gossips to a random person, who in turn tells the story to another random person, and so on.

Using simulation, calculate the approximate probability that before the story is told 150 times, it is returned to the first person.

6. A city has n taxis numbered 1 through n. A statistician takes taxis numbered 31, 50, and 112 at three random occasions. Based on this information, determine, using simulation, which of the numbers 112 through 120 is a better estimate for n.

   HINT: For each n ($112 \le n \le 120$), repeat the following experiment a large number of times: Choose three random numbers from $\{1, 2, \ldots, n\}$ and check to see if they are 31, 50, and 112.

7. Suppose that an airplane passenger whose itinerary requires a change, of airplanes in Ankara, Turkey has a 4% chance, independently, of losing each piece of his or her luggage. Suppose that the probability of losing each piece of luggage in this way is 5% at Da Vinci airport in Rome, 5% at Kennedy airport in New York, and 4% at O'Hare airport in Chicago. Dr. May travels from Bombay to San Francisco with three suitcases. He changes airplanes in Ankara, Rome, New York, and Chicago. Using simulation, find the approximate probability that one of Dr. May's suitcases does not reach his destination with him.

12.2   Simulation of Combinatorial Problems

Suppose that $X(1), X(2), X(3), \ldots, X(n)$ are n objects numbered from 1 to n in a convenient way. For example, suppose that $X(1), X(2), \ldots, X(52)$ are the cards of an ordinary deck of 52 cards and they are assigned numbers 1 to 52 in some convenient order. One of the most common problems in simulation of combinatorial problems is to find an efficient procedure for choosing m ($m \le n$) distinct objects randomly from $X(1), X(2), X(3), \ldots, X(n)$. To solve this problem, we choose a random integer from $\{1, 2, 3, \ldots, n\}$ and call it n_1. Clearly, $X(n_1)$ is a random object from $\{X(1), X(2), \ldots, X(n)\}$. To choose a second random object distinct from $X(n_1)$, we exchange $X(n)$ and $X(n_1)$ and then choose a random object from $\{X(1), X(2), X(3), \ldots, X(n-1)\}$, as before. That is, we choose a random integer, n_2, from $\{1, 2, 3, \ldots, n-1\}$ and then exchange $X(n_2)$ and $X(n-1)$. At this point, $X(n)$ and $X(n-1)$ are two distinct random objects from the original set of objects. Continuing this procedure, after m selections the set $\{X(n), X(n-1), X(n-2), \ldots, X(n-m+1)\}$ consists of m distinct random objects from $\{X(1), X(2), \ldots, X(n)\}$. In particular, if $m = n$, the *ordered set* $\{X(n), X(n-1), X(n-2), \ldots, X(1)\}$ is a permutation of $\{X(1), X(2), \ldots, X(n)\}$. An algorithm follows.

Algorithm 12.1

STEP 1:   *Set i = 1;*
STEP 2:   *While i ≤ m, do steps 3 to 5.*
STEP 3:   *Choose a random integer k from {1, 2, 3, ..., n − i + 1}.*
STEP 4:   *Set Z = X(k);*
          Set X(k) = X(n − i + 1);
          Set X(n − i + 1) = Z;
STEP 5:   *Set i = i + 1;*
STEP 6:   *Output {X(n), X(n − 1), X(n − 2), ..., X(n − m + 1)}.*
   STOP

Example 12.3   A nurse in a hospital mixes the 10 pills of 10 patients accidentally. Suppose that she gives a pill to each patient at random from the mixed-up batch. Write an algorithm to approximate by simulation the probability that at least one person gets his or her own pill. Assume that none of the 10 pills are of the same type.

SOLUTION: Let the pill of patient i be numbered i, $1 \le i \le 10$. If $X(i)$ denotes the pill that the nurse gives to patient i, then $X(i)$ is a random integer between 1 and 10. Thus $X(1), X(2), \ldots, X(10)$ are the numbers $1, 2, 3, \ldots, 10$ in some random order, and patient i gets his or her pill if $X(i) = i$. One way to simulate this problem is that for a large n, we generate n random orders for 1 through 10, find m, the number of those in which there is at least one i with $X(i) = i$, and then divide m by n. To do so, we present the following algorithm, which puts the numbers 1 through 10 in random order n times and calculates m and $p = m/n$.

STEP 1:   *Set m = 0;*
STEP 2:   *Set i = 1;*
STEP 3:   *While i ≤ n, do steps 4 to 10.*
STEP 4:   *Generate a random permutation {X(10), X(9), ..., X(1)} of*
          {1, 2, 3, ..., 10} (see Algorithm 12.1).
STEP 5:   *Set k = 0;*
STEP 6:   *Set j = 1;*
STEP 7:   *While j ≤ 10, do step 8.*
STEP 8:   *If X(j) = j, then*
                Set j = 11;
                Set k = 1;
          else
                Set j = j + 1.

STEP 9:   *Set m = m + k;*
STEP 10:   *Set i = i + 1;*
STEP 11:   *Set p = m/n;*
STEP 12:   *Output(p).*
  STOP

A Pascal program based on this algorithm was executed for several values of *n*. The corresponding values of *p* were obtained as follows:

n	p
10	0.7000
1,000	0.6090
10,000	0.6340
100,000	0.6326

For comparison, it is worthwhile to mention that the mathematical solution of this problem, up to four decimal points, gives 0.6321 (see Example 2.21).   ♦

EXERCISES

WARNING: In some combinatorial probability problems, the desired quantity is very small. For this reason, in simulations, the number of experiments to be performed should be very large to get good approximations.

1. Suppose that 18 customers stand in a line at a box office, nine with $5 bills and nine with $10 bills. Each ticket costs $5 and the box office has no money initially. Write a simulation program to calculate an approximate value for the probability that none of the customers has to wait for change.

2. A cereal company puts exactly one of its 20 prizes in every box of its cereals at random.
   (a) Julie has bought 50 boxes of cereals from this company. Write a simulation program to calculate an approximate value of the probability that she gets all 20 prizes.
   (b) Suppose that Jim wants to have at least a 50% chance of getting all 20 prizes. Write a simulation program to calculate the approximate value of the minimum number of cereal boxes that he should buy from this company.

3. Nine students lined up in some order and had their picture taken. One year later, the same students lined up again in random order and had their picture taken. Using simulation, find an approximate probability that the second time, no student was standing next to the one next to whom he or she was standing the previous year.

4. Use simulation to find an approximate value for the probability that in a class of 50, exactly four students have the same birthday and the other 46 students all have different birthdays. Assume that the birth rates are constant throughout the year and that each year has 365 days.

5. Using simulation, find the approximate probability that at least four students of a class of 87 have the same birthday. Assume that the birth rates are constant throughout the year and that each year has 365 days. ANSWER: Approximately 0.4998.

6. Suppose that two pairs of the vertices of a regular 10-gon are selected at random and connected. Using simulations, find the approximate probability that they do not intersect.
HINT: The exact value of the desired probability is

$$\frac{\dfrac{1}{11}\dbinom{20}{10}}{\dbinom{20}{2}\dbinom{18}{2}} \approx 0.58.$$

7. In the last 10 days, everyday an executive has written seven letters and his secretary has prepared seven envelopes for the letters. The secretary has kept envelopes of each day in chronological order in a packet, but she has forgotten to insert the letters in the envelopes. Today the secretary finds out that all 70 letters and all 10 packets of envelopes are mixed up. If she selects packets of envelopes at random and inserts randomly chosen letters into the envelopes of each packet, what is the approximate probability, obtained from using simulation, that none of the letters will be addressed correctly? Note that the envelopes within a packet are left in chronological order but the packets themselves are mixed up.
ANSWER: Approximately 0.53416. For an analytic discussion of this problem, see the article by Steve Fisk in the April 1988 issue of *Mathematics Magazine*.

12.3   Simulation of Conditional Probabilities

Let S be the sample space of an experiment and A and B be two events of S with $P(B) > 0$. To find approximate values for $P(A \mid B)$ using computer simulation, either we will reduce the sample space to B and then calculate $P(AB)$ in the reduced sample space or we use the formula $P(A \mid B) = P(AB)/P(B)$. In the latter case, we perform the experiment a large number of times, find n and m, the number of times in which B and AB occur, respectively, and then divide m by n. In the former case, $P(AB)$ is estimated simply by performing the experiment in the reduced sample space n times, for a large n, finding m, the number of those in which AB occurs, and dividing m by n. We illustrate this method by Example 12.4, and the other method by Example 12.5.

Example 12.4   From an ordinary deck of 52 cards, 10 cards are drawn at random. If exactly four of them are hearts, write an algorithm to approximate by simulation the probability of at least two spades.

SOLUTION: The problem in the reduced sample space is as follows: A deck of cards consists of 9 hearts, 13 spades, 13 clubs, and 13 diamonds. If six cards are drawn at random, what is the probability that none of them is a heart and that two or more are spades? To simulate this problem, suppose that hearts are numbered 1 to 9, spades 10 to 22, clubs 23 to 35, and diamonds 36 to 48. For a large number of times, say n, each time choose six distinct random integers between 1 and 48 and check to see if there are no numbers between 1 and 9 and at least two numbers between 10 and 22. Let m be the number of those draws satisfying both conditions; then m/n is the desired approximation. For sufficiently large n, the final answer of any accurate simulation is close to 0.42. An algorithm follows:

STEP 1:   *Set $k = 1$;*

STEP 2:   *Set $m = 0$;*

STEP 3:   *While $k \le n$, do steps 4 to 13.*

STEP 4:   *Choose six distinct numbers $X(1), X(2), \ldots, X(6)$ randomly from $\{1, 2, 3, \ldots, 48\}$ (see Algorithm 12.1 of Section 12.2).*

STEP 5:   *For $j = 1$ to 6, do steps 6 and 7.*

STEP 6:   *For $l = 1$ to 9, do step 7.*

STEP 7:   *If $X(j) = l$, then*
        Set $k = k + 1$;
        Goto step 3.

STEP 8:   *Set $s = 0$;*

STEP 9:   *For $j = 1$ to 6, do steps 10 to 12.*

STEP 10: *For l = 10 to 22, do steps 11 and 12.*
STEP 11: *If X(j) = l, then*
            Set s = s + 1;
STEP 12: *If s = 2, then*
            Set m = m + 1;
            Set k = k + 1;
            Goto step 3.
STEP 13: *Set k = k + 1; Goto step 3;*
STEP 14: *Set p = m/n.*
STEP 15: *Output(p).*
    STOP

Example 12.5 **(Laplace's Law of Succession)** Suppose that u urns are numbered 0 through $u - 1$, and that the ith urn contains i red and $u - 1 - i$ white balls, $0 \le i \le u - 1$. An urn is selected at random and then its balls are removed one by one, at random, and with replacement. If the first m balls are all red, write an algorithm to find by simulation an approximate value for the probability that the $(m + 1)$st ball removed is also red.

SOLUTION: Let A be the event that the first m balls drawn are all red, and let B be the event that the $(m + 1)$st ball drawn is red. We are interested in $P(B \mid A) = P(BA)/P(A)$. To find an approximate value for $P(B \mid A)$ by simulation, for a large n, perform this experiment n times. Let a be the number of those experiments in which the first m draws are all reds, and let b be the number of those in which the first $m + 1$ draws are all reds. Then to obtain the desired approximation, divide b by a. To write an algorithm, introduce i to count the number of experiments and a and b to count, respectively, the numbers of those in which the first m and the first $m + 1$ draws are all reds. Initially, set $i = 1$, $a = 0$, and $b = 0$. Each experiment begins with choosing t, a random number from $(0, 1)$. If $t \in (0, 1/u]$, set $k = 0$, meaning that urn number 0 is selected; if $t \in (1/u, 2/u]$, set $k = 1$, meaning that urn number 1 is selected; and so on. If $k = 0$, urn 0 is selected. Since urn 0 has no red balls, and hence neither A nor AB occurs, do not change the values of a and b; simply change i to $i + 1$ and start a new experiment. If $k = u - 1$, urn number $u - 1$, which contains only red balls, is selected. In such a case both A and AB will occur; therefore, set $i = i + 1$, $a = a + 1$, and $b = b + 1$ and start a new experiment. For $k = 1, 2, 3, \ldots, u - 2$, the urn that is selected has $u - 1$ balls, of which exactly k are red. Start choosing random numbers from $(0, 1)$ to represent drawing balls from this urn. If a random number is less than $k/(u - 1)$, a red ball is drawn. Otherwise, a white ball is drawn. If all the first m draws are reds, set $a = a + 1$. In this case draw another ball, if it is red, then set $b = b + 1$ as well. If among the first m draws, a white ball is removed at any draw, then only change i to $i + 1$ and start a new experiment. If the first m draws are red but the $(m + 1)$st one is not, change i to $i + 1$ and a to $a + 1$ but keep b unchanged and start a new experiment. When $i = n$, all the experiments are complete. The approximate value of the desired probability is then equal to $p = b/a$. An algorithm follows:

STEP 1:   *Set $i = 1$;*

STEP 2:   *Set $a = 0$;*

STEP 3:   *Set $b = 0$;*

STEP 4:   *While $i \leq n$, do steps 5 to 20.*

STEP 5:   *Generate a random number t from $(0, 1)$.*

STEP 6:   *Set $r = 1$;*

STEP 7:   *While $r \leq u$, do step 8.*

STEP 8:   *If $t \leq r/u$, then*
        Set $k = r - 1$;
        Set $r = u + 1$;
    else
        Set $r = r + 1$;

STEP 9:   *If $k = 0$, then*
        Set $i = i + 1$;
        Goto step 4.

STEP 10: *If $k = u - 1$, then*
        Set $i = i + 1$;
        Set $a = a + 1$;
        Set $b = b + 1$;
        Goto step 4.

STEP 11: *Set $j = 0$;*

STEP 12: *While $j < m$, do steps 13 and 14.*

STEP 13: *Generate a random number r from $(0, 1)$.*

STEP 14: *If $r < k/(u - 1)$, then*
        Set $j = j + 1$;
    else
        Set $j = m + 1$;

STEP 15: *If $j \neq m$, then*
        Set $i = i + 1$;
        Goto step 4.

STEP 16: *If $j = m$, then do steps 17 to 19.*

STEP 17: *Set $a = a + 1$;*

STEP 18: *Generate a random number r from (0, 1).*
STEP 19: *If r < k/(u − 1), then*
$$Set\ b = b + 1;$$
STEP 20: *Set i = i + 1;*
STEP 21: *Set p = b/a.*
STEP 22: *Output(p).*
    STOP

A sample run of a Pascal program based on this algorithm gave us the following results for $n = 1000$ and $m = 5$:

u	2	10	50	100	200	500	1000	2000	5000
p	1	0.893	0.886	0.852	0.852	0.848	0.859	0.854	0.853

It can be shown that if the number of urns is large, the answer is approximately $(m + 1)/(m + 2) = 6/7 \approx 0.857$ (see Exercise 37 of Section 3.4). Hence the results of these simulations are quite good. ◆

EXERCISES

1. The cards of an ordinary deck of 52 cards are dealt among A, B, C, and D, 13 each. If A and B have a total of six hearts and five spades, using computer simulation find an approximate value for the probability that C has two of the remaining seven hearts and three of the remaining eight spades.

2. From families with five children, a *family* is selected at random and found to have a boy. Using computer simulation, find the approximate value of the probability that the family has three boys and two girls and that the middle child is a boy. Assume that in a five-child family all gender distributions have equal probabilities.

3. In Example 12.5, Laplace's law of succession, suppose that among the first m balls drawn r are red and $m - r$ are white. Using computer simulation, find the approximate probability that the $(m + 1)$st ball is red.

4. Targets A and B are placed on a wall. It is known that in every shot the probabilities of hitting A and hitting B by a missile are, respectively, 0.3 and 0.4. If in an experiment target A was not hit, use simulation to calculate an approximate value for the probability that target B was hit.

(The exact answer is 4/7. Hence the result of simulation should be close to $4/7 \approx 0.571$.)

12.4   Simulation of Random Variables

Let X be a random variable over the sample space S. To approximate $E(X)$ by simulation, we use the strong law of large numbers. That is, for a large n, we repeat the experiment over which X is defined independently n times. Each time we will find the value of X. We then add the n values obtained and divide the result by n. For example, in the experiment of rolling a die, let X be the following random variable:

$$X = \begin{cases} 1 & \text{if the outcome is 6} \\ 0 & \text{if the outcome is not 6.} \end{cases}$$

To find $E(X)$ by simulation, for a large number n, we generate n random numbers from $(0, 1)$ and find m, the number of those that are in $(0, 1/6]$. The sum of all the values of X that are obtained in these n experiments is m. Hence m/n is the desired approximation. A more complicated example is the following:

Example 12.6   A fair coin is flipped successively until for the first time four consecutive heads are obtained. Write an algorithm for a computer simulation to determine the approximate value of the average number of trials that are required.

SOLUTION: Let H stand for the outcome heads. We present an algorithm that repeats the experiment of flipping the coin until the first HHHH, n times. In the algorithm, j is the variable that counts the number of experiments. It is initially 1; each time that an experiment is finished, the value of j is updated to $j + 1$. When $j = n$, the last experiment is performed. For every experiment we begin choosing random numbers from $(0, 1)$. If an outcome is in $(0, 1/2)$, heads is obtained; else, tails is obtained. The variable i counts the number of successive heads. Initially, it is 0; every time that heads is obtained, one unit is added to i, and every time that the outcome is tails, i becomes 0. Therefore, an experiment is over when for the first time i becomes 4. The variable k counts the number of trials until the first HHHH, m is the sum of all the k's for the n experiments, and a is the average number of trials until the first HHHH. Therefore, $a = m/n$. An algorithm follows.

STEP 1:   *Set $m = 0$;*
STEP 2:   *Set $j = 0$;*

STEP 3:   *While j < n, do steps 4 to 11.*

STEP 4:   *Set k = 0;*

STEP 5:   *Set i = 0;*

STEP 6:   *While i < 4, do steps 7 to 9.*

STEP 7:   *Set k = k + 1;*

STEP 8:   *Generate a random number r from (0, 1).*

STEP 9:   *If r < 0.5, then*
           Set i = i + 1;
      else
           Set i = 0;

STEP 10: *Set m = m + k;*

STEP 11: *Set j = j + 1;*

STEP 12: *Set a = m/n;*

STEP 13: *Output(a).*

   STOP

We executed a Pascal program based on this algorithm for several values of n. The corresponding values obtained for a were as follows:

n	a	n	a
5	23.80	5,000	29.90
50	27.22	50,000	30.00
500	29.28	500,000	30.03

It can be shown that the exact answer to this problem is 30. Therefore, these simulation results for $n \geq 5000$ are quite good. ◆

To simulate a Bernoulli random variable X with parameter p, generate a random number from $(0, 1)$. If it is in $(0, p]$, let $X = 1$; otherwise, let $X = 0$. To simulate a binomial random variable Y with parameters (n, p), choose n independent random numbers from $(0, 1)$ and let Y be the number of those that lie in $(0, p]$. To simulate a geometric random variable T with parameter p, we may keep choosing random points from $(0, 1)$ until for the first time a number selected lies in $(0, p]$. The number of random points selected is a simulation of T. However, this method of simulating T is very inefficient. By the following theorem, T can be simulated most efficiently, namely, by just choosing one random point from $(0, 1)$.

THEOREM 12.1   *Let r be a random number from the interval* (0, 1) *and let*

$$T = 1 + \left[\frac{\ln r}{\ln(1 - p)}\right], \qquad 0 < p < 1,$$

where by [x] *we mean the largest integer less than or equal to x. Then T is a geometric random variable with parameter p.*

PROOF: For all $n \geq 1$,

$$P(T = n)$$

$$= P\left\{1 + \left[\frac{\ln r}{\ln(1 - p)}\right] = n\right\}$$

$$= P\left\{\left[\frac{\ln r}{\ln(1 - p)}\right] = n - 1\right\} = P\left\{n - 1 \leq \frac{\ln r}{\ln(1 - p)} < n\right\}$$

$$= P\{(n - 1)\ln(1 - p) \geq \ln r > n\ln(1 - p)\}$$

$$= P\{\ln(1 - p)^{n-1} \geq \ln r > \ln(1 - p)^n\}$$

$$= P\{(1 - p)^{n-1} \geq r > (1 - p)^n\} = (1 - p)^{n-1} - (1 - p)^n$$

$$= (1 - p)^{n-1}[1 - (1 - p)] = (1 - p)^{n-1}p.$$

This shows that T is geometric with parameter p.   ◆

Therefore, to simulate a geometric random variable T with parameter p, $0 < p < 1$, all we got to do is to choose a point r at random from $(0, 1)$ and let $T = 1 + [\ln r / \ln(1 - p)]$.

Let X be a negative binomial random variable with parameters (n, p), $0 < p < 1$. By Example 9.17, X is a sum of n independent geometric random variables. Therefore, to simulate X, it suffices to choose n independent random numbers $r_1, r_2, \ldots, r_n$ from $(0, 1)$ and let

$$X = \sum_{i=1}^{n}\left(1 + \left[\frac{\ln r_i}{\ln(1 - p)}\right]\right).$$

We now explain simulation of continuous random variables. The following theorem is a key to simulation of many special continuous random variables.

THEOREM 12.2   *Let X be a continuous random variable with probability distribution function F. Then $F(X)$ is a uniform random variable over $(0, 1)$.*

PROOF: Let S be the sample space over which X is defined. The functions $X: S \to \mathbf{R}$ and $F: \mathbf{R} \to [0, 1]$ can be composed to obtain the random variable $F(X): S \to [0, 1]$. Clearly,

$$P\{F(X) \le t\} = \begin{cases} 1 & \text{if } t \ge 1 \\ 0 & \text{if } t \le 0. \end{cases}$$

Let $t \in (0, 1)$; it remains to prove that $P\{F(X) \le t\} = t$. To show this, note that since F is continuous, $F(-\infty) = 0$, and $F(\infty) = 1$, the inverse image of t, $F^{-1}(\{t\})$, is nonempty. We know that F is nondecreasing; since F is not necessarily strictly increasing, $F^{-1}(\{t\})$ might have more than one element. For example, if F is the constant t on some interval $(a, b) \subseteq (0, 1)$, then $F(x) = t$ for all $x \in (a, b)$, implying that (a, b) is contained in $F^{-1}(\{t\})$. Let

$$x_0 = \inf\{x : F(x) > t\}.$$

Then $F(x_0) = t$ and

$$F(x) \le t \qquad \text{if and only if} \qquad x \le x_0.$$

Therefore,

$$P\{F(X) \le t\} = P\{X \le x_0\} = F(x_0) = t.$$

We have shown that

$$P\{F(X) \le t\} = \begin{cases} 0 & \text{if } t \le 0 \\ t & \text{if } 0 \le t \le 1 \\ 1 & \text{if } t \ge 1, \end{cases}$$

meaning that $F(X)$ is uniform over $(0, 1)$.   ◆

Based on this theorem, to simulate a continuous random variable X with *strictly increasing* probability distribution function F, it suffices to generate a random number u from $(0, 1)$ and then solve the equation $F(t) = u$ for t. The solution t to this equation, $F^{-1}(u)$, is unique because F is strictly increasing. It is a simulation of X. For example, let X be an exponential random variable with parameter λ. Since $F(t) = 1 - e^{-\lambda t}$ and the solution to $1 - e^{-\lambda t} = u$ is $t = -1/\lambda \ln(1 - u)$, we have that for any random number u from $(0, 1)$, the quantity $-1/\lambda \ln(1 - u)$ is a simulation of

X. But if u is a random number from $(0, 1)$, then $(1 - u)$ is also a random number from $(0, 1)$. Therefore,

$$-\frac{1}{\lambda}\ln[1 - (1 - u)] = -\frac{1}{\lambda}\ln u$$

is also a simulation of X. So, *to simulate an exponential random variable X with parameter λ, it suffices to generate a random number u from $(0, 1)$ and then calculate $-(1/\lambda)\ln u$.* The method described is called *the inverse transformation method* and is good whenever the equation $F(t) = u$ can be solved without complications.

As we know, there are close relations between Poisson processes and exponential random variables and also between exponential and gamma random variables. These relations enable us to use the results obtained for simulation of the exponential case [e.g., to simulate Poisson processes and gamma random variables with parameters (n, λ), n being a positive integer]. Let X be a gamma random variable with such parameters. Then $X = X_1 + X_2 + \cdots + X_n$, where X_i's $(i = 1, 2, \ldots, n)$ are independent exponential random variables each with parameter λ. Hence *if $u_1, u_2, \ldots, u_n$ are n independent random numbers generated from $(0, 1)$, then*

$$\sum_{i=1}^{n} -\frac{1}{\lambda}\ln u_i = -\frac{1}{\lambda}\ln(u_1 u_2 \cdots u_n)$$

is a simulation of a gamma random variable with parameters (n, λ).

To simulate a Poisson process $\{N(t) : t \geq 0\}$ with parameter λ, note that $N(t)$ is the number of "events" that have occurred at or prior to t. Let X_1 be the time of the first event, X_2 be the elapsed time between the first and second events, X_3 be the elapsed time between the second and third events, and so on. Then the sequence $\{X_1, X_2, X_3, \ldots\}$ is an independent sequence of exponential random variables each with parameter λ. Clearly,

$$N(t) = \max\{n : X_1 + X_2 + \cdots + X_n \leq t\}.$$

Let $U_1, U_2, U_3, \ldots$ be independent uniform random variables from $(0, 1)$; then

$$N(t) = \max\left\{n : -\frac{1}{\lambda}\ln U_1 - \frac{1}{\lambda}\ln U_2 - \cdots - \frac{1}{\lambda}\ln U_n \leq t\right\}$$

$$= \max\{n : -\ln(U_1 U_2 \cdots U_n) \leq \lambda t\}$$

$$= \max\{n : \ln(U_1 U_2 \cdots U_n) \geq -\lambda t\}$$

$$= \max\{n : U_1 U_2 \cdots U_n \geq e^{-\lambda t}\}.$$

This implies that

$$N(t) + 1 = \min\{n : U_1 U_2 \cdots U_n < e^{-\lambda t}\}.$$

Therefore, *to simulate $N(t)$, we keep generating random numbers u_1, u_2, u_3, ... from $(0, 1)$ until for the first time $u_1 u_2 \cdots u_n$ is less than $e^{-\lambda t}$. The number of random numbers generated minus 1 is a simulation of $N(t)$.*

Let N be a Poisson random variable with parameter λ. Since N has the same distribution as $N(1)$, to simulate N all we have to do is to simulate $N(1)$ as explained above.

The inverse transformation method is not appropriate for simulation of normal random variables. This is because there is no closed form for F, the distribution function of a normal random variable, and hence, in general, $F(t) = u$ cannot be solved for t. However, other methods can be used to simulate normal random variables. Among them the following, introduced by Box and Muller, is perhaps the simplest. It is based on the following theorem proved in Example 8.29.

THEOREM 12.3 *Let V and W be two independent uniform random variables over $(0, 1)$. Then the random variables*

$$Z_1 = \cos(2\pi V)\sqrt{-2\ln W}$$

and

$$Z_2 = \sin(2\pi V)\sqrt{-2\ln W}$$

are independent standard normal random variables.

Based on this theorem, *to simulate the standard normal random variable Z, we may generate two independent random numbers v and w from $(0, 1)$ and let $z = \cos(2\pi v)\sqrt{-2\ln w}$. The number z is then a simulation of Z.* The advantage of Box and Muller's method is that it can generate two independent standard normal random variables at the same time. Its disadvantage is that it is not very efficient.

As for an arbitrary normal random variable, suppose that $X \sim N(\mu, \sigma^2)$. Then since $Z = (X - \mu)/\sigma$, is standard normal, we have that $X = \sigma Z + \mu$. Thus, *to simulate X, we may generate two independent random numbers v and w from $(0, 1)$ and let $t = \sigma \cos(2\pi v)\sqrt{-2\ln w} + \mu$. The quantity t is then a simulation of X.*

Example 12.7 Passengers arrive at a train station according to a Poisson process with parameter λ. If the arrival time of the next train is uniformly distributed over the interval $(0, T)$, write an algorithm to approximate by simulation the probability that by the time the next train arrives there are at least N passengers at the station.

SOLUTION: For a large positive integer n, we generate n independent random numbers $t_1, t_2, \ldots, t_n$ from $(0, T)$. Then for each t_i we simulate a Poisson random variable X_i with parameter λt_i. Finally, we will find m, the number of X_i's that are greater than or equal to N. The desired approximation is $p = m/n$. An algorithm follows.

> STEP 1: *Set $m = 0$.*
>
> STEP 2: *Do steps 3 to 5 n times.*
>
> STEP 3: *Generate a random number t from $(0, T)$.*
>
> STEP 4: *Generate a Poisson random variable X with parameter λt.*
>
> STEP 5: *If $X \geq N$, then set $m = m + 1$.*
>
> STEP 6: *Set $p = m/n$.*
>
> STEP 7: *Output (p).*
>
> **STOP** ◆

Example 12.8 The time between any two earthquakes in San Francisco and the time between any two earthquakes in Los Angeles are exponentially distributed with means $1/\lambda_1$ and $1/\lambda_2$, respectively. Write an algorithm to approximate by simulation the probability that the next earthquake in Los Angeles occurs prior to the next earthquake in San Francisco.

SOLUTION: Let X and Y denote the times between now and the next earthquakes in San Francisco and Los Angeles, respectively. For a large n, let n be the total number of simulations of the times of the next earthquakes in San Francisco and Los Angeles. If m is the number of those simulations in which $X > Y$, then $p = P(X > Y)$ is approximately m/n. An algorithm follows.

> STEP 1: *$m = 0$.*
>
> STEP 2: *Do steps 3 to 5 n times.*
>
> STEP 3: *Generate an exponential random variable X with parameter λ_1.*
>
> STEP 4: *Generate an exponential random variable Y with parameter λ_2.*
>
> STEP 5: *If $X > Y$, then set $m = m + 1$.*
>
> STEP 6: *Set $p = m/n$.*
>
> STEP 7: *Output (p).*
>
> **STOP** ◆

Example 12.9 The distributions of the grades of students in calculus, statistics, music, computer programming, and physical education are, respectively, $N(70, 400)$, $N(75, 225)$, $N(80, 400)$, $N(75, 400)$, and $N(85, 100)$. Write an algorithm to approximate by simulation the probability that the median of the grades of a randomly selected student taking these five courses is at least 75.

SOLUTION: Let $X(1)$, $X(2)$, $X(3)$, $X(4)$, and $X(5)$ be the grades of the randomly selected student in calculus, statistics, music, computer programming, and physical education, respectively. Let X be the median of these grades. We will simulate $X(1)$ through $X(5)$ n times. Each time we calculate X by sorting $X(1)$, $X(2)$, $X(3)$, $X(4)$, and $X(5)$ and letting $X = X(3)$. Then we check to see if X is at least 75. In all these n simulations, let m be the total number of X's that are at least 75; m/n is the desired approximation. An algorithm follows.

STEP 1: *Set $m = 0$.*

STEP 2: *Do steps 3 to 12 n times.*

STEP 3: *Generate $X(1) \sim N(70, 400)$.*

STEP 4: *Generate $X(2) \sim N(75, 225)$.*

STEP 5: *Generate $X(3) \sim N(80, 400)$.*

STEP 6: *Generate $X(4) \sim N(75, 400)$.*

STEP 7: *Generate $X(5) \sim N(85, 100)$.*

STEP 8: *Do steps 9 and 10 for $L = 2$ to 4.*

STEP 9: *Do step 10 for $j = 1, 6 - L$.*

STEP 10: *If $X(j) > X(j + 1)$, then*
$$\text{Set } K = X(j + 1);$$
$$\text{Set } X(j + 1) = X(j);$$
$$\text{Set } X(j) = K.$$

STEP 11: *Set $X = X(3)$.*

STEP 12: *If $X \geq 75$, set $m = m + 1$.*

STEP 13: *Set $p = m/n$.*

STEP 14: *Output (p).*

   STOP ♦

EXERCISES

1. Let X be a random variable with probability distribution function

$$F(x) = \begin{cases} \dfrac{x-3}{x-2} & \text{if } x \geq 3 \\ 0 & \text{elsewhere.} \end{cases}$$

Develop a method to simulate X.

2. Explain how a random variable X with the following probability density function can be simulated:

$$f(x) = e^{-2|x|}, \qquad -\infty < x < +\infty.$$

3. Explain a procedure for simulation of lognormal random variables. A random variable X is called lognormal with parameters μ and σ^2 if $\ln X \sim N(\mu, \sigma^2)$

4. Use the result of Example 9.10 to explain how a gamma random variable with parameters $(n/2, 1/2)$ (n being a positive integer) can be simulated.

5. It can be shown that the median of $(2n + 1)$ random numbers from the interval $(0, 1)$ is a beta random variable with parameters $(n + 1, n + 1)$. Use this property to simulate a beta random variable with parameters $(n + 1, n + 1)$, n being a positive integer.

6. Suppose that in a community the distributions of the heights of men and women, in centimeters, are $N(173, 40)$ and $N(160, 20)$, respectively. Write an algorithm to calculate by simulation the approximate value of the probability that (a) a wife is taller than her husband; (b) a husband is at least 10 centimeters taller than his wife.

7. Mr. Jones is at a train station waiting to make a phone call. There are two public telephone booths next to each other and occupied by two persons, say A and B. If the duration of each telephone call is an exponential random variable with $\lambda = 1/8$, using simulation, approximate the probability that among Mr. Jones, A, and B, Mr. Jones is not the last person to finish his call.

8. The distributions of the grades of the students of probability and calculus in a certain university are, respectively, $N(65, 400)$ and $N(72, 450)$. Dr. Olwell teaches a calculus class with 28 and a probability class with 22 students. Write an algorithm to simulate the probability that the difference between the averages of the final grades of the classes of Dr. Olwell is at least 2.

HINT: Note that if $X_1, X_2, \ldots, X_n$ are all $N(\mu, \sigma^2)$, then

$$\overline{X} = \frac{X_1 + X_2 + \cdots + X_n}{n} \sim N\left(\mu, \frac{\sigma^2}{n}\right).$$

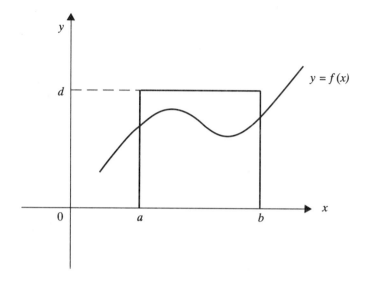

Figure 12.1   Area under f to be calculated using the Monte Carlo procedure.

12.5   Monte Carlo Procedure

As explained in Section 8.1, if S is a subset of the plane with area $A(S)$ and R is a subset of S with area $A(R)$, the probability that a random point from S falls in R is equal to $A(R)/A(S)$. This important fact gives an excellent algorithm, called *the Monte Carlo procedure*, for finding the area under a bounded curve $y = f(x)$ by simulation. Suppose that we want to estimate, I, the area under $y = f(x)$ from $x = a$ to $x = b$ of Figure 12.1. To do so, we first construct a rectangle $[a, b] \times [0, d]$ that includes the region under $y = f(x)$ from a to b as a subset. Then for a large integer n, we choose n random points from $[a, b] \times [0, d]$ and count m, the number of those that lie below the curve $y = f(x)$. Now the probability that a random point from $[a, b] \times [0, d]$ is under the curve $y = f(x)$ is approximately m/n. Thus

$$\frac{m}{n} \approx \frac{\text{area under } f(x) \text{ from } a \text{ to } b}{\text{area of the rectangle } [a, b] \times [0, d]}$$

$$= \frac{I}{(b - a)(d - 0)} = \frac{I}{(b - a)d},$$

and hence

$$I \approx \frac{md(b - a)}{n}.$$

This method of simulation was introduced by the Hungarian-born American mathematician John von Neumann (1903–1957), a great genius of this century, and Stanislaw Ulam (1909–1984). They used it during World War II to study the extent that neutrons can travel through various materials. Since their studies were classified, von Neumann gave it the code name *Monte Carlo method*. An algorithm for this procedure is as follows.

STEP 1:   *Set m = 0;*

STEP 2:   *Set i = 1;*

STEP 3:   *While i ≤ n, do steps 4 to 6.*

STEP 4:   *Generate a random number x from [a, b].*

STEP 5:   *Generate a random number y from [0, d].*

STEP 6:   *If y < f(x), then*
>>   *Set i = i + 1;*
>>   *Set m = m + 1;*
> *else*
>>   *Set i = i + 1.*

STEP 7:   *Set I = [md(b − a)]/n.*

STEP 8:   *Output(I).*
   STOP

We now explain one of the most interesting problems of geometric probability, *Buffon's needle problem.* Then we show how the solution of Buffon's needle problem and the Monte Carlo method can be used to find estimations for π by simulation. Georges Louis Buffon (1707–1784), who proposed and solved this famous problem, was a French naturalist who in the eighteenth century used probability to study many natural phenomena. In addition to the needle problem, his studies of the distribution and expectation of the remaining lifetimes of human beings are famous among mathematicians. These works, together with many more, are published in his gigantic 44-volume *Histoire Naturelle* (Natural History; 1749–1804).

Example 12.10   (Buffon's Needle Problem)   A plane is ruled with parallel lines a distance d apart. A needle of length ℓ, $\ell < d$, is tossed at random onto the plane. What is the probability that the needle intersects one of the parallel lines?

SOLUTION: Let X denote the distance from the center of the needle to the closest line, and let Φ denote the angle between the needle and the line. The position of the needle is completely determined by the coordinates X and Φ. As Figure 12.2 shows, the needle intersects the line if and only if the length of the hypotenuse of the triangle, $X/\sin \Phi$, is less than $\ell/2$. Since X

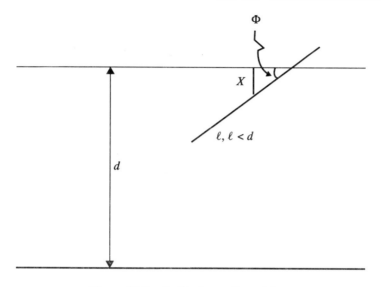

Figure 12.2 Buffon's needle problem.

and Φ are independent uniform random variables over the intervals $(0, d/2)$ and $(0, \pi)$, respectively, the probability that the needle intersects a line is $\iint_R f(x, \phi) \, dx \, d\phi$, where $f(x, \phi)$ is the joint probability density function of X and Φ and is given by

$$f(x, \phi) = f_X(x) f_\Phi(\phi) = \begin{cases} \dfrac{2}{\pi d} & \text{if } 0 \le \phi \le \pi, \ 0 \le x \le \dfrac{d}{2}. \\ 0 & \text{elsewhere} \end{cases}$$

and

$$R = \left\{ (x, \phi) : \frac{x}{\sin \phi} < \frac{\ell}{2} \right\} = \left\{ (x, \phi) : x < \frac{\ell}{2} \sin \phi \right\}.$$

Thus the desired probability is

$$\iint_R f(x, \phi) \, dx \, d\phi = \int_0^\pi \int_0^{(\ell/2) \sin \phi} \frac{2}{\pi d} \, dx \, d\phi$$

$$= \frac{\ell}{\pi d} \int_0^\pi \sin \phi \, d\phi = \frac{2\ell}{\pi d}. \quad \blacklozenge$$

The formula obtained in Example 12.10 has been used throughout the history of probability to determine approximate values for π. For fixed and predetermined values of d and ℓ, $\ell < d$, rule a plane with parallel lines a distant d apart and toss a needle of length ℓ, n times, and count m, the number of times that the needle intersects a line. Then m/n determines an approximate value for the probability that the needle intersects a line, provided that n is a large number. Thus $2\ell/\pi d \approx m/n$ and hence $\pi \approx 2n\ell/md$. In 1850, Rudolf Wolf, a scientist from Switzerland, conducted such an experiment. With 5000 tosses of a needle he got 3.1596 for π. In 1855, another mathematician, Smith, did the same experiment with 3204 tosses and got 3.1553. In 1860, De Morgan repeated the experiment; with 600 tosses he got 3.137 for π. In 1894 and 1901, Fox and Lazzarini conducted similar experiments; with 1120 and 3408 tosses, respectively, they got the values 3.1419 and 3.1415929. It should be noted that the latter results are subject to skepticism, because it can be shown that the probability of getting such close approximations to π in such low numbers of tosses of a needle is extremely small. For details, see N. T. Gridgeman's paper "Geometric Probability and Number π", *Scripta Mathematika*, 1960, Volume 25, Number 3.

To find more accurate estimates for π using the result of Buffon's needle problem, we may use the Monte Carlo procedure. As we explained above, every time that the needle is tossed, a point (Φ, X) is selected from the square $[0, \pi] \times [0, d/2]$, and the needle intersects the line if and only if $X/\sin \Phi < \ell/2$. Therefore, if for a large number n, we select n

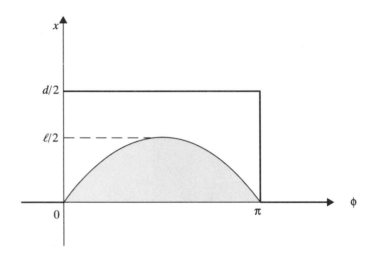

Figure 12.3   The curve $x = (\ell/2) \sin \phi$.

random points from the square $[0, \pi] \times [0, d/2]$ and find m, the number of those that fall below the curve $x = (\ell/2) \sin \phi$ (see Figure 12.3), then m/n is approximately the probability that the needle intersects a line, that is, $m/n \approx 2\ell/\pi d$, which gives $\pi \approx 2n\ell/md$.

EXERCISES

1. Three concentric circles of radii 2, 3, and 4 are the boundaries of the regions that form a circular target. If a person fires a shot at random at the target, use simulation to approximate the probability that it lands in the middle region.

   HINT: To simulate a random point located inside a circle of radius r centered at the origin, choose random points (X, Y) inside the square $[-r, r] \times [-r, r]$ until you obtain one that satisfies $X^2 + Y^2 < r^2$.

2. Using Monte Carlo procedure, write a program to estimate

$$\Phi(3) - \Phi(-1) = \int_{-1}^{3} \frac{1}{\sqrt{2\pi}} e^{-x^2/2} \, dx.$$

   Then run your program for $n = 10,000$ and compare the answer with that obtained from Table 1 of the Appendix.

3. In Buffon's needle problem, let $\ell = 1$ and $d = 2$ and use Monte Carlo procedure to estimate π with $n = 10,000$. Then do the same for $\ell = 1$ and $d = 3/2$, then for $\ell = 1$ and $d = 5/4$.

APPENDIX

Table 1   Area under the Standard Normal Distribution to the Left of x

$$\Phi(x) = \frac{1}{\sqrt{2\pi}} \int_{-\infty}^{x} e^{-y^2/2}\, dy, \qquad \Phi(-x) = 1 - \Phi(x)$$

x	0	1	2	3	4	5	6	7	8	9
.0	.5000	.5040	.5080	.5120	.5160	.5199	.5239	.5279	.5319	.5359
.1	.5398	.5438	.5478	.5517	.5557	.5596	.5636	.5675	.5714	.5753
.2	.5793	.5832	.5871	.5910	.5948	.5987	.6026	.6064	.6103	.6141
.3	.6179	.6217	.6255	.6293	.6331	.6368	.6406	.6443	.6480	.6517
.4	.6554	.6591	.6628	.6664	.6700	.6736	.6772	.6808	.6844	.6879
.5	.6915	.6950	.6985	.7019	.7054	.7088	.7123	.7157	.7190	.7224
.6	.7257	.7291	.7324	.7357	.7389	.7422	.7454	.7486	.7517	.7549
.7	.7580	.7611	.7642	.7673	.7703	.7734	.7764	.7794	.7823	.7852
.8	.7881	.7910	.7939	.7967	.7995	.8023	.8051	.8078	.8106	.8133
.9	.8159	.8186	.8212	.8238	.8264	.8289	.8315	.8340	.8365	.8389
1.0	.8413	.8438	.8461	.8485	.8508	.8531	.8554	.8577	.8599	.8621
1.1	.8643	.8665	.8686	.8708	.8729	.8749	.8770	.8790	.8810	.8830
1.2	.8849	.8869	.8888	.8907	.8925	.8944	.8962	.8980	.8997	.9015
1.3	.9032	.9049	.9066	.9082	.9099	.9115	.9131	.9147	.9162	.9177
1.4	.9192	.9207	.9222	.9236	.9251	.9265	.9279	.9292	.9306	.9319
1.5	.9332	.9345	.9357	.9370	.9382	.9394	.9406	.9418	.9429	.9441
1.6	.9452	.9463	.9474	.9484	.9495	.9505	.9515	.9525	.9535	.9545
1.7	.9554	.9564	.9573	.9582	.9591	.9599	.9608	.9616	.9625	.9633
1.8	.9641	.9649	.9656	.9664	.9671	.9678	.9686	.9693	.9699	.9706
1.9	.9713	.9719	.9726	.9732	.9738	.9744	.9750	.9756	.9761	.9767
2.0	.9772	.9778	.9783	.9788	.9793	.9798	.9803	.9808	.9812	.9817
2.1	.9821	.9826	.9830	.9834	.9838	.9842	.9846	.9850	.9854	.9857
2.2	.9861	.9864	.9868	.9871	.9875	.9878	.9881	.9884	.9887	.9890
2.3	.9893	.9896	.9898	.9901	.9904	.9906	.9909	.9911	.9913	.9916
2.4	.9918	.9920	.9922	.9925	.9927	.9929	.9931	.9932	.9934	.9936
2.5	.9938	.9940	.9941	.9943	.9945	.9946	.9948	.9949	.9951	.9952
2.6	.9953	.9955	.9956	.9957	.9959	.9960	.9961	.9962	.9963	.9964
2.7	.9965	.9966	.9967	.9968	.9969	.9970	.9971	.9972	.9973	.9974
2.8	.9974	.9975	.9976	.9977	.9977	.9978	.9979	.9979	.9980	.9981
2.9	.9981	.9982	.9982	.9983	.9984	.9984	.9985	.9985	.9986	.9986
3.0	.9987	.9987	.9987	.9988	.9988	.9889	.9889	.9889	.9990	.9990
3.1	.9990	.9991	.9991	.9991	.9992	.9992	.9992	.9992	.9993	.9993
3.2	.9993	.9993	.9994	.9994	.9994	.9994	.9994	.9995	.9995	.9995
3.3	.9995	.9995	.9995	.9996	.9996	.9996	.9996	.9996	.9996	.9997
3.4	.9997	.9997	.9997	.9997	.9997	.9997	.9997	.9997	.9997	.9998
3.5	.9998	.9998	.9998	.9998	.9998	.9998	.9998	.9998	.9998	.9998
3.6	.9998	.9998	.9999	.9999	.9999	.9999	.9999	.9999	.9999	.9999

Table 2 Expectations, Variances, and Moment-Generating Functions of Important Distributions

Distribution of X	Probability Function of X	Range of X	$E(X)$	$\text{Var}(X)$	$M_X(t)$
Binomial with parameters n and p	$\binom{n}{x} p^x (1-p)^{n-x}$	$x = 0, 1, \ldots, n$	np	$np(1-p)$	$(pe^t + 1 - p)^n$
Poisson with parameter $\lambda > 0$	$\dfrac{e^{-\lambda}\lambda^x}{x!}$	$x = 0, 1, 2, \ldots$	λ	λ	$\exp\left[\lambda\left(e^t - 1\right)\right]$
Geometric with parameter p	$p(1-p)^{x-1}$	$x = 1, 2, \ldots$	$\dfrac{1}{p}$	$\dfrac{1-p}{p^2}$	$\dfrac{pe^t}{1 - (1-p)e^t}$
Negative binomial with parameters r and p	$\binom{x-1}{r-1} p^r (1-p)^{x-r}$	$x = r, r+1, r+2, \ldots$	$\dfrac{r}{p}$	$\dfrac{r(1-p)}{p^2}$	$\left[\dfrac{pe^t}{1 - (1-p)e^t}\right]^r$
Uniform over (a, b)	$\dfrac{1}{b-a}$	$a < x < b$	$\dfrac{a+b}{2}$	$\dfrac{(b-a)^2}{12}$	$\dfrac{e^{tb} - e^{ta}}{t(b-a)}$
Normal with parameters μ and σ^2	$\dfrac{1}{\sqrt{2\pi}}\exp\left[-\dfrac{(x-\mu)^2}{2\sigma^2}\right]$	$-\infty < x < \infty$	μ	σ^2	$\exp\left(\mu t + \dfrac{\sigma^2 t^2}{2}\right)$
Exponential with parameter λ	$\lambda e^{-\lambda x}$	$0 \leq x < \infty$	$\dfrac{1}{\lambda}$	$\dfrac{1}{\lambda^2}$	$\dfrac{\lambda}{\lambda - t}$
Gamma with parameters r and λ	$\dfrac{\lambda e^{-\lambda x}(\lambda x)^{r-1}}{\Gamma(r)}$	$x \geq 0$	$\dfrac{r}{\lambda}$	$\dfrac{r}{\lambda^2}$	$\left(\dfrac{\lambda}{\lambda - t}\right)^r$

ANSWERS
TO ODD-NUMBERED
EXERCISES

Section 1.2

1. $\{RRR, RRB, RBR, RBB,$
$BRR, BRB, BBR, BBB\}$.

3. Denote the dictionaries by d_1,
d_2; the third book by a. The
answers are: $\{d_1d_2a, d_1ad_2,$
$d_2d_1a, d_2ad_1, ad_1d_2, ad_2d_1\}$ and
$\{d_1d_2a, ad_1d_2\}$.

5. $\{QQ, QN, QP, QD, DN, DP,$
$NP, NN, PP\}, \{QP\},$
$\{DN, DP, NN\}, \emptyset$.

7. If E or F occurs, then G occurs.
If G occurs, then E and F occur.

9. (a) $F \cup EG$, (b) FE.

17. $\bigcap_{m=1}^{\infty} \bigcup_{n=m}^{\infty} A_n$.

Section 1.4

1. No!

3. 7/15.

5. Use $P(A \cup B) \le 1$.

7. (a) False, (b) False.

11. $\sum_{i=1}^{n} p_{ij}$.

13. (a) 1/8, (b) 5/24, (c) 5/24.

15. 4/7.

19. 0.172.

21. 99/200.

Section 1.7

1. 2/3.

3. (a) False, (b) False.

5. 0.

Review Problems for Chapter 1

1. 0.54.

3. (a) False, (b) True.

5. $\{x_1 x_2 \cdots x_n : n \ge 1, x_i \in$
$\{H, T\}; x_i \ne x_{i+1}, 1 \le i \le$
$n - 2; x_{n-1} = x_n\}$.

9. Denote a box of books from
publisher i by $a_i, i = 1, 2, 3$.

Then the sample space is
$S = \{x_1 x_2 x_3 x_4 x_5 x_6 : \text{two } x_i\text{'s are}$
a_1, two are a_2, and the remaining
two are $a_3\}$. The desired event is
$\{x_1 x_2 x_3 x_4 x_5 x_6 \in S : x_5 = x_6\}$.

11. 0.571.

13. (a) $U_i^c D_i^c$,
   (c) $(U_1^c D_1^c) \cup \cdots \cup (U_n^c D_n^c)$,
   (e) $D_1^c D_2^c \cdots D_n^c$.

15. 173/216.

Section 2.2

1. 427,608.

3. Yes.

5. 0.00000024.

7. 1/6.

9. 0.00000000093.

11. Yes, there are.

13. 36.

15. 625 and 505, respectively.

17. 0.274.

19. $1 - [(N - 1)^n / N^n]$.

21. 30.43%.

23. 0.469.

25. 0.868.

27. 0.031.

29. 0.067.

Section 2.3

1. 6.

3. 0.875.

5. 30.

7. 50,400.

9. 92,400.

11. 34,650.

13. 6.043×10^{-13}.

15. 0.6.

17. 0.00068.

19. 0.0079.

21. 0.985.

23. (a) m^n, (b) $_m P_n$, (c) $n!$.

25. 7.61×10^{-6} and 3.13×10^{-24},
   respectively.

27. 0.385.

Section 2.4

1. 38,760.

3. 6,864,396,000.

5. n/N.

7. 560.

9. 0.318.

11. 71,680.

13. (a) 0.246, (b) 0.623.

15. (a) 0.228, (b) 0.00084.

17. 3^n and $(x + 1)^n$, respectively.

19. 0.00151.

21. 6.

23. -7560.

25. 4.47×10^{-28}.

27. 36.

31. (a) 0.347, (b) 0.520, (c) 0.130,
   (d) 0.0031.

33. (a) $(1/n^m) \sum_{i=0}^{n} (-1)^i \binom{n}{i}$
   $(n - i)^m$. (b) $(1/n^m) \binom{n}{r} \sum_{i=0}^{n-r}$
   $(-1)^i \binom{n-r}{i} (n - r - i)^m$.

39. 0.264.

41. 16,435,440.

45. $[1/(10 \times 2^{2n})][(5 + 3\sqrt{5})(1 + \sqrt{5})^n + (5 - 3\sqrt{5})(1 - \sqrt{5})^n]$.

Section 2.5

1. $1/\sqrt{\pi n}$, $\sqrt{2}/4^n$, respectively.

Review Problems for Chapter 2

1. 127.

3. 0.278.

5. 7560.

7. 82,944.

9. In one way.

11. 34.62%.

13. 43,046,721.

15. 0.467.

17. 0.252.

19. 531,441.

21. 0.379.

23. $(N - n + 1)/\binom{N}{n}$.

25. 369,600.

Section 3.1

1. 60%.

3. 0.625.

5. 2/3.

7. 2/5.

9. 0.239.

11. (a) 1/7, (b) 4/7, (c) 3/7.

15. 0.269.

17. (a) 0.0144, (b) 0.344.

19. (a) 0.3334, (b) 0.5332.

21. 0.0712.

23. 0.883.

25. 0.059.

Section 3.2

1. 0.02625.

3. 0.633.

5. 1/4.

7. 0.7055.

9. 0.034.

11. 0.4. (Even without calculations you may argue that the answer is 0.4.)

15. 7/10.

17. 5/12.

19. He should distribute one green and 0 red balls in one urn, $N - 1$ green and N red balls in the other urn.

Section 3.3

1. 3/5.

3. 14.63%.

5. 0.084.

7. 0.056.

9. 0.21.

11. 205/297.

13. 0.61.

Section 3.4

1. You should not agree!

3. 0.00503

5. 0.526 and 0.0756, respectively.

7. Let A be the event of heads and B be the event of tails in the experiment of flipping a coin once.

11. 0.9917.

13. 0.226

17. $1 - [(n - 1)/n]^n$; approaches $1 - (1/e)$.

19. No!

21. 0.2.

23. $2p^4 - p^6$.

27. (a)0.168, (b) 0.286.

31. 0.495.

33. $2p^2 + 2p^3 - 5p^4 + p^5$.

35. $1/q$, p/q, p^2/q, where the denominator is $q = (1 + p)(1 + p + p^2)$.

Review Problems for Chapter 3

1. 13/75.

3. 65%.

5. (a) 0.783, (b) 0.999775, (c) 0.217
(d) 0.000225.

7. 0.0796.

9. 1/6.

11. 0.35.

13. 0.03.

15. 0.36.

17. Either way gives them the same odds.

19. (a) 0.3696, (b) 0.5612. Knowing that Adam has the king of diamonds reduces the sample space to a size considerably smaller than the case in which we are given that he has a king.

Section 4.2

1. Possible values are 0, 1, 2, 3, 4, and 5. Probabilities associated with these values are 6/36, 10/36, 8/36, 6/36, 4/36, and 2/36, respectively.

3. $(1-p)^{i-1}p$ if $1 \le i \le N$, $(1-p)^N$ if $i = 0$.

5. The answers are 1/2, 1/6, 1/4, 1/2, 0, and 1/3, respectively.

7. (a) 1/33, (b) 25/33, (c) 9/33, (d) 5/6.

11. It is a distribution function.

13. 0.277.

15. $F(t) = 0$ if $t < 0$, $t/(1-t)$ if $0 \le t < 1/2$, 1 if $t \ge 1/2$.

17. The distribution function of Y is $F(t)$ if $t < 5$ and 1 if $t \ge 5$.

Section 4.3

1. $F(x) = 0$ if $x < 1$, 1/15 if $1 \le x < 2$, 3/15 if $2 \le x < 3$,

6/15 if $3 \le x < 4$, 10/15 if $4 \le x < 5$, and 1 if $x \ge 5$.

3. $p(-2) = 1/2$, $p(2) = 1/10$, $p(4) = 13/45$, $p(6) = 1/9$.

5. Mode of $p = 1$, mode of $q = 1$.

7. $[\binom{18}{i}\binom{28}{12-i}]/\binom{46}{12}$, $0 \le i \le 12$.

9. $p(i) = (5/6)^{i-1}(1/6)$, $i \ge 1$. $F(x)$ is 0 if $x < 1$; $1 - (5/6)^n$ if $n \le x < n+1$, $n \ge 1$.

Section 4.4

1. He should park at a lot.

3. -0.38.

5. 0.75

9. 11π.

11. 22.

13. $\sum_{k=1}^{c} \sum_{j=k}^{c} (k\alpha_j/j)$.

15. 10.6

17. $E(X)$ does not exist.

Section 4.5

1. The first business.

3. 3.

5. $(N+1)/2$, $(N^2-1)/12$, $\sqrt{(N^2-1)/12}$.

7. 36.

Section 4.6

1. Mr. Norton should hire the salesperson who worked in store 2.

Review Problems for Chapter 4

1. It is 1/45 for $i = 1, 2, 16, 17$; 2/45 for $i = 3, 4, 14, 15$; 3/45 for $i = 5, 6, 12, 13$; 4/45 for $i = 7, 8, 10, 11$; and 5/45 for $i = 9$.

3. 35.

5. 1.09.

7. They are 16, 16, 0.013, and 0.008, respectively.

9. (a) $k = e^{-2t}$. (b) They are $k \sum_{i=0}^{3}(2t)^i / i!$ and $1 - k - 2tk$, respectively.

Section 5.1

1. 0.087.

3. 0.054.

5. 0.33.

7. $p(x) = \binom{4}{x}(0.60)^x(0.4)^{4-x}$, $x = 0, 1, 2, 3, 4$; $q(y) = \binom{4}{(y-1)/2}(0.60)^{(y-1)/2}$ $(0.4)^{(9-y)/2}$, $y = 1, 3, 5, 7, 9$.

9. 0.108.

11. They are $np + p$ and $np + p - 1$.

13. 0.219.

15. $k = 4$ and the maximum probability is 0.238.

17. 0.995.

19. It is preferable to send in a single parcel.

21. $(1/2)^n \sum_{i=0}^{[\frac{n+(b-a)}{2}]} \binom{n}{i}$, where $[\frac{n+(b-a)}{2}]$ is the greatest integer $\leq \frac{n+(b-a)}{2}$.

25. 90,072.

Section 5.2

Poisson as an Approximation to Binomial

1. 0.9502.

3. 0.594.

5. 0.823.

7. 0.063.

9. 83%.

11. $-N \ln(1 - \alpha)/M$.

Poisson Process

1. 0.21.

3. 0.87.

5. 0.13.

7. (a) $\frac{1}{2}(1 + e^{-2\lambda\alpha})$, (b) $\frac{1}{2}(1 - e^{-2\lambda\alpha})$.

9. 0.035.

Section 5.3

1. (a) 12, (b) 0.07.

3. 0.055.

5. 0.42.

9. 24.27.

11. $p(i) = \binom{i+9}{9}(0.15)^{10}(0.85)^i$, $i \geq 0$.

13. 0.987.

15. 0.6346.

17. 254.80.

19. Approximately 9.015.

21. $\binom{2N-1}{N}(1/2)^{2N}$.

23. (a) D/N, (b) $(D - 1)/(N - 1)$.

Review Problems for Chapter 5

1. 0.0009.

3. 1.067.

5. 0.179.

7. 0.244.

9. 0.285.

11. They are both 2.

13. 0.772.

15. 0.91.

17. No, it is not.

19. 0.0989.

21. 7.

Section 6.1

1. (a) 3, (b) 0.78.

3. (a) 6; (b) 0 if $x < 1$,
$-2x^3 + 9x^2 - 12x + 5$ if
$1 \leq x < 2$, and 1 if $x \geq 2$;
(c) 5/32 and 1/2, respectively.

5. (a) $1/\pi$; (b) 0 if $x < -1$,
$\frac{1}{\pi} \arcsin x + \frac{1}{2}$ if $-1 \leq x < 1$, 1
if $x \geq 1$.

7. (b) They are symmetric about 3
and 1, respectively.

9. 0.3327.

11. $\alpha = 1/2$; $\beta = 1/\pi$,
$f(x) = 2\left[\pi(4+x^2)\right]$,
$-\infty < x < \infty$.

Section 6.2

1. The probability density function
of Y is $(1/12)y^{-2/3}$ if
$-8 < y < 8$; 0, otherwise. The
probability density function of Z
is $(1/8)z^{-3/4}$ if $0 < z < 16$; 0
otherwise.

3. The probability density function
of Y is $[2/(3\sqrt[3]{y})]\exp(-y^{2/3})$ if
$y \in (0, \infty)$; 0, elsewhere. The
probability density function of Z
is 1 if $Z \in (0, 1)$; 0, otherwise.

5. $(3\lambda/2)\sqrt{y}$ if $-\lambda y\sqrt{y}$ if $y \geq 0$;
0 otherwise.

7. $1/\pi$ if $-\pi/2 \leq z \leq \pi/2$; 0,
elsewhere.

Section 6.3

1. (a) 8.

3. The muffler of company B.

5. 0.

7. 0.3069.

9. 2.

11. $(\alpha/\lambda) + (\beta/\mu)$.

Review Problems for Chapter 6

1. $1/y2$ if $y \in [1, \infty)$; 0, elsewhere.

3. $11/(5\sqrt{5})$.

5. No!

7. $[15(1 - \sqrt[4]{y})^2]/(2\sqrt[4]{y})$ if
$y \in (0, 1)$; 0, elsewhere.

11. 0.3.

Section 7.1

1. 3/7.

3. a is 1:54 P.M. and b is 2:06 P.M.

5. The answers are 40π and 1/2,
respectively.

7. 1/3.

9. a/b.

11. $P([nX] = i) = 1/n$ if
$i = 0, 1, 2, \cdots, n - 1$;
$P([nX] = i) = 0$, otherwise.

Section 7.2

1. 0.5948.

5. They are all equal to 0.

7. 0.2%, 4.96%, 30.04%, 45.86%,
and 18.94%, respectively.

9. 0.4796.

11. 0.388.

13. The median is μ.

15. No, it is not correct.

17. They are 1 and 1/4, respectively.

19. $k = \pi$.

21. 0.742.

23. $\frac{1}{\sigma t\sqrt{2\pi}} \exp\left[-(\ln t - \mu)^2/(2\sigma^2)\right]$,
$t \geq 0$; 0, otherwise.

25. 10 copies.

27. Yes, absolutely!

31. 15.93 and 0.29, respectively.

33. $1/(2\sqrt{\lambda})$.

Section 7.3

1. 0.0001234.

3. $\exp(-y - e^{-y})$, $-\infty < y < \infty$.

5. (a) 0.1535, (b) 0.1535.

7. (a) $e^{-\lambda t}$, (b) $e^{-\lambda t} - e^{-\lambda s}$.

9. \$323.33.

11. (a) 1/2.

Section 7.4

3. 0.3208.

5. 2 hours and 35 minutes.

7. Gamma with parameters 2 and $n\lambda$.

Section 7.5

1. 0.7712.

Review Problems for Chapter 7

1. 5/12.

3. 0.248.

5. 0.999999927.

7. $(1/\sqrt{2\pi}) \int_{-\infty}^{x/2.5} \exp[-(t-4)^2/2] \, dt$.

9. 0.99966.

11. 0.51%, 12.2%, 48.7%, 34.23%, and 4.36%, respectively.

13. It is uniform over the interval $(-5/10^{k+1}, 5/10^{k+1})$.

15. 0.89.

Section 8.1

Joint Probability Functions

1. (a) 2/9. (b) $p_X(x) = x/3$, $x = 1, 2$; $p_Y(y) = 2/(3y)$, $y = 1, 2$. (c) 2/3.

3. (a) 1/25, (b) $p_X(x)$ is 12/25 if $x = 1$, 13/25 if $x = 2$; $p_Y(y)$ is 2/25 if $y = 1$, 23/25 if $y = 3$.

5. $[\binom{7}{x}\binom{8}{y}\binom{5}{4-x-y}]/\binom{20}{4}$, $0 \le x \le 4$, $0 \le y \le 4$, $0 \le x + y \le 4$.

7. $p(1,1) = 0$, $p(1,0) = 0.30$, $p(0,1) = 0.50$, $p(0,0) = 0.20$.

Joint Probability Density Functions

1. (a) $f_X(x) = 2x$, $0 \le x \le 1$; $f_Y(y) = 2(1-y)$, $0 \le y \le 1$. (b) They are 1/4, 1/2, and 0, respectively.

3. $f_X(x) = e^{-x}$, $x > 0$; $f_Y(y) = (1/2)y$, $0 < y < 2$.

5. $f(x, y) = 6$ if $(x, y) \in R$; 0, elsewhere.

7. 3/4.

11. 3/4.

13. $a/2b$.

15. $f(x) = 1 - |x|$, $-1 \le x \le 1$.

Section 8.2

1. Yes, they are.

3. 4/81 and 4/27, respectively.

5. 0.0179.

7. $(1/2)^{2n} \binom{2n}{n}$.

9. No, they are not.

11. No, they are not.

13. 1.

15. $\ell/3$.

17. 1/3.

19. They are not.

21. 11/23.

23. 0.12.

25. The distribution function is 0 if $t < 1$, $(t-1)/t$ if $t \ge 1$. The density function is $1/t^2$ if $t \ge 1$; 0, otherwise.

Section 8.3

1. $p_{X|Y}(x|y) = (x^2 + y^2)/(2y^2 + 5)$, $x = 1, 2$, $y = 0, 1, 2$. The desired conditional probability and conditional expectation are 5/7 and 12/7, respectively.

3. It is $\binom{x-6}{2}(1/2)^{x-5}$, $x = 8, 9, 10, \cdots$.

7. $1/(y+1)$.

9. 10/33.

11. The joint probability density function of X and Y is $1/(20y)$ if $20 < x < 20 + (2y)/3$ and $0 < y < 30$. It is 0 otherwise.

13. Given that $X = x$, $Y - x$ is binomial with parameters p and $n - m$.

15. Given that $N(s) = k$, $N(t) - k$ is Poisson with parameter $\lambda(t - s)$.

19. (a) $f(x, y) = 2$ if $x \geq 0$, $y \geq 0$, $x + y \geq 1$; 0, otherwise.
(b) $f_{X|Y}(x|y) = 1/(y - 1)$ if $0 \leq x \leq 1 - y$, $0 \leq y < 1$.
(c) $E(X|Y = y) = (1 - y)/2$, $0 < y < 1$.

Section 8.4

1. $p(h, d, c, s) = [\binom{13}{h}\binom{13}{d}\binom{13}{c}\binom{13}{s}]/\binom{52}{13}$, $h + d + c + s = 13, 0 \leq h, d, c, s \leq 13$.

3. $p_{X,Y}(x, y) = xy/54$, $x = 4, 5$, $y = 1, 2, 3$. $p_{Y,Z}(y, z) = yz/18$, $y = 1, 2, 3$, $z = 1, 2$. $p_{X,Z}(x, z) = xz/27$, $x = 4, 5$, $z = 1, 2$.

5. They are not independent.

7. (a) They are independent.
(b) $f(x, y, z) =$

$\lambda_1\lambda_2\lambda_3 \exp(-\lambda_1 x - \lambda_2 y - \lambda_3 z)$.
(c) $\lambda_1\lambda_2/[(\lambda_2+\lambda_3)(\lambda_1+\lambda_2+\lambda_3)]$.

9. $[1 - (r^2/R^2)]^n$.

11. Yes, it is.

13. X is exponential with parameter $\lambda_1 + \lambda_2 + \cdots + \lambda_n$.

15. They are $n/(n + 1)$ and $1/(n + 1)$, respectively.

19. They are not independent.

Section 8.5

1. 0.028.

3. 0.171.

5. (a) 0.8125, (b) 0.135, (c) 0.3046.

7. $p = 1/2$.

Section 8.6

1. $l(u, v) = 1/(2uv^2)$, $v \geq 1$, $u \geq 0$.

3. $l(u, v) = 1/(ve^u)$, $v > 1$, $u > 0$.

7. (a) $l(u, v) = (u\lambda^{r_1+r_2})/[\Gamma(r_1)\Gamma(r_2)]$ $e^{-\lambda u}(uv)^{r_1-1}(u - uv)^{r_2-1}$.

Review Problems for Chapter 8

1. 0.54.

3. $[\binom{13}{x}\binom{26}{9-x}]/\binom{39}{9}$, $0 \leq x \leq 9$.

5. $[\binom{13}{x}\binom{13}{6-x}]/\binom{26}{6}$, $0 \leq x \leq 6$.

7. (a) 1/2, (b) $f_X(x) = 1/2$, $0 < x < 1/2$; $f_Y(y) = (1/2) \ln(2/y)$, $0 < y < 2$.

11. 6.

13. $f_X(x) = (-3/2)x^2 + 3/2$, $0 < x < 1$; $f_Y(y) = (-3/2)y^2 + 3/2$, $0 < y < 1$. The desired probability is 29/64.

15. $p(x, y) =$
$\binom{10}{x}\binom{15}{y}(1/4)^{x+y}(3/4)^{25-x-y}$,
$0 \le x \le 10, \; 0 \le y \le 15$.

17. (a) $c = 12$, (b) They are not independent.

19. $[1 - (\pi/6)(r/a)^3]^n$.

21. (a) 1/2, (b) 2/9.

23. 0.00135.

25. 0.2775.

27. $E(Y|X = x) = 0$,
$E(X|Y = y) = (1 - y)/2$ if
$-1 < y < 0$ and $(1 + y)/2$ if
$0 < y < 1$.

Section 9.1

1. $M_X(t) =$
$(1/5)(e^t + e^{2t} + e^{3t} + e^{4t} + e^{5t})$.

3. $M_X(t) =$
$2 \exp(t - \ln 3)/[1 - \exp(t - \ln 3)]$,
$t < \ln 3. \; E(X) = 3/2$.

5. (a) $M_X(t) = (12/t^3)(1 - e^t) +$
$(6/t^2)(1 + e^t), \; t \ne 0$;
$M_X(t) = 1$ if $t = 0$.
(b) $E(X) = 1/2$.

7. (a) $M_X(t) = \exp[\lambda(e^t - 1)]$.
(b) $E(X) = \text{Var}(X) = \lambda$.

9. $E(X) = 1/p$,
$\text{Var}(X) = (1 - p)/p^2$.

11. $p(1) = 5/15, \; p(3) = 4/15$,
$p(4) = 2/15, \; p(5) = 4/15$;
$p(i) = 0, \; i \notin \{1, 3, 4, 5\}$.

13. $E(X) = 3/2, \; \text{Var}(X) = 3/4$.

17. 8/9.

21. $M_X(t) = [\lambda/(\lambda - t)]^r$,
$E(X) = r/\lambda, \; \text{Var}(X) = r/\lambda^2$.

Section 9.2

3. The sum is gamma with parameters n and λ.

5. The sum is gamma with parameters $r_1 + r_2 + \cdots + r_n$ and λ.

7. The answer is $\binom{n}{i}\binom{m}{j-i}/\binom{n+m}{j}$.

9. 0.558.

11. 0.0571

13. 1/2.

15. (a) 0.289, (b) 0.0836.

17. $X_1 + X_2 + \cdots + X_k$ is binomial
with parameters n and
$p_1 + p_2 + \cdots + p_k$.

Section 9.3

1. 13/6.

3. 26.

5. 11.42.

7. \$28.49.

9. $n/32$.

11. 23.41

15. 9.001.

Review Problems for Chapter 9

1. 1.

3. 0.0262

5. 20/9 and 0.299.

7. $1 + na_{n-1}$.

9. X is uniform over $(-1/2, 1/2)$.

11. The nth moment is
$(-1)^{n+1}(n + 1)!$.

13. $1/(1 - t^2), \; -1 < t < 1$.

Section 10.1

1. (a) 0.4545, (b) 0.472.

3. (a) 0.4, (b) 0.053, (c) 0.222.

5. 16 days earlier.

7. It is $\le 1/\mu$.

Section 10.2

1. The limit is 1/3 with probability 1.

Section 10.3

1. 0.6046.
3. 0.6826.
5. 0.0174.
7. 0.0793.
9. 50,000.

Review Problems for Chapter 10

1. 23,073.
3. 1/9.
5. 0.4844.

Section 11.1

5. They are $n(n-1)p(1-p)$ and $-np(1-p)$, respectively.
7. They are dependent but uncorrelated.
11. $n/12$.
17. $-n/36$.
19. $\text{Var}(X) = \frac{nD(N-D)}{N^2}(1 - \frac{n-1}{N-1})$.

Section 11.2

1. 112.
3. They are -1 and $-1/12$, respectively.
5. No! it is not.
7. Show that $\rho(X, Y) = -0.248 \neq \pm 1$.

Section 11.3

1. 0.5987.
3. $-\rho(X, Y)(\sigma_Y/\sigma_X)$.
5. They are $(1 + y)/2, 0 < y < 1$; $x/2, 0 < x < 1$; and 1/2, respectively.

Section 11.4

1. 5.
3. (a) $b - d[1 - (1/b)]^n$. (b) $1 - (d/b)[1 - (1/b)]^n$.
5. 3.786.

Review Problems for Chapter 11

1. 1554.
3. 204.
5. $2/\pi$.
7. $\alpha = \mu_Y - \rho(\sigma_Y/\sigma_X)\mu_X$, $\beta = \rho(\sigma_Y/\sigma_X)$.
9. 0.6256.
11. The distribution function of $\sum_{i=1}^{N} X_i$ is exponential with parameter λp.

INDEX

The following *abbreviations* are used in this index: r.v. for random variable, p.d.f. for probability density function, p.f. for probability function, and m.g.f. for moment-generating function.